AF692612

Integrated Aquaculture and Monoculture of Low-Trophic Species

Integrated Aquaculture and Monoculture of Low-Trophic Species

Editors

Wagner C. Valenti
Eduardo Luis Ballester

MDPI

Basel • Beijing • Wuhan • Barcelona • Belgrade • Novi Sad • Cluj • Manchester

Editors
Wagner C. Valenti
São Paulo State University (UNESP)
Jaboticabal
Brazil

Eduardo Luis Ballester
Federal University of Paraná
Palotina
Brazil

Editorial Office
MDPI AG
Grosspeteranlage 5
4052 Basel, Switzerland

This is a reprint of articles from the Special Issue published online in the open access journal *Fishes* (ISSN 2410-3888) (available at: https://www.mdpi.com/journal/fishes/special_issues/500Z3E5HT4).

For citation purposes, cite each article independently as indicated on the article page online and as indicated below:

Lastname, A.A.; Lastname, B.B. Article Title. *Journal Name* **Year**, *Volume Number*, Page Range.

ISBN 978-3-7258-2585-1 (Hbk)
ISBN 978-3-7258-2586-8 (PDF)
doi.org/10.3390/books978-3-7258-2586-8

© 2024 by the authors. Articles in this book are Open Access and distributed under the Creative Commons Attribution (CC BY) license. The book as a whole is distributed by MDPI under the terms and conditions of the Creative Commons Attribution-NonCommercial-NoDerivs (CC BY-NC-ND) license.

Contents

About the Editors

Wagner C. Valenti

Wagner C. Valenti, a fellow of the World Aquaculture Society, is now a Professor and Senior Research Scientist at the Aquaculture Center, São Paulo State University (UNESP), Brazil. He is also the Coordinator of the Brazilian Sustainable Aquaculture Network, Executive Editor of *Aquaculture Reports* (Elsevier), and Section Editor-in-Chief of the *Fishes* journal (MDPI). Wagner has served the aquaculture profession through teaching, R&D, and developing national policies over the past 40 years. He has created and directed new institutions and graduate programs, led research networks, held international positions, and published over 200 peer-reviewed articles, books, and book chapters on aquaculture and aquatic biology. He also holds patents and has founded technological startups. He has organized numerous national and international conferences and has been selected as a keynote or plenary speaker several times. His research work includes the assessment of sustainability, integrated aquaculture, nutrient budget, and bioeconomy.

Eduardo Luis Ballester

Eduardo Luis Ballester is a Professor and Research Scientist at the Federal University of Paraná (UFPR), Brazil. He also serves as the Director of the Research Center of Aquaculture and Sustainable Development—UFPR and the Head of the Prawn/Shrimp Culture Laboratory at UFPR. With around 25 years of experience in aquaculture research projects, he created the Aquaculture and Sustainable Development Post-Graduation Program at UFPR. He has published more than 100 peer-reviewed articles and book chapters on aquaculture and supervised over 100 undergraduate and graduate degree students studying aquaculture. His main current projects involve the use of essential oils as anesthetics and food additives for prawns and the use of protein hydrolysates as a source of protein for shrimp and prawns. He has also developed research projects with Biofloc Technology (BFT), shrimp and prawn nutrition, and the toxicity of nitrogenous compounds to prawns.

fishes

Editorial

Integrated Aquaculture and Monoculture of Low-Trophic Species

Wagner C. Valenti [1,*] and Eduardo Luis Ballester [2,*]

1 Aquaculture Center, São Paulo State University (UNESP), Jaboticabal 14884-700, SP, Brazil
2 Aquaculture and Sustainable Development Post-Graduation Program, Federal University of Paraná, Palotina 85950-000, PR, Brazil
* Correspondence: w.valenti@unesp.br (W.C.V.); elcballester@ufpr.br (E.L.B.)

Aquaculture is undoubtedly a critical sector for satisfying the needs of a growing human population and meeting the Sustainable Development Goals of Agenda 2030. Nevertheless, some current paradigms should be changed to match the sustainable production of aquatic organisms. To improve sustainability, the linear economy model should move to a circular economy, and low-trophic-level species should replace high-trophic-level species monoculture.

Combining species with complementary ecosystemic functions and requirements in the same culture allows for more efficient production systems. In this way, developing integrated multi-spatial, multi-niche, or multi-trophic aquaculture (IMTA) may be a solution. Innovative systems associating autotrophic organisms, microbes, and suspension and deposit feeders with manufactured-diet-fed species may improve the efficiency of using natural resources and the circularity. In integrated systems, the co-products, frequently classified as worthless wastes in monoculture, are used as supplies to produce other species instead of being discarded into the environment.

On the other hand, farming low-trophic species (LTS) in monoculture and integrated culture is more environmentally efficient than farming high-trophic-level ones. Low-trophic species in aquaculture are defined as the primary producers (algae and aquatic plants), primary consumers (herbivorous and detritus feeders), and other animals that feed mainly on plankton and small benthic invertebrates. Many detritivorous species use more nutrients from bacteria, fungi, algae, and invertebrates than from dead particulate organic matter.

This Special Issue introduces different insights into how aquaculture can benefit from integrated systems, as well as the culture of low-trophic species. Freshwater prawns and bivalve filtering mollusks have emerged as a great alternative to produce human food using natural or cheap feed sources. However, the availability of seeds may be a constraint. Vetorelli et al. (article 1) evaluated the intensification of hatchery systems for the Amazon river prawn, *Macrobrachium amazonicum*, which has shown great potential for monoculture or IMTA systems. The results showed that *M. amazonicum* tolerates high intensification in the recirculating hatchery, which can boost seed production. Sühnel et al. (article 2) evaluated different stocking densities and different algal diets in a fluidized bed bottle nursery system for the oysters *Crassostrea gasar* and *Crassostrea gigas*. The results demonstrated that the production costs and time to oyster growth can be improved in this type of system. In addition, regarding seed supplies, Crisóstomo et al. (article 3) evaluated the conditioning of *Argopecten purpuratus* broodstock; here, the seed quality indicators used by the authors were slightly lower compared to the broodstock kept in the natural environment. However, the study showed the importance of further optimizing the broodstock conditioning aspects, allowing more predictable and sustainable production to ensure permanent seed supplies.

The biofloc technology (BFT) used for LTS in both monoculture and integrated culture was primarily investigated. In their first paper, Carvalho et al. (article 4) explored how

Citation: Valenti, W.C.; Ballester, E.L. Integrated Aquaculture and Monoculture of Low-Trophic Species. *Fishes* **2024**, *9*, 450. https://doi.org/10.3390/fishes9110450

Received: 22 October 2024
Accepted: 25 October 2024
Published: 4 November 2024

Copyright: © 2024 by the authors. Licensee MDPI, Basel, Switzerland. This article is an open access article distributed under the terms and conditions of the Creative Commons Attribution (CC BY) license (https://creativecommons.org/licenses/by/4.0/).

the integration of different species in BFT can lead to more sustainable aquaculture; they evaluated the growth and nutrient absorption of the macroalgae *Ulva lactuca* when cultivated in an integrated system with *Penaeus vannamei* and *Oreochromis niloticus*. The results revealed an increase in macroalgae biomass, increased nitrate and phosphate removal, and superior water-quality parameter maintenance, demonstrating the viability of macroalgae cultivation integrated into BFT systems. These results were confirmed in a second study conducted by the same team (Carvalho et al., article 5), in which they determined that the concentrations of total suspended solids in an integrated culture of *U. lactuta* and *P. vannamei* did not affect the macroalgae growth or shrimp performance, showing that macroalgae may be used for nitrate absorption. In another study, Rind et al. (article 6) evaluated using different carbon sources in BFT for Nile tilapia and determined that tapioca flour improved the water quality, fish growth, the status of hematology, immunity, and the antioxidants in fish juveniles. In addition, in a BFT system, Costa et al. (article 7) demonstrated that the oyster *Crassostrea gasar* can act as a suspensivorous feeder; however, the suspended solids must be kept at concentrations below 200 mg/L to achieve the best results. All these studies demonstrated that even in intensive systems, such as BFT, combining species with different functions can bring benefits and improve production sustainability.

Accessing circular economy practices, Checa et al. (article 8) used bioremediation and efficiency indicators to evaluate the circularity performance of four IMTA trials in three aquaculture facilities established in Ireland, Brazil, and South Africa. Salmon, white shrimp, tilapia, abalone, and sea urchins were studied and cultivated together in various combinations with several low-trophic species in these IMTA trials to evaluate the improvement in circularity compared with the corresponding monoculture conditions. The results showed an increase in the circularity of up to 90% in terms of water recirculation, as well as bioremediation, which was improved by 80–90%, providing evidence for the potential role of IMTA in the circularity transition. To control biofouling in the cages of IMTA systems, Montgomery et al. (article 9) examined the use of California sea cucumbers (*Apostichopus californicus*) on cages containing adult Chinook salmon (*Oncorhynchus tshawytscha*) at a commercial farming operation. The results showed that the sea cucumbers actively fed on the biofouling but preferred to consume uneaten feed/feces at the bottom of the cages. Biofouling control in cages would likely be possible with a higher density of sea cucumbers. These results may contribute to developing a management framework for sea cucumber/salmon integrated multi-tropic aquaculture. Studying another bioremediator species in integrated systems, Hou et al. (article 10) investigated the effects of farming the snail *Bellamya purificata* at different stocking densities on the algal and fungal communities in sediment. The results showed that *B. purificata* at a low stocking density might enhance the resource utilization efficiency and minimize the environmental pollution.

Most of the freshwater fish farmed worldwide have a low-trophic status. Lima et al. (article 11) explored the use of fertilization to supplement the diet of tambaqui (*Colossoma macropomum*), the principal native fish farmed in South America. Their results showed that tambaqui can thrive by alternating between natural food sources and commercial feed without any growth change. This suggests that tambaqui is well-suited for farming in restorative and integrated aquaculture systems as well as in intensive systems that rely on commercial feed.

The culture of filtering mollusks is expanding in the West Hemisphere. Iitembu et al. (article 12) explored regional variations in oyster production techniques, market dynamics, and consumption patterns across seven Atlantic regions. The results showed that Crassostrea gigas is farmed in most Atlantic regions, except the USA, and that the financial challenges for small businesses, the ecological implications of seed production techniques, the biosecurity risks, and public health considerations are critical areas for attention. This study introduces valuable information for policymakers, aquaculture practitioners, and stakeholders in optimizing global shellfish industry strategies.

We are sure that the contents of this Special Issue represent an essential contribution to the culture of low-trophic species, mainly in integrated systems. This entails changes in the current paradigms, moving aquaculture toward a more conservative and restorative production process. In addition, it allows aquaculture development according to the circular economy model. In this way, aquaculture can contribute to reaching the Sustainable Development Goals of Agenda 2030.

Conflicts of Interest: The authors declare no conflicts of interest.

List of Contributions

1. Vetorelli, M.; Rodrigues, L.; Kimpara, J.; Valenti, W. Intensification of Amazon River Prawn Hatchery. *Fishes* **2024**, *9*, 82. https://doi.org/10.3390/fishes9030082.
2. Sühnel, S.; Lagreze-Squella, F.; Corrêa, G.; Araújo, J.; de Souza, G.; Ferreira, J.; da Silva, F.; de Miranda Gomes, C.; de Melo, C. Stocking Density and Diet of Two Oyster (*Crassostrea gasar* and *Crassostrea gigas*) Seeds in Fluidized Bed Bottle Nursery System. *Fishes* **2024**, *9*, 183. https://doi.org/10.3390/fishes9050183.
3. Crisóstomo, R.; Pepe-Victoriano, R.; Méndez-Ancca, S.; Zambrano-Cabanillas, A.; Marín-Machuca, O.; Perez, H.; Yana-Mamani, V.; Ruiz-Choque, M. Reproductive Conditioning of the Peruvian Scallop *Argopecten purpuratus* in Different Environments. *Fishes* **2024**, *9*, 9. https://doi.org/10.3390/fishes9010009.
4. Carvalho, A.; Brandão, H.; Zemor, J.; Cardozo, A.; Vieira, F.; Okamoto, M.; Turan, G.; Poersch, L. Effect of Organic or Inorganic Fertilization on Microbial Flocs Production in Integrated Cultivation of *Ulva lactuca* with *Oreochromis niloticus* and *Penaeus vannamei*. *Fishes* **2024**, *9*, 191. https://doi.org/10.3390/fishes9060191.
5. Carvalho, A.; Costa, L.; Holanda, M.; Poersch, L.; Turan, G. Influence of Total Suspended Solids on the Growth of the Sea Lettuce Ulva lactuca Integrated with the Pacific White Shrimp *Litopenaeus vannamei* in a Biofloc System. *Fishes* **2023**, *8*, 163. https://doi.org/10.3390/fishes8030163.
6. Rind, K.; Habib, S.; Ujan, J.; Fazio, F.; Naz, S.; Batool, A.; Ullah, M.; Attaullah, S.; Khayyam, K.; Khan, K. The Effects of Different Carbon Sources on Water Quality, Growth Performance, Hematology, Immune, and Antioxidant Status in Cultured Nile Tilapia with Biofloc Technology. *Fishes* **2023**, *8*, 512. https://doi.org/10.3390/fishes8100512.
7. Costa, L.; Carvalho, A.; Holanda, M.; Santos, J.; Borges, L.; Guterres, B.; Nam Junior, J.; Fonseca, V.; Muller, L.; Romano, L.; et al. Biological Responses of Oyster *Crassostrea gasar* Exposed to Different Concentrations of Biofloc. *Fishes* **2023**, *8*, 586. https://doi.org/10.3390/fishes8120586.
8. Checa, D.; Macey, B.; Bolton, J.; Brink-Hull, M.; O'Donohoe, P.; Cardozo, A.; Poersch, L.; Sánchez, I. Circularity Assessment in Aquaculture: The Case of Integrated Multi-Trophic Aquaculture (IMTA) Systems. *Fishes* **2024**, *9*, 165. https://doi.org/10.3390/fishes9050165.
9. Montgomery, E.; Cannon, B.; Pearce, C. Exploring Biofouling Control by the California Sea Cucumber (*Apostichopus californicus*) in Integrated Multi-Trophic Aquaculture (IMTA) with Organic Chinook Salmon (*Oncorhynchus tshawytscha*). *Fishes* **2023**, *8*, 430. https://doi.org/10.3390/fishes8090430.
10. Hou, Y.; Zhou, M.; Jia, R.; Sun, W.; Yang, Y.; Huang, X.; Li, B.; Zhu, J. Effects of Snail *Bellamya purificata* Farming at Different Stocking Densities on the Algal and Fungal Communities in Sediment. *Fishes* **2023**, *8*, 488. https://doi.org/10.3390/fishes8100488.
11. Lima, A.; dos Reis, A.; Costa, V.; Valenti, W. Natural Food Intake and Its Contribution to Tambaqui Growth in Fertilized and Unfertilized Ponds. *Fishes* 2024, *9*, 139. https://doi.org/10.3390/fishes9040139.
12. Iitembu, J.; Fitzgerald, D.; Altintzoglou, T.; Boudry, P.; Britz, P.; Byron, C.; Delago, D.; Girard, S.; Hannon, C.; Kafensztok, M.; et al. Comparative Description and Analysis of Oyster Aquaculture in Selected Atlantic Regions: Production, Market Dynamics, and Consumption Patterns. *Fishes* **2023**, *8*, 584. https://doi.org/10.3390/fishes8120584.

Disclaimer/Publisher's Note: The statements, opinions and data contained in all publications are solely those of the individual author(s) and contributor(s) and not of MDPI and/or the editor(s). MDPI and/or the editor(s) disclaim responsibility for any injury to people or property resulting from any ideas, methods, instructions or products referred to in the content.

fishes

Article

Intensification of Amazon River Prawn Hatchery

Michelle Pinheiro Vetorelli [1,*], Laurindo André Rodrigues [2], Janaina Mitsue Kimpara [3] and Wagner C. Valenti [4]

[1] Faculty of Agricultural Sciences, Federal University of Grande Dourados, Dourados 79804-970, MS, Brazil
[2] Embrapa Western Agriculture Research Center, Brazilian Agricultural Research Corporation, Dourados 79804-970, MS, Brazil; laurindo.rodrigues@embrapa.br
[3] Embrapa Digital Agriculture Research Center, Brazilian Agricultural Research Corporation, Campinas 13083-886, SP, Brazil; janaina.kimpara@embrapa.br
[4] Aquaculture Center, São Paulo State University, Jaboticabal 14884-900, SP, Brazil; wcvalenti@gmail.com
* Correspondence: michellevetorelli@ufgd.edu.br

Abstract: The effects of the intensification on the performance of the Amazon River prawn hatchery carried out in a simple recirculation system were investigated. Newly hatched larvae were stocked in 120 L tanks at 80, 100, 120 and 140 larvae L^{-1} in a closed recirculating system. The experiment used a randomized block design with five replicates. An exponential equation was adjusted to express the relationship between the stocking density and productivity (postlarvae L^{-1}). The development, larval quality, survival and postlarval (PL) dry weight did not significantly differ among the treatments ($p > 0.05$). When 80 larvae were stocked, the productivity (54 ± 11 PL L^{-1}) was lower than those at higher densities ($p < 0.05$). Stocking 120 and 140 larvae L^{-1} resulted in higher productivities (75 ± 18 and 80 ± 17 PL L^{-1}, respectively) with a lower use of *Artemia* nauplii to produce each postlarvae (~1200 *Artemia* nauplii PL^{-1}). The maximum mean *M. amazonicum* postlarval production estimated by the exponential model was 93 PL L^{-1}. This means that despite the increase in stocking density, productivity tends to stabilize. The results showed that *M. amazonicum* tolerates high intensification in recirculating hatchery systems based on a crushed shell bed biofilter, and the intensification optimizes *Artemia* use.

Keywords: stocking density; productivity; recirculation system; larviculture; *Macrobrachium amazonicum*; *Artemia* nauplii

Key Contribution: *M. amazonicum* postlarvae can be produced in simple RAS systems stocked at 100 to 140 larvae L^{-1}, and productivity may reach around 93 PL L^{-1} in about 20 days. As the culture intensifies, the use of *Artemia* nauplii per production unity is reduced by ~20%.

Citation: Vetorelli, M.P.; Rodrigues, L.A.; Kimpara, J.M.; Valenti, W.C. Intensification of Amazon River Prawn Hatchery. *Fishes* **2024**, *9*, 82. https://doi.org/10.3390/fishes9030082

Academic Editor: Domitília Matias

Received: 8 February 2024
Revised: 15 February 2024
Accepted: 20 February 2024
Published: 22 February 2024

Copyright: © 2024 by the authors. Licensee MDPI, Basel, Switzerland. This article is an open access article distributed under the terms and conditions of the Creative Commons Attribution (CC BY) license (https://creativecommons.org/licenses/by/4.0/).

1. Introduction

Aquaculture may play an essential role in reaching the Sustainable Development Goals defined in Agenda 2030 [1]. Thus, more sustainable aquaculture systems should be implemented as a form of nature-positive food production that serves the people and the planet. The use of native low trophic species (LTSs) and integrated multitrophic aquaculture (IMTA) systems has been considered more environmentally sustainable than the culture of exotic or high trophic level species and monocultures [2]. Native species showed a lower impact on surrounding biodiversity; LTSs generally fed on detritus and small natural biota, recovering organic matter to trophic webs; and IMTAs use the wastes of one species to feed others, according to the principles of the circular economy. The Amazon River prawn, *Macrobrachium amazonicum*, is a LTS widely distributed in rivers and lakes of South America [3–5]. This species has been primarily exploited by artisanal fisheries [6,7] and has great potential for use in aquaculture [8,9]. The Amazon River prawn has been demonstrated to adapt very well to IMTA systems [10–12]. There is a large local market for *M. amazonicum*, mainly in the Amazon and northeastern Brazil [3,6].

An essential constraint to producing native LTS organisms is the lack of seed availability [13]. Therefore, developing hatchery technology to provide the postlarvae of *M. amazonicum* for sustainable grow-out farms is relevant. Prawn larviculture is generally performed in recirculating aquaculture systems (RASs) [8]. These systems are more conservative of water and heat, generate fewer effluents than flow-through systems and are less exposed to climate changes. However, they are more expensive to set up and manage than most aquaculture systems.

An RAS is an intensive aquaculture that uses about 1 to 10% of the water used in conventional aquaculture systems and allows for the total control of water variables and effluents [14]. Generally, intensification has been associated with unsustainable systems; however, intensification may be a way to save financial and natural resources and decrease the harmful effluents per unit of product. Additionally, job positions may be created if planned accordingly. High productivity is essential to reduce unit production costs. The challenge is finding the intensification level that maximizes productivity and increases job positions while minimizing costs and the environmental impact.

Increasing intensification leads to an increasing stocking density. It depends on the carrying capacity of the system [15] and the intrinsic characteristics of the cultured species. High densities may increase the levels of ammonia and intraspecific competition, which may result in low animal welfare, poor growth, survival and production. Low stocking densities may result in lower productivity. Therefore, it is essential to know the effect of the levels of intensification on the water quality and prawn development to define the best stocking densities. Papers that have focused on the stocking densities of the freshwater prawn *M. rosenbergii* in the larviculture phase were found in the literature [16–18]. However, for *M. amazonicum*, intensification has only been studied in the nursery [19–21] and grow-out [6] phases. Considering the above rationale, the objective of this paper was to evaluate the effect of the intensification of an *M. amazonicum* hatchery performed using simple and cheap recirculating aquaculture systems.

2. Materials and Methods

2.1. Experimental Conditions

Intensification was evaluated based on the increase in the larval stocking density in the rearing tanks. This experiment was set up according to a randomized block design with four treatments (stocking densities) and five replicates. The tested stocking densities were 80, 100, 120 and 140 larvae L^{-1}. The evaluated variables were the larval development, larval quality, *Artemia* nauplii utilization, survival, metamorphosis rate, productivity, dry weight and average maximum productivity.

The *Macrobrachium amazonicum* larvae were obtained from females maintained in earthen ponds in a semi-intensive system. This stock was generated by animals from a diadromous population captured in Pará State (01°13′25″ S, 48°17′40″ W) in 2001, and it has been maintained for research purposes since then. Females with transparent eggs were collected and placed into larval hatching tanks (70 female m^{-2}). The water was maintained at a salinity of ~5, a temperature of ~29 °C and with constant aeration.

After hatching, all the larvae were counted and transferred to rearing tanks at densities of 80, 100, 120 and 140 larvae L^{-1}. Larvae were reared in 120 L cylindrical tanks with conical bottoms and under a closed recirculation aquaculture system (RAS) [8]. Crushed shell bed biofilters with 30 L (25% of that of the rearing tanks), provided with a heater and intense aeration were used for each tank. The water was moved via an air-lift pump and the recirculation rate was about 24 times per day. Larvae were fed beginning on the 2nd day with newly hatched *Artemia* nauplii supplied "ad libitum" in the afternoon (17:00 h). A moist inert diet (egg-custard-based; see Mallasen and Valenti [22] for composition) was supplied from the 9th day after stocking twice a day (at 8:00 h and 11:00 h). The feeding rate was adjusted daily and corresponded to consumption. The temperature (°C), water recirculation rate (% day^{-1}), total ammonia nitrogen (TAN, mg L^{-1}) and nitrite concentration (N-NO_2 mg L^{-1}) were monitored daily. The dissolved oxygen (DO, mg L^{-1}) and DO saturation (%), salinity and pH were monitored weekly. Nitrogen compounds were analyzed using colorimetric tests. The dissolved oxygen was determined using a YSI Model 55 oxygen meter (Yellow Springs Instruments Co., Inc., Yellow Springs, OH, USA), and the salinity and pH were measured using a YSI Model 63 digital pH meter (Yellow Springs Instruments Co., Inc. Yellow Springs, OH, USA). The mean and standard deviation of the variables for each treatment are presented in Table 1. The photoperiods were kept constant at 12:12 h (light:dark) with ~1000 lux.

Table 1. Water data (means ± standard deviations) obtained during *Macrobrachium amazonicum* larviculture at different stocking density treatments. The means did not significantly differ ($p > 0.05$). TAN = total ammonia nitrogen; DO = dissolved oxygen.

Variables	Stocking Density (Larvae L^{-1})			
	80	100	120	140
Temperature (°C)	29.8 ± 0.1	29.8 ± 0.1	29.9 ± 0.1	29.9 ± 0.1
Recirculation rate (% day^{-1})	22.4 ± 0.4	25.0 ± 6.0	21.5 ± 2.9	22.5 ± 3.5
TAN (mg L^{-1})	0.23 ± 0.02	0.23 ± 0.02	0.24 ± 0.07	0.25 ± 0.05
N-NO_2 (mg L^{-1})	0.05 ± 0.01	0.05 ± 0.01	0.05 ± 0.03	0.04 ± 0.01
DO (mg L^{-1})	7.84 ± 0.17	8.08 ± 0.15	7.83 ± 0.60	7.84 ± 0.13
DO Saturation (%)	103.3 ± 4.0	106.3 ± 1.1	103.1 ± 4.6	106.8 ± 4.4
Salinity	10.2 ± 0.4	10.4 ± 0.5	10.1 ± 0.2	9.9 ± 0.4
pH	7.71 ± 0.56	7.71 ± 0.58	7.76 ± 0.56	7.75 ± 0.53

2.2. Larval Stage Index and Larval Condition Index

Samples containing 10 larvae were taken from each tank, and analyses of the larval stage index (LSI) and larval condition index (LCI) were performed using stereomicroscopy every two days. Larval stages were identified according to Guest [23]. The LSI was determined using the weighted average method described by Manzi et al. [24]:

$$LSI = \left(\sum Si \times ni\right)N^{-1}, \quad (1)$$

in which: Si is the larvae PL^{-1} stage (I = 1–10), ni = number of animals in stage Si, and N = total number of animals observed.

The LCI was determined following the criteria developed by Tayamen and Brown [25] for *M. rosenbergii* adapted to *M. amazonicum*. The criteria for determining the condition index for evaluating larval quality were the gut fullness, gut lipid content, pigmentation, body coloration, setation, muscle-to-gut ratio, abdominal muscle appearance, melanization, fouling organisms, the photopositive response between stage I and V, and, between stage VI and IX, swimming behavior was added. Each criterion was given a score, where 0 = poor, 1 = fair, and 2 = excellent.

2.3. Use of *Artemia nauplii*

The feed ratios are expressed as the *Artemia* nauplii density in the rearing water (nauplii mL^{-1} day^{-1}). The estimated use of nauplii mL^{-1} was obtained daily. Feeding was monitored by estimating the concentration of nauplii mL^{-1} using a 5 mL pipette (mean of five replicates). After 24 h, another estimation was conducted to quantify the amount of remaining nauplii inside the tank. The daily *Artemia* consumption, in nauplii mL^{-1} concentration, was obtained by subtracting both these values for each treatment. The mean value of the five samples was then multiplied by the volume of water in the tank to estimate the total number of nauplii in the tank.

As we had a single cohort inside the tanks, no mass mortality was observed during the experiment, and survival was high, it is reasonable to suppose that the mortality rate was almost constant [26]. Thus, we can compute the instantaneous mortality rate (m), using the number of prawns stocked (N_0) and harvested (N_T), and the culture duration (T, in days) as Equation (2):

$$m = ln(N_T/N_0)/T, \quad (2)$$

Then, based on Wineliller and Dailey [27], we determined the populational decline curve for each tank as Equation (3):

$$N_t = N_0 e^{-mt}, \quad (3)$$

in which: N_t = number of prawns inside the tank at time t (in days). This equation was determined for each replicate and used to estimate the daily larvae quantity inside the tanks.

The individual use of prey (nauplii larval day) was obtained by dividing the total number of nauplii consumed per day by the number of larvae in the tank on that day. The number of *Artemia* nauplii per postlarvae (PL) was obtained by dividing the total number of nauplii used per tank during the entire culture by the number of PLs produced.

2.4. Survival, Metamorphosis Rate, Productivity and Dry Weight

The end of the experiment occurred 21, 19, 20, 20 and 20 days after stocking in blocks I, II, III, IV and V, respectively. All larvae and postlarvae (PL) were collected and counted individually. The variables determined were the survival of the larvae and postlarvae (%), metamorphosis rate in postlarvae (% PL), productivity in postlarvae (PL L^{-1}) and postlarvae dry weight (mg).

To determine the postlarvae dry weight, PL samples from all replicates in each treatment were taken, rinsed in distilled water, dried on filter paper and transferred to aluminum cartridges at predetermined weights. The cartridges containing the PL were dried (60 °C) for 24 h and kept inside the desiccator for at least two hours. Then, the cartridges were weighed on an analytical scale (Mettler Toledo AT21, accuracy 1 µg). Ten replicates for each tank were weighted.

To express the relationship between the stocking density and productivity, an exponential Equation (4) was adjusted:

$$P = P_{max}\left[1 - e^{-k(D-D_0)}\right], \quad (4)$$

where P = the productivity of postlarvae L^{-1}, P_{max} = the maximum average productivity of postlarvae L^{-1} that can be obtained in this system, e = base of natural logarithms, D = the stocking density, K and D_0 = constants.

2.5. Statistical Analyses

The larval stage index, larval condition index, nauplii *Artemia* used by tank (nauplii mL^{-1} day^{-1}), nauplii *Artemia* used by larvae and postlarvae (nauplii $larvae^{-1}$ and nauplii $postlarvae^{-1}$), survival, metamorphosis rate, productivity (PL L^{-1}) and weight were expressed as the mean ± standard deviation. Percentage data were normalized via arcsine transformation before statistical analyses. All data were subjected to normality (Cramer–von Mises) and variance (Levene's homoscedasticity test) tests. As no deviations were observed, data were subjected to an analysis of variance (one-way ANOVA). When significant differences ($p < 0.05$) were detected, treatment means were compared by the Tukey–Kramer test. All the statistical analyses were performed based on Sheskin [28], using a Statistical Analysis System (SAS Institute, Inc., Cary, NC, USA, version 9.0). The regression between stocking density and postlarvae productivity was determined using Excel® Version 2401/2023 (Microsoft Corporation, Redmond, WA, USA).

3. Results

The intensification of the *M. amazonicum* hatchery did not affect the larval development or quality ($p > 0.05$). The LSI did not differ among the treatments with different stocking densities during the rearing cycle (Table 2). The larval condition index (LCI) showed random variation throughout the rearing cycle above 1.6 and did not differ among stocking densities. The mean LCIs were 1.70 ± 0.12, 1.71 ± 0.09, 1.72 ± 0.09 and 1.70 ± 0.09 for the 80, 100, 120 and 140 larvae L^{-1} treatments, respectively.

Table 2. Larval stage indices (means ± standard deviations) for larviculture of *M. amazonicum* at different stocking densities. CV = coefficient of variation.

Rearing Time (Days)	Stocking Density (Larvae L^{-1})				*F*-Value	*p*-Value	CV (%)
	80	100	120	140			
2	1.1 ± 0.1	1.4 ± 0.4	1.3 ± 0.5	1.1 ± 0.0	0.54	0.66	27.5
4	2.8 ± 0.2	2.8 ± 0.5	2.7 ± 0.1	2.7 ± 0.5	0.04	0.98	14.2
6	4.5 ± 0.3	4.4 ± 0.4	4.7 ± 0.2	4.5 ± 0.2	0.64	0.60	6.2
8	5.5 ± 0.3	5.2 ± 0.2	5.4 ± 0.4	5.5 ± 0.3	0.86	0.49	6.0
10	6.3 ± 1.0	6.5 ± 0.6	6.8 ± 0.4	6.6 ± 0.2	0.40	0.75	9.6
12	7.9 ± 0.3	7.6 ± 0.2	8.0 ± 0.3	7.8 ± 0.5	0.76	0.54	4.7
14	8.4 ± 0.2	8.3 ± 0.3	8.3 ± 0.4	8.3 ± 0.2	0.18	0.91	3.35
16	8.6 ± 0.3	8.6 ± 0.1	8.7 ± 0.4	8.5 ± 0.3	0.28	0.83	3.8
18	8.7 ± 0.2	9.0 ± 0.1	8.6 ± 0.2	8.6 ± 0.5	0.52	0.68	2.8

The average daily *Artemia* use ranged from 2 to 9 nauplii mL^{-1} and did not differ ($p > 0.05$) among the treatments. However, the amount of *Artemia* used per larvae was lower at high densities ($p < 0.05$), and the *Artemia* used per postlarvae produced was greater at a density of 80 larvae L^{-1} ($p < 0.05$) (Table 3). Survival and metamorphosis rates were not affected by culture intensification (Table 4). The productivity (PL L^{-1}) was lower for the 80 larvae L^{-1} density than for the other stocking densities ($p < 0.05$) (Table 4). The postlarval dry weight was not significantly different among the tested stocking densities ($p > 0.05$) (Table 4). The productivity of postlarvae L^{-1} curve as a function of the stocking density showed that the average maximum productivity reached in this system was 93 PL L^{-1} (Figure 1).

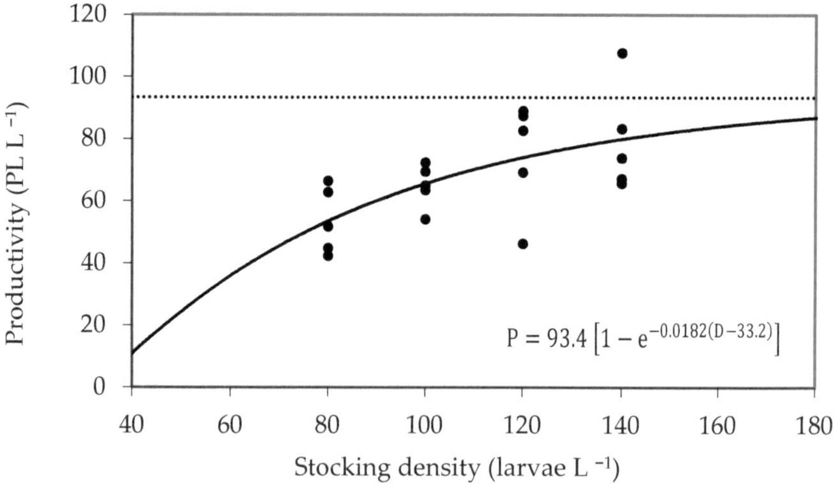

Figure 1. Relationships between stocking density and productivity of *Macrobrachium amazonicum* hatchery in recirculating aquaculture system. PL = postlarvae. The dotted line represents the asymptotic maximum productivity, which is estimated by the exponential model.

Table 3. Means ± standard deviations of the daily consumption of *Artemia*, measured in nauplii per milliliter (mL) of water and in nauplii per larvae, and the total number of nauplii ingested per larvae during culture to reach the postlarvae (PL) stage.

Rearing Time (Days)	Stocking Densities (Larvae L^{-1})							
	80		100		120		140	
	Nauplii mL^{-1}	Nauplii $Larvae^{-1}$	Nauplii mL^{-1}	Nauplii $Larvae^{-1}$	Nauplii mL^{-1}	Nauplii $Larvae^{-1}$	Nauplii mL^{-1}	Nauplii $Larvae^{-1}$
2	1.6 ± 2.3	20 ± 29	1.3 ± 1.1	14 ± 11	1.9 ± 2.1	16 ± 19	3.0 ± 1.0	22 ± 7
3	3.6 ± 0.7	46 ± 14	1.8 ± 2.7	19 ± 29	3.7 ± 1.1	32 ± 10	4.3 ± 1.8	32 ± 14
4	4.5 ± 1.3	59 ± 26	4.6 ± 1.0	49 ± 11	5.0 ± 1.2	45 ± 11	4.2 ± 1.2	32 ± 9
5	5.0 ± 1.5	66 ± 15	5.1 ± 1.5	55 ± 16	5.7 ± 1.8	52 ± 15	5.5 ± 1.3	42 ± 10
6	5.1 ± 0.8	68 ± 8	5.0 ± 1.8	55 ± 20	5.5 ± 0.8	52 ± 12	5.7 ± 1.7	45 ± 13
7	5.9 ±1.0	79 ± 13	5.1 ± 1.1	57 ± 12	6.4 ± 1.7	61 ± 18	5.8 ± 1.6	47 ± 14
8	4.9 ± 1.9	67 ± 27	5.5 ± 0.6	63 ± 6	6.0 ± 2.7	57 ± 24	5.2 ± 0.9	42 ± 8
9	5.4 ± 2.1	74 ± 23	4.3 ± 1.9	49 ± 21	6.0 ± 1.5	58 ± 10	6.1 ± 0.9	50 ± 7
10	5.7 ± 2.9	80 ± 31	6.0 ± 1.6	71 ± 17	6.1 ± 0.9	61 ± 7	6.9 ± 1.9	57 ± 16
11	6.3 ± 4.5	91 ± 53	6.4 ± 1.3	77 ± 18	5.6 ± 1.2	57 ± 12	5.2 ± 1.1	44 ± 9
12	5.1 ± 2.3	75 ± 59	5.1 ± 2.3	62 ± 29	4.8 ± 1.1	51 ± 16	5.5 ± 0.7	47 ± 7
13	3.9 ± 0.8	56 ± 16	5.9 ± 3.1	74 ± 41	3.9 ± 1.8	42 ± 20	6.6 ± 2.3	57 ± 20
14	4.3 ± 1.9	63 ± 23	5.0 ± 1.3	63 ± 18	5.3 ± 2.4	60 ± 36	6.1 ± 2.2	54 ± 19
15	5.3 ± 2.0	79 ± 27	5.0 ± 1.4	65 ± 19	5.4 ± 2.5	59 ± 27	6.8 ± 1.9	60 ± 15
16	7.3 ± 5.0	110 ± 59	6.5 ± 1.6	85 ± 21	7.4 ± 2.5	80 ± 21	7.3 ± 3.0	67 ± 29
17	7.2 ± 1.6	108 ± 25	7.7 ± 1.6	101 ± 17	7.6 ± 1.6	88 ± 30	8.7 ± 1.9	80 ± 15
18	6.5 ± 2.0		9.4 ± 2.5		8.6 ± 2.7		6.1 ± 0.6	
Nauplii PL^{-1}	1559 ± 497 a		1237 ± 185 b		1230 ± 423 b		1206 ± 261 b	

Letters indicate significant differences ($p < 0.05$) among treatments in the same line by an ANOVA followed by the Tukey's test.

Table 4. Variables of production (mean ± standard deviations) obtained in the larviculture of *M. amazonicum* at different stocking densities. PL = postlarvae.

Variables	Stocking Densities (Larvae L^{-1})			
	80	100	120	140
Survival (%)	80 ± 5	72 ± 6	72 ± 15	75 ± 5
Metamorphosis rate (%)	67 ± 13	65 ± 7	62 ± 15	57 ± 12
Productivity (PL L^{-1})	54 ± 11 a	65 ± 7 ab	75 ± 18 b	80 ± 17 b
PL dry weight (mg)	1.29 ± 0.10	1.21 ± 0.20	1.15 ± 0.20	1.23 ± 0.30

Letters indicate significant differences ($p < 0.05$) among treatments in the same line by an ANOVA followed by the Tukey's test.

4. Discussion

Intensifying the *M. amazonicum* hatchery up to 140 larvae L^{-1} in a simple RAS did not affect the water quality or larval development, well-being, growth or survival. On the other hand, the intensification increased productivity, while decreasing the quantity of *Artemia* used to produce each postlarvae. Consequently, it improved the efficiency of the rearing system. Producing more PL per unit of water, using the same infrastructure may reduce production costs and increase profitability and sustainability because it optimizes the use of resources.

Water variables were kept within the range recommended for *M. amazonicum* larvae at all stocking densities [3,6,29]. Therefore, intensifying the *M. amazonicum* hatchery up to 140 larvae L^{-1} does not interfere with the water quality in an RAS comprised of crushed shell bed biofilters, dimensioned at 25% of the total rearing tank volume. This simple system was effective in converting ammonia and nitrite to nitrate and maintaining the temperature, dissolved oxygen, salinity and pH suitable for the culture of *M. amazonicum*.

Larvae presented similar development indices when cultured from 80 to 140 individuals L^{-1} throughout the rearing cycle. In addition, the media larval condition index obtained in this study was greater than 1.7 at all stocking densities. Therefore, larvae reared at stocking densities up to 140 larvae L^{-1} exhibit suitable development and satisfactory larval conditions, suggesting that larvae welfare was effective in accordance with McKay et al., 2023 [30].

The survival and metamorphosis rates were similar for stocking up to 140 larvae L^{-1} Barreto and Soares [31] reported that no correlation was observed between the stocking density and survival or metamorphosis when *M. amazonicum* is stocked from 10 to 75 larvae L^{-1} However, a negative correlation between the larvae stocking density and survival was found for the prawn *M. rosenbergii* [16,17,32] and for fish larvae production systems [33] The authors attribute the survival reduction at higher stocking densities to cannibalism In other studies, *M. rosenbergii* larvae cannibalism was related to the stocking density and feeding regime [16,34,35]. Authors recommend that the increase in stocking density must be concomitant with an increase in feeding supply. Nhan et al. [16] suggest that increasing the stocking density to 200 larvae per liter and the feeding frequency to three times a day led to the highest production efficiency of *M. rosenbergii*. Following this management, productivity was 48 PL L^{-1}, and the consumption of *Artemia* per postlarvae produced was 7100 nauplii. David et al. [17] recommend stocking hatchery tanks with 80 to 100 larvae L^{-1} to ensure an optimal production of about 50 PL mL^{-1}. Therefore, the performance of *M. amazonicum* obtained in the present study was higher than that obtained for *M. rosenbergii* in previous studies.

In the present study, larvae were fed "ad libitum"; therefore, the feed was not limiting. There was no significant difference in the consumption of *Artemia*, measured as nauplii mL^{-1}, among the different stocking densities throughout the rearing cycle. However, nauplii consumption per larvae to reach the stage of postlarvae was higher at 80 larvae L^{-1} than in higher densities. Thus, larvae stocked at higher densities have a lower predator–prey relationships. According to Barros and Valenti [36], the predator–prey relationship influences consumption because it is associated with the highest number of encounter opportunities. These findings were consistent with those of Maciel et al. [26] for *M. amazonicum*, Gomes et al. [37] for *Macrobrachium equidens* and David et al. [17] for *M. rosenbergii*, respectively.

The highest feeding of *Artemia* at the lowest density did not improve the larval development, survival, metamorphosis rate or final dry weight. Therefore, the efficiency of feed use increased at densities from 100 larvae L^{-1} on. This suggests the occurrence of superfluous feeding in *M. amazonicum* larvae similar to *M. rosenbergii* (see David et al. [17]). The intensification of the *M. amazonicum* hatchery optimized the *Artemia* cyst use by ~20%. This may represent a substantial expense reduction because *Artemia* is up to 24% of variable costs in freshwater prawn hatcheries [18].

Agonistic behavior and cannibalism were not quantified, but they were observed especially at high densities (120 and 140 larvae L^{-1}) during the last days of culture. This is a common behavior in *Macrobrachium* species, as evidenced by Coyle [16,38], and may have occurred due to the competition for space and a lack of shelters. However, the increase in cannibalism was not enough to decrease survival at the highest stocking densities. Cannibalism in a *Macrobrachium amazonicum* hatchery was also reported by Araujo and Valenti [39] at a density of 80 larvae L^{-1}. Therefore, this behavior does not seem to depend on the farming density.

The maximum mean productivity found in this study was 80 PL L^{-1} when the stocking density was 140 larvae L^{-1}. Experimental studies on *M. amazonicum* hatcheries showed mean productivities of 70–75 PL L^{-1} [26], 64 PL L^{-1} [40] and 59 PL L^{-1} [41], for a stocking density of 80 larvae L^{-1} cultured during ~20 days, like in the present study. The most farmed freshwater prawn in the world is the *Macrobrachium rosenbergii*. New ref. [42] reported that commercial hatcheries of this species, operating in RASs, showed a mean productivity of 50 PL L^{-1}. Nhan et al. [16] observed a productivity of 48 PL L^{-1}, stocking

200 larvae L^{-1}. David et al. [17] stocked 80 to 140 larvae L^{-1} of *M. rosenbergii* and obtained a productivity from 42 to 52 PL L^{-1}. The cycle of larviculture is about 22 to 35 days [17,42]. These data show that an *M. amazonicum* hatchery operating in an RAS may be more productive than the hatcheries of *M. rosenbergii*, which have the solid hatchery technologies used in different countries [8].

The model adjusted showed that the potential maximum mean productivity in the studied system was approximately 93 PL L^{-1}. Productivity stabilizes around this value even with increasing stocking density. A similar study conducted with *M. rosenbergii* showed the maximum theoretical mean productivity of 51 PL L^{-1} [17]. Changes in the production system design or feeding regime may alter the carrying capacity of the system and therefore affect the potential maximum productivity. Nevertheless, no limitation from the feed or water quality was observed in the maximum stocking density tested in the present study as well as in the study of David et al. [17], which was 140 larvae L^{-1}. Therefore, the results suggest that these limits are due to the intrinsic characteristics of the species and probably that the space is the principal limiting factor.

5. Conclusions

In conclusion, an *M. amazonicum* hatchery may be intensified by at least 140 larvae L^{-1} using simple RASs. Productivity may reach around 93 PL L^{-1} in about 20 days of culture. The use of *Artemia* nauplii per PL produced is reduced by ~20% as the culture is intensified. Intensification may increase the profitability and sustainability of the system because more PL is produced using the same quantity of water, space, energy and *Artemia*. These experimental results should be confirmed in large commercial tanks. Sustainability and economic studies should be performed to determine the best level of intensification for a hatchery of *M. amazonicum*.

6. Patents

This study partially supports the *Macrobrachium amazonicum* hatchery technology patented on the Brazilian patent basis #BR 10 2019 027248 1; 19 December 2019.

Author Contributions: Conceptualization, M.P.V. and W.C.V.; methodology, M.P.V.; software, M.P.V.; validation, M.P.V., J.M.K., L.A.R. and W.C.V.; formal analysis, M.P.V.; investigation, M.P.V. and L.A.R.; resources, W.C.V.; data curation, M.P.V.; writing—original draft preparation, M.P.V.; writing—review and editing, L.A.R., J.M.K. and W.C.V.; visualization, J.M.K. and L.A.R.; supervision, W.C.V.; project administration, W.C.V.; funding acquisition, W.C.V. All authors have read and agreed to the published version of the manuscript.

Funding: This study was financed in part by the Coordination of Superior Level Staff Improvement—CAPES Foundation, code 001 and in part by the São Paulo Research Foundation (FAPESP), code 03/12570-4.

Institutional Review Board Statement: The National Council for the Control of Animal Experimentation (Concea) regulates the Production, Maintenance, or Use of Animals belonging to the phylum Chordata, subphylum Vertebrata, for Teaching or Scientific Research Activities in Brazil. The institutional animal ethics committee accredited by CONCEA therefore exempts the approval of lower-order animals (invertebrates).

Data Availability Statement: Data are contained within the article.

Conflicts of Interest: Laurindo André Rodrigues and Janaina Mitsue Kimpara are employed at the Brazilian Agricultural Research Corporation (Embrapa). The authors declare that the conflict of interest did not influence the content and results of the study. The other authors declare no conflicts of interest.

References

1. Troell, M.; Costa-Pierce, B.; Stead, S.; Cottrell, R.S.; Brugere, C.; Farmery, A.K.; Little, D.C.; Strand, Å.; Pullin, R.; Soto, D.; et al. Perspectives on aquaculture's contribution to the improved human and planetary health. *J. World Aquac. Soc.* **2023**, *54*, 251–342. [CrossRef]
2. Boyd, C.E.; D'Abramo, L.R.; Glencross, B.D.; Huyben, D.C.; Juarez, L.M.; Lockwood, G.S.; Mcnevin, A.A.; Tacon, A.G.J.; Teletchea, F.; Tomasso, J.R.; et al. Achieving sustainable aquaculture: Historical and current perspectives and future needs and challenges. *J. World Aquac. Soc.* **2020**, *51*, 578–633. [CrossRef]
3. Maciel, C.R.; Valenti, W.C. Biology, fisheries and aquaculture of the Amazon River Prawn *Macrobrachium amazonicum*: A review. *Nauplius* **2009**, *17*, 61–79.
4. Da Costa, T.V.; de Mattos, L.A.; Machado, N.d.J.B. Estrutura populacional de *Macrobrachium amazonicum* em dois lagos de várzea da Amazônia. *Bol. Inst. Pesca* **2016**, *42*, 281–293. [CrossRef]
5. Costa e Silva, R.; Cunha, M.C.; Mossolin, E.C.; Jacobucci, G.B. Population structure of *Macrobrachium amazonicum* (Heller, 1862) (Decapoda: Palaemonidae) in Miranda Hydroelectric Plant Reservoir, Araguari river, Minas Gerais, Brazil. *Acta Limnol. Bras.* **2019**, *31*, e14. [CrossRef]
6. Moraes-Valenti, P.; Valenti, W.C. Culture of the Amazon River Prawn *Macrobrachium amazonicum*. In *Freshwater Prawns: Biology and Farming*; New, M.B., Valenti, W.C., Tidwell, J.H., D'abramo, L.R., Kutty, M.N., Eds.; Wiley-Blackwell: Oxford, UK, 2010; pp. 485–501.
7. Taddei, F.G.; Reis, S.D.S.; David, F.S.; Silva, T.E.D.; Fransozo, V.; Fransozo, A. Population structure, mortality, and recruitment of *Macrobrachium amazonicum* (Heller, 1862) (Caridea: Palaemonidae) in the eastern Amazon region, Brazil. *J. Crust. Biol.* **2017**, *37*, 131–141. [CrossRef]
8. Valenti, W.C.; Daniels, W.H.; New, M.B.; Correia, E. Hatchery systems and management. In *Freshwater Prawns: Biology and Farming*; New, M.B., Valenti, W.C., Tidwell, J.H., D'Abramo, L.R., Kutty, M.N., Eds.; Wiley-Blackwell: Oxford, UK, 2010; pp. 55–85.
9. Marques, H.L.A.; Barros, H.P.B.; Mallasen, M.; Boock, M.V.; Moraes-Valenti, P.M.C. Influence of stocking densities in the nursery phase on the growth of *Macrobrachium amazonicum* reared in net pens. *Aquaculture* **2012**, *358–359*, 240–245. [CrossRef]
10. Rodrigues, C.G.; Garcia, B.F.; Verdegem, M.; Santos, M.R.; Amorim, R.V.; Valenti, W.C. Integrated culture of Nile tilapia and Amazon river prawn in stagnant ponds, using nutrient-rich water and substrates. *Aquaculture* **2019**, *503*, 111–117. [CrossRef]
11. Dantas, D.P.; Flickinger, D.L.; Costa, G.A.; Batlouni, S.R.; Moraes-Valenti, P.; Valenti, W.C. Technical feasibility of integrating Amazon river prawn culture during the first phase of tambaqui grow-out in stagnant ponds, using nutrient-rich water. *Aquaculture* **2020**, *516*, 734611. [CrossRef]
12. Marques, A.M.; Boaratti, A.Z.; Belmudes, D.; Ferreira, J.R.C.; Mantoan, P.V.L.; Moraes-Valenti, P.; Valenti, W.C. Improving the Efficiency of Lambari Production and Diet Assimilation Using Integrated Aquaculture with Benthic Species. *Sustainability* **2021**, *13*, 10196. [CrossRef]
13. Iitembu, J.A.; Fitzgerald, D.; Altintzoglou, T.; Boudry, P.; Britz, P.; Byron, C.J.; Delago, D.; Girard, S.; Hannon, C.; Kafensztok, M.; et al. Comparative Description and Analysis of Oyster Aquaculture in Selected Atlantic Regions: Production, Market Dynamics, and Consumption Patterns. *Fishes* **2023**, *8*, 584. [CrossRef]
14. Ebeling, J.M.; Timmons, M.B. Recirculating Aquaculture Systems. In *Aquaculture Production Systems*; Tidwell, J.H., Ed.; John Wiley & Sons, Inc.: Hoboken, NJ, USA, 2012; pp. 245–277.
15. Tierney, T.W.; Fleckenstein, L.J.; Ray, A.J. The effects of density and artificial substrate on intensive shrimp *Litopenaeus vannamei* nursery production. *Aquac. Eng.* **2020**, *89*, 102063. [CrossRef]
16. Nhan, D.T.; Wille, M.; Hung, L.T.; Sorgeloos, P. Effects of larval stocking density and feeding regime on larval rearing of giant freshwater prawn (*Macrobrachium rosenbergii*). *Aquaculture* **2010**, *300*, 80–86. [CrossRef]
17. David, F.S.; Cohen, F.P.A.; Valenti, W.C. Intensification of the Giant river prawn *Macrobrachium rosenbergii* hatchery production. *Aquac. Res.* **2016**, *47*, 3747–3752. [CrossRef]
18. David, F.S.; Fonseca, T.; Bueno, G.W.; Valenti, W.C. Economic feasibility of intensification of *Macrobrachium rosenbergii* hatchery. *Aquac. Res.* **2018**, *49*, 3769–3776. [CrossRef]
19. Marques, H.L.; New, M.B.; Boock, M.V.; Barros, H.P.; Mallasen, M.; Valenti, W.C. Integrated freshwater prawn farming: State-of-the-art and future potential. *Rev. Fish. Sci. Aquac.* **2016**, *24*, 264–293. [CrossRef]
20. Medeiros, M.V.; Aubin, J.; Camargo, A.F. Life cycle assessment of fish and prawn production: Comparison of monoculture and polyculture freshwater systems in Brazil. *J. Clean. Prod.* **2017**, *156*, 528–537. [CrossRef]
21. Peña-Herrejón, G.A.; Sánchez-Velázquez, J.; García-Trejo, J.F.; Soto-Zarazúa, G.M.; Rico-García, E. Effect of stocking density on growth and survival of the prawn *Macrobrachium tenellum*, cultured in a recirculating aquaculture system. *Lat. Am. J. Aquat. Res.* **2019**, *47*, 342–348. [CrossRef]
22. Mallasen, M.; Valenti, W.C. Comparison of artificial and natural, new and reused brackish water for the larviculture of the freshwater prawn *Macrobrachium rosenbergii* in a recirculating system. *J. World Aquac. Soc.* **1998**, *29*, 345–350. [CrossRef]
23. Guest, W.C. Laboratory Life History of the palaemonid shrimp *Macrobrachium amazonicum* (Heller) (Decapoda, Palaemonidae). *Crustaceana* **1979**, *37*, 141–152. [CrossRef]
24. Manzi, J.J.; Maddox, M.B.; Sandifer, P.A. Algal supplement enhancement of *Macrobrachium rosenbergii* (De Man) larviculture. *Proc. World Maric. Soc.* **1977**, *8*, 207–223. [CrossRef]

25. Tayamen, M.; Brown, J.H. A condition index for evaluating larval quality of *Macrobrachium rosenbergii* (De Man, 1879). *Aquac. Res.* **1999**, *30*, 917–922. [CrossRef]
26. Maciel, C.R.; New, M.B.; Valenti, W.C. The predation of *Artemia* nauplii by the larvae of the Amazon River Prawn, *Macrobrachium amazonicum* (Heller, 1862), is affected by prey density, time of day, and *ontogenetic* development. *J. World Aquac. Soc.* **2012**, *43*, 659–669. [CrossRef]
27. Winemiller, K.O.; Dailey, W.H. Life history strategies, population dynamics, and consequences for supplemental stocking of tarpon. *Contrib. Mar. Sci.* **2002**, *35*, 81–94.
28. Sheskin, D.J. *Handbook of Parametric and Nonparametric Statistical Procedures*, 5th ed.; CRC Press: Boca Raton, FL, USA, 2011; 1886p.
29. Hayd, L.A.; Lemos, D.; Valenti, W.C. Effects of ambient nitrite on Amazon river prawn, *Macrobrachium amazonicum*, larvae. *J. World Aquac. Soc.* **2014**, *45*, 55–64. [CrossRef]
30. McKay, H.; McAuliffe, W.; Waldhorn, D.R. *Welfare Considerations for Farmed Shrimp*; Rethink Priorities: San Francisco, CA, USA, 2023. [CrossRef]
31. Barreto, A.V.; Soares, C.M.A. Produção de pós-larvas de *Macrobrachium amazonicum* (Heller, 1862) (Decapoda; Palaemonidae), sob condições controladas de laboratório. *Rev. Bras. Zool.* **1982**, *1*, 51–53. [CrossRef]
32. Rahman, M.M.; Salin, K.R.; Tsusaka, T.W.; Anal, A.K.; Rahi, M.L.; Yakupitiyage, A. Effect of stocking density on growth performance and gonadal maturity of all-female giant freshwater prawn, *Macrobrachium rosenbergii*. *J. World Aquac. Soc.* **2022**, *53*, 1120–1133. [CrossRef]
33. Arifin, O.Z.; Prakoso, V.A.; Subagja, J.; Kristanto, A.H.; Pouil, S.; Slembrouck, J. Effects of stocking density on survival, food intake and growth of giant gourami (*Osphronemus goramy*) larvae reared in a recirculating aquaculture system. *Aquaculture* **2019**, *509*, 159–166. [CrossRef]
34. Coelho-Filho, P.A.; Gonçalvez, A.P.; Barros, H.P. *Artemia* nauplii intake by *Macrobrachium carcinus* at different larval stages in laboratory. *Aquaculture* **2018**, *484*, 333–337. [CrossRef]
35. Nhan, D.T. Evaluation of different diets to replace *Artemia* nauplii for larval rearing of giant freshwater prawn (*Macrobrachium rosenbergii*). *Agric. Dev. Mag.* **2018**, *17*, 35–43. [CrossRef]
36. Barros, H.P.; Valenti, W.C. Food intake of *Macrobrachium rosenbergii* during larval development. *Aquaculture* **2003**, *216*, 165–176. [CrossRef]
37. Gomes, J.N.; Abrunhosa, F.A.; Costa, A.K.; Maciel, C.R. Feeding and larval growth of an exotic freshwater prawn *Macrobrachium equidens* (Decapoda: Palaemonidae), from Northeastern Pará, Amazon Region. *An. Acad. Bras. Ciên.* **2014**, *86*, 1525–1536. [CrossRef]
38. Coyle, S.D.; Tidwell, J.H.; Danaher, J.; Yasharian, D.K.; Bright, L.A. The Effect of Biomass Density, Salinity, and Substrate on Transport Survival of Juvenile Freshwater Prawns *Macrobrachium rosenbergii* in Continuously Oxygenated, Vented Containers. *N. Am. J. Aquac.* **2006**, *68*, 271–275. [CrossRef]
39. Araujo, M.C.; Valenti, W.C. Effects of feeding strategy on larval development of the Amazon River prawn *Macrobrachium amazonicum*. *Rev. Bras. Zoot.* **2017**, *46*, 85–90. [CrossRef]
40. Araujo, M.C.; Valenti, W.C. Efeito da intensidade luminosa no desenvolvimento larval do *Macrobrachium amazonicum*. *Bol. Inst. Pesca* **2011**, *37*, 155–164.
41. Maciel, C.R.; Valenti, W.C. Assessing the potential of partial replacing of *Artemia* by practical inert diet in the larviculture of the Amazon River Prawn. *Bol. Inst. Pesca* **2014**, *40*, 69–78.
42. New, M.W. Freshwater prawn farming: Global status, recent research and a glance at the future. *Aquac. Res.* **2005**, *36*, 210–230. [CrossRef]

Disclaimer/Publisher's Note: The statements, opinions and data contained in all publications are solely those of the individual author(s) and contributor(s) and not of MDPI and/or the editor(s). MDPI and/or the editor(s) disclaim responsibility for any injury to people or property resulting from any ideas, methods, instructions or products referred to in the content.

fishes

MDPI

Article

Stocking Density and Diet of Two Oyster (*Crassostrea gasar* and *Crassostrea gigas*) Seeds in Fluidized Bed Bottle Nursery System

Simone Sühnel [1,*], Francisco José Lagreze-Squella [2], Gabriel Nandi Corrêa [1], Jaqueline Araújo [1], Glauber de Souza [1], João Paulo Ramos Ferreira [1], Francisco Carlos da Silva [1], Carlos Henrique Araújo de Miranda Gomes [1] and Claudio Manoel Rodrigues de Melo [1]

[1] Laboratory of Marine Mollusks, Department of Aquaculture, Federal University of Santa Catarina, Florianopolis, SC 88061-600, Brazil; gabriel.xops@hotmail.com (G.N.C.); jaqueline.araujo@ufsc.br (J.A.); glauber.souza@posgrad.ufsc.br (G.d.S.); joao.ramos.f@ufsc.br (J.P.R.F.); francisco.cs@ufsc.br (F.C.d.S.); carlos.henrique.miranda@ufsc.br (C.H.A.d.M.G.); claudio.melo@ufsc.br (C.M.R.d.M.)

[2] Laboratory of Shellfish Aquaculture Engineering, Centre of Marine Studies, Federal University of Paraná, Pontal do Paraná, PR 83255-000, Brazil; lagreze@ufpr.br

* Correspondence: ssuhnel@gmail.com

Abstract: *Crassostrea* is the most farmed oyster genus worldwide and has significant economic and social impacts with environmental benefits. Hatchery oyster seed production is a highly costly phase, and a fluidized nursery system can help reduce this cost and reduce seed production time. The present study evaluated the survival and growth of two oyster species (*Crassostrea gasar* and *Crassostrea gigas*) in a fluidized bed bottle nursery system. With *C. gasar*, two experiments were performed; one tested three stocking densities and the other three bialgae diets. With *C. gigas*, one experiment with a bialgae and monoalgae in an initial bottle occupation of 8.8% produced more seeds per bottle, but an initial bottle occupation of 2.2% produced bigger seeds. Also, the experiment with *C. gasar* and with *C. gigas* tested diets did not affect seed survival, but the diets with bialgae *I. galbana* and *N. oculate* promoted more seed growth. The fluidized bed bottle nursery system developed for this study was adequate for the seeds of the oysters *C. gasar* and *C. gigas* in the nursery phase.

Keywords: mangrove oyster; Pacific oyster; microalgae; shellfish; spat

Key Contribution: The fluidized bed bottle nursery system developed for this study was adequate for the oyster seeds of *C. gasar* and *C. gigas* in the nursery phase in the hatchery. All initial tested stocking densities can be used with the *C. gasar* and *C. gigas* seeds. The oyster seeds of *C. gasar* and *C. gigas* fed with a bialgae diet (*I. galbana* and *N. oculate*) showed high growth. The tested microalgae in monoalgae or bialgae diet affect more seed growth than survival. The fluidization velocity used in the present study for *C. gasar* and *C. gigas* showed a relationship with biomass in the bottle.

Citation: Sühnel, S.; Lagreze-Squella, F.J.; Corrêa, G.N.; Araújo, J.; de Souza, G.; Ferreira, J.P.R.; da Silva, F.C.; de Miranda Gomes, C.H.A.; de Melo, C.M.R. Stocking Density and Diet of Two Oyster (*Crassostrea gasar* and *Crassostrea gigas*) Seeds in Fluidized Bed Bottle Nursery System. *Fishes* **2024**, *9*, 183. https://doi.org/10.3390/fishes9050183

Academic Editor: Yuanzi Huo

Received: 10 April 2024
Revised: 8 May 2024
Accepted: 14 May 2024
Published: 17 May 2024

Copyright: © 2024 by the authors. Licensee MDPI, Basel, Switzerland. This article is an open access article distributed under the terms and conditions of the Creative Commons Attribution (CC BY) license (https://creativecommons.org/licenses/by/4.0/).

1. Introduction

The oyster genus *Crassostrea* (Sacco 1897) is the most cultured oyster group, with economic, social, and environmental importance. Around 26 species are in the genus *Crassostrea* [1] and according to FAO [2], *Crassostrea gigas* (Thunberg 1793) is the most cultured oyster species worldwide. In 2021, *Crassostrea* spp. production was 6.0 million tons, representing 33.1% of the world mollusk production in quantity [3] and 21.4% of the world mollusk production in value [4]. Seed oyster production in hatcheries has an important contribution to the large-scale culture of *Crassostrea*, especially of *C. gigas*, as in China [5] and Brazil [6]. In Brazil, besides Pacific oyster production, another oyster species farmed is the native oyster *Crassostrea gasar* (according to Ferreira et al. [7] the descriptor for *C. gasar* is Dillwyn 1817; syn. *Crassostrea brasiliana and Crassostrea tulipa*), commonly known as mangrove oyster, gasar oyster, and bottom oyster, among others.

Oyster hatchery production involves at least six key points: water supply, broodstock obtention and conditioning, larviculture, settlement and nursery, microalgae production for feed larvae, seed (spat), broodstock, and skilled staff. Settlement, and especially the nursery for early seeds, can be expensive. Depending on the culture system used, it requires a large quantity of microalgae, large areas, and a lot of human labor effort.

The hatcheries' seed culture system (nursery) starts after larvae metamorphosis [7]. Larvae metamorphosis is generally induced with neurotransmitters, such as epinephrine, for single oyster hatchery production. When the metamorphosis is completed, the seeds are transferred to the nursery systems, which include the upwelling or downwelling systems [7]. Generally, the upweller and downweller use silos, which are containers with mesh in the bottom to hold the seeds and to permit water flow ascendant (upweller) or descendent (downweller), suspended in tanks. A combination of flow and seston concentration (i.e., microalgae) can affect the optimal stocking density used in the culture units [8,9], affecting oxygen and food availability and waste removal due to the water movement through the oyster bed. The upweller-using bottles are known as fluidized bed nurseries, generally used with oysters, and are capable of holding higher stocking densities than standard upwelling systems (i.e., silos). In this fluidized bed nursery system, oyster seeds are lifted by the flow [7] due to the fluidization of the bed. In high superficial flow velocity, the oyster's seeds can be transported by the fluid and washed out from the culture unit. However, in intermediate velocities, each individual oyster is suspended in the fluid [10], inside of the culture unit. Many commercial hatcheries are using a fluidized bed system, but little has been published about the zootechnical performance of oyster seeds in this system. The fluidized bed nursery system was evaluated with the oyster *Crassostrea virginica* (Gmelin 1791) [10] and the clam *Mercenaria mercenaria* (Linnaeus 1758) [11].

Microalgae diets are another important aspect of the seed's zootechnical performance during the nursery phase. The biochemical composition of microalgae can vary among species [12], where a combination of microalgae species in the diet needs to be considered to provide more nutrients for the oyster seeds. Monoalgae (single-species diets) and multialgae (mixed-species diets) diet have been studied for the oyster nursery phase, as for example monoalgae diets for *C. gigas* [13,14] and *Ostrea edulis* (Linnaeus 1758) [15] and bialgae diets for *C. gigas* [14], *O. edulis* [15,16], *Crassostrea corteziensis* (Hertlein 1951) [17], and *Saccostrea commercialis* (Iredale & Roughley 1933) [18].

In this sense, the zootechnical performance, measured by survival and growth, of oysters (*C. gasar* and *C. gigas*) in a fluidized bed bottle nursery system was evaluated. Two experiments with *C. gasar* were conducted: one testing stocking density and the other with bialgae diet; and one experiment with *C. gigas* testing bialgae and monoalgae diet in a fluidized bed bottle nursery system.

2. Materials and Methods

This study was performed at the Laboratory of Marine Mollusks of the Federal University of Santa Catarina (LMM-UFSC), located in Florianópolis, Brazil (27°35′06.35″ S, 48°26′27.05″ W). Single oyster seeds of the mangrove oyster *C. gasar* and the Pacific oyster *C. gigas* were obtained from the LMM-UFSC hatchery.

Three experiments were performed to evaluate oyster survival and growth (described below) in a fluidized bed bottle nursery system in a closed aquaculture system (FBBN-CAS). Two experiments were performed with the native mangrove oysters (*C. gasar*): the first (experiment I) tested stocking density and the second (experiment II) tested diet. A third experiment (experiment III) was performed with the Pacific oysters (*C. gigas*) to test the diet.

2.1. Fluidized Bed Bottle Nursery System in Closed Aquaculture System (FBBN-CAS)

Three identical fluidized bed bottle nursery (FBBN) systems in a closed aquaculture system (CAS) were developed for this study. Each FBBN system (Figure 1) was composed of cylinder-conical acrylic bottles (transparent; a total volume of 1.26 L with 0.9 L useful volume; experimental unit = EU), each EU with a shut-off valve (SOV; glass marble in the bottom of the bottle), two ball valve (BV) and one flowmeter (FW) coupled in the bottom, and a drain in the top of each bottle (BDR); a wood rack to fix the bottles and a PVC water drainer; a sump tank (SUMP; fiberglass; 664 L) with a magnetic pump (MP; Boyu, 5500 L·h^{-1}); a distribution tank (Dtank; carboy; 20 L; steady head); and a feeding tank (Ftank; fiberglass; 250 L) with a peristaltic pump (PP; Seko Tekna EVO803; 20–54 L·h^{-1}). The total volume of the system was 685 L. For the stocking density experiment (I), one FBBN-CAS system with three EUs (n = 3) for each treatment was used, totaling nine EUs randomly distributed in the FBBN-CAS system. For experiments with diet (II and III), three identical FBBN-CAS systems were used, each with four EUs (n = 4) randomly distributed in the system.

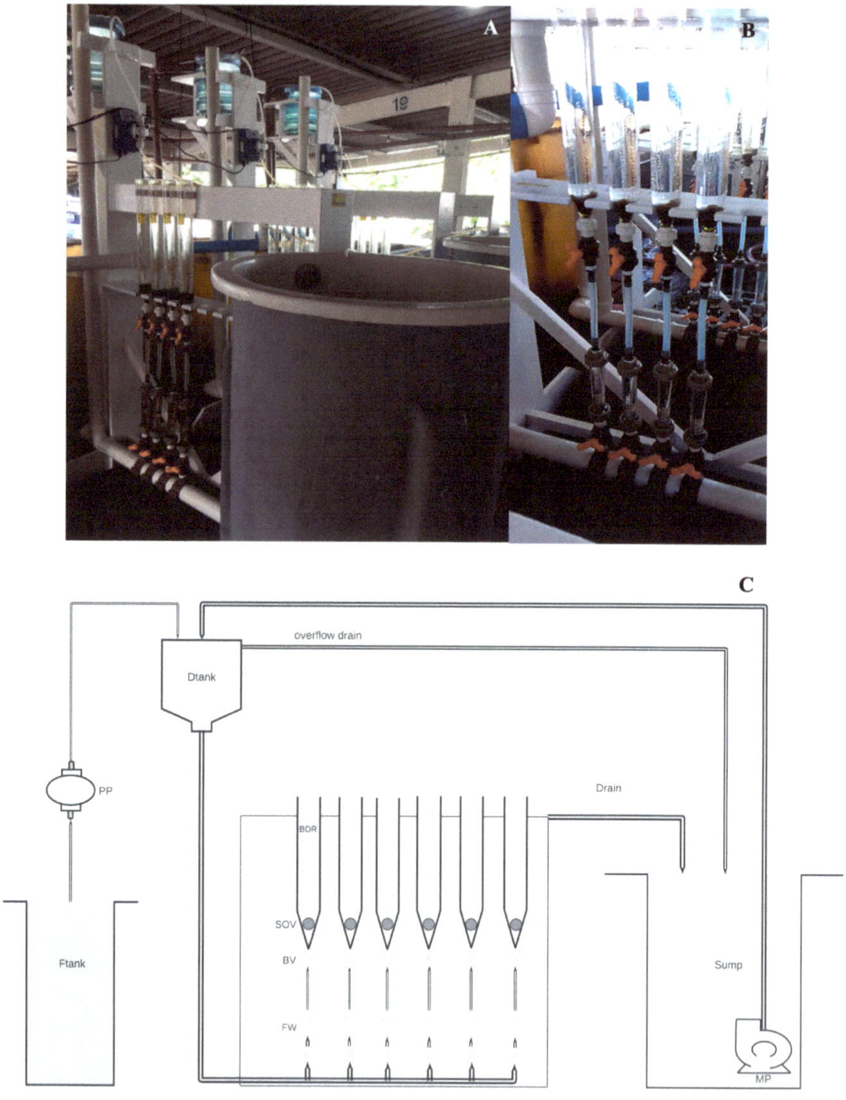

Figure 1. Fluidized bed bottle nursery in closed aquaculture system (FBBN-CAS) system used in the experiments I, II, and III. (**A**), an overview of the FBBN; (**B**), a detail of the acrylic bottle and flow control system used; and (**C**), a schematic drawing showing water direction movement in the FBBN-CAS.

Seawater (filtered at 1 μm and sterilized with UV) was pumped from the sump tank to the distribution tank (highest level in the system) by gravity distribution through a pipe (lowest level in the system) to the bottom of each bottle, creating an ascendant flow, going from the bottom of the bottle, through the seeds, to the top drainers and returning to the sump tank by gravity. The distribution tank has an overflow. Drain microalgae were pumped (140 mL.min^{-1}) from the feed tank by the peristaltic pump to the distribution tank. The flow was regulated through a flowmeter to maintain the seeds in suspension in the bottle. The flow of the inlet seawater in the bottles was fixed to be the same for all treatments in each experiment and regulated according to the seed's growth (Table 1).

Table 1. Mean and standard deviation of the temperature, salinity, pH, and residual algae; minimum and maximum temperature; and the experiment period in each experiment (I, II, and III), and the range of inlet seawater flow in the bottle in the sump of experiment I and in each treatment in experiments II and III, where Exp. = experiment; C = *Crassostrea;* D15 = 15 seeds·mL^{-1}; D30 = 30 seeds·mL^{-1}; D60 = 60 seeds·mL^{-1}; IC = 50% of *Isochrysis galbana* and 50% of *Chaetoceros muelleri;* IN = 50% of *I. galbana* and 50% of *Nannochloropsis oculata;* RC = 50% of *Rhodomonas salina* and of 50% of *C. muelleri;* I = 100% of *I. galbana;* N = 100% of *N. oculata;* Min = minimum; and Max = maximum.

Exp.	Specie	Tested Parameter	Temperature (°C)			Salinity (ppt)	pH	Exp. Period (Days)	Flow (L·h^{-1})	Fluidization Velocity (cm·s^{-1})	Residual Algae (×10^4 cell·mL^{-1})
			Mean	Min	Max						
I	*C. gasar*	Stocking density	26.9 ± 1.5	22.0	30.0	35.0 ± 0.0	8.4 ± 0.3	29	D15: 75 to 250 D30: 75 to 250 D60: 75 to 250	D15: 0.7 to 2.5 D30: 0.7 to 2.5 D60: 0.7 to 2.5	SUMP: 1.8 ± 1.2
II	*C. gasar*	Diet	21.0 ± 1.5	12.7	23.8	35.0 ± 0.2	7.8 ± 0.3	29	IC: 50 to 225 IN: 50 to 250 RC: 50 to 225	IC: 0.5 to 2.5 IN: 0.5 to 2.5 RC: 0.5 to 2.2	IC: 9.4 ± 5.7 IN: 37.2 ± 33.8 RC: 9.6 ± 5.4
III	*C. gigas*	Diet	18.8 ± 2.7	9.5	22.4	35.1 ± 0.3	8.1 ± 0.2	21	I: 50 to 100 IN: 50 to 150 N: 50 to 125	I: 0.5 to 1.0 IN: 0.5 to 1.5 N: 0.5 to 1.2	I: 10.4 ± 10.6 IN: 97.4 ± 81.8 N: 157.6 ± 119.7

FBBS-CAS daily handling consisted of seawater (sump tank) and feed (feeding tank) exchange. For that, first, the shut-off valve of each bottle was closed to avoid seed escape, and then the pumps (magnetic and peristaltic) were turned off. The seawater from the sump tank and the feeding tank were drained, and both tanks were cleaned with fresh water before being refilled with seawater (sump tank) and microalgae (feeding tank). Before and after seawater exchange, the temperature (infrared sensor; Figure 2), salinity (refractometer: Kasvi), and pH (pHmeter; Alfakit AT-350) of the sump tank for each experiment were registered (Table 1).

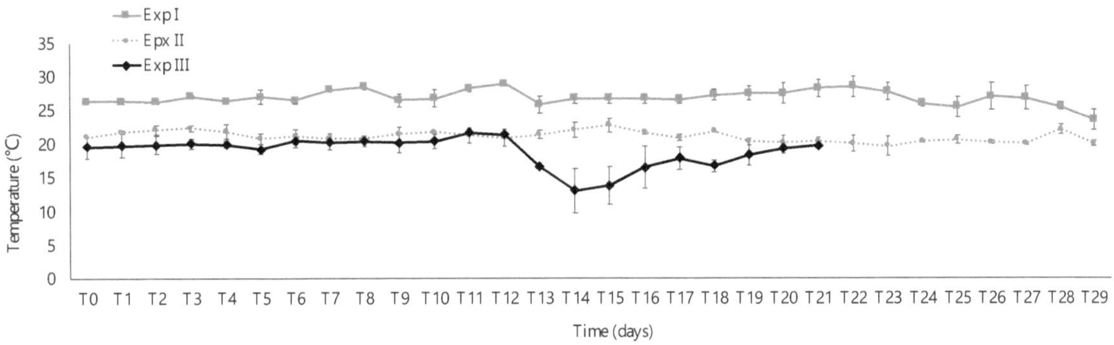

Figure 2. Temperature (mean ± standard deviation) during the experimental period in experiments I, II, and III where Exp. = experiment and T = time.

2.2. Experiment I: Stocking Density (Crassostrea Gasar)

In experiment I, three initial stocking densities (D15, a stocking density of 15 seeds·mL^{-1} (20 mL of seeds; 2.2% of the total bottle volume occupation); D30, a stocking density of 30 seeds·mL^{-1} (40 mL of seeds; 4.4% of the total bottle volume occupation); and D60, a stocking density of 60 seeds·mL^{-1} (80 mL of seeds; 8.8% of the total bottle volume occupation)) of *C. gasar* seeds (1.98 ± 0.31 mm of shell height) in FBBS-CAS were tested. Each stocking density treatment was calculated as the percentage of the useful bottle volume filled with seeds. Experiment I lasted for 29 days, being T0 at the beginning of the experiment (planting day) and 29 days at the end (T29).

The seed diet during the experimental period was composed of two microalgae species, *Isochrysis galbana* (Parke 1949) (Iso) and *Chaetocheros muelleri* (Lemmermann 1898) (Cm), with a ratio of 30: 70 (Iso: Cm). The microalgae concentration started at 12 × 10^4 cells·mL^{-1} (1 to 5 days of the experiment), and increased to 15 × 10^4 cells·mL^{-1} (6 to 17 days), 20 × 10^4 cells·mL^{-1} (18 to 22 days), and 22 × 10^4 cells·mL^{-1} (23 to 30 days) according to the seed growth. The microalgae concentration was calculated using the total volume system (685 L). The microalgae concentration used in this experiment was chosen for the LMM protocol for *C. gasar* seeds maintenance. Daily, the residual microalgae concentration in the outlet seawater from the bottles was monitored (Table 1).

2.3. Experiment II: Diet (Crassostrea Gasar)

In experiment II, *C. gasar* seeds (1.77 ± 0.12 mm of shell height) were used to test three bialgae diets: IC, Iso, and Cm; IN, Iso, and *Nannochloropsis oculata* (Hibberd 1981) (N); and RC, *Rhodomonas salina* (Hill & Wetherbee 1989) (R), and Cm; with the proportion 1:1 of each microalgae species. The microalgae quantities in each treatment were calculated for 10% of the total seed whole weight (fresh weight) [16]. Weekly, the total microalgae quantities were adjusted to the total seed whole weight, maintaining the feed in 10% of the biomass. Experiment II lasted for 29 days. Each experimental unit was planted with 10 mL of seeds (7.5 seeds·mL^{-1}). Daily, the residual microalgae concentration in the outlet seawater from the bottles was monitored (Table 1).

2.4. Experiment III: Diet (Crassostrea Gigas)

In experiment III, *C. gigas* seeds (1.68 ± 0.14 mm of shell height) were used to test three diets, two monoalgae diets (I: 100% of Iso; and N: 100% of N) and one bialgae diet (IN: 50% of Iso and 50% of N). The microalgae quantities in each treatment were calculated for 10% of the total whole weight of the seeds [16]. Weekly, the total microalgae quantities were adjusted to the total seed fresh weight, maintaining the feed at 10% of the total seed fresh weight. Experiment III lasted for 21 days. Each experimental unit was planted with 10 mL of seeds (7.5 seeds·mL^{-1}). Daily, the residual microalgae concentration in the outlet seawater from the bottles was monitored (Table 1).

The dried microalgae weight for *I. galbana* and *R. salina* used in the present study is cited by Brown [14]; for *N. oculata*, we considered the dried weight of *Nannochloropsis*-like sp. CS-246 cited by Brown [14], and for *C. muelleri*, the dried weight is cited by Helm et al. [7].

2.5. Data Collection and Statistical Analysis

For data collection (Table 2), the seed number (total live seeds in each EU of each treatment) and survival (survival in each time in each EU of each treatment) in experiments I and II were measured after 14 and 29 days (T14 and T29, respectively) and in experiment III after 14 and 21 days (T14 and T21, respectively). For seed number quantification, the total volume of seeds from each EU for each treatment was measured, and after that, the seeds were sieved in four screen mesh (1, 2, 3, and 4 mm). Three samples (n = 3) from each screen (0.5 mL for the seeds retained in the screen 1 mm and below 1 mm, and 2.5 mL for the seeds retained in the screens 2, 3, and 4 mm) were taken. The sum of the number of live seeds from each screen was calculated to obtain the total number of seeds per EU. The seed survival (%) was quantified in each time sample (i.e., seed survival in T14 in relation to T0, seed survival in T29 in relation to T14 for experiments I and II, seed survival in T14 in relation to T0, and seed survival in T21 in relation to T14 for experiment III).

Table 2. Analyzed parameters in each experiment (I, II, and III) in each time sampling (T), where C = *Crassostrea*; T = time; FV = fluidization velocity; TV = total volume of live seeds; IV = increase in volume of seeds; PSS = percentage of seeds by screen; TWW = total seed whole weight; and IWW = individual seed whole weight.

Exp.	Specie	Tested Parameter	Analyzed Parameter			
			T7	T14	T21	T29
I	*C. gasar*	Stocking density	FV, survival, seed number, TV, IV, and PSS	FV and shell height	FV	FV, survival, seed number, TV, IV, PSS, and shell height
II	*C. gasar*	Diet	FV, survival, seed number, TV, IV, TWW, IWW, and specific weight	FV	FV	FV, survival, seed number, TV, IV, TWW, IWW, and specific weight
III	*C. gigas*	Diet	FV, survival, seed number, TV, IV, TWW, IWW, shell height, shell length, and specific weight	FV, shell height, and shell length	FV, survival, seed number, TV, IV, TWW, IWW, shell height, shell length, and specific weight	

For seed growth analysis in the experiment with stocking densities (experiment I), the total volume (TV) of live seeds, the increase in the volume (IV) of seeds, and the percentage of seeds by screen (PSS) were quantified. IV was defined as the increase in the total volume of seeds in a container as a measurement of oyster growth. In the *C. gasar* diet experiment (experiment II), the seed growth per treatment was analyzed with TV, total seed whole weight (TWW) in each EU, and individual seed whole weight (IWW) at T14 and T29. For *C. gigas* diet experiment (experiment III), the seed growth per treatment was analyzed with TV, TWW, and IWW at T14 and T21, and shell height and shell length at T7, T14, and T21.

For TV per each EU in each treatment measurement graduated test tubes were used, and IV, the increase in the total volume (times more volume) of seeds compared to T0, was calculated according to Equation (1).

$$IV = TV \div iV \quad (1)$$

where

IV: increase in volume (times more volume of seeds than in T0)

TV: total volume of seeds in each EU at each time (T14 and T29 for experiments I and II and at T14 and T21 for experiment III; mL) for each treatment;

iV: initial volume of seeds planted at T0 (mL)

To analyze PSS, the following five size classes were used: (i) <#1: the seeds not retained in the screen of 1 mm (seeds < 1 mm); (ii) #1–2: the seeds retained the screens of 1 mm (seeds $\geq$ 1 mm and < 2 mm); (iii) #2–3: the seeds retained in between the screens of 2 mm (seeds $\geq$ 2 mm and < 3 mm); (iv) #3–4: the seeds retained in between the screens of 3 mm (seeds $\geq$ 3 mm and < 4 mm); and (v) $\geq$#4: the seeds retained in the screen of 4 mm (seeds $\geq$ 4 mm). These five size classes were used to calculate seed growth in total volume (TV). The percentage of seeds in each size class (<#1, #1–2, #2–3, #3–4, and $\geq$#4) was calculated with the total seed number in each size class and the total seed number per EU.

For TWW quantification, all animals from each EU were retained in a sieve (230 μm) and dried for 1 h over napkins (napkins were changed every 15 min to continue retaining water). They were then weighted (total whole weight) in an analytical balance (with a resolution of 0.001 g; Shimadzu, UX4200H). After weighing, seeds from each EU returned to their own bottle to continue the experiment.

For seed shell height and shell length, a sample (n = 30) from each EU for each treatment was tacked, and measurement was performed in microscopy (Leica; using software LAZ EZ 3.0.0) and stereomicroscope (Leica 4D) according to the size of the seeds.

Fluidization velocity (FV; $cm{\cdot}s^{-1}$), defined as the velocity that promotes oyster seed fluidization, was calculated for experiments II and III. According to Equation (2), the fluidization velocity was calculated for each flow in each treatment at T0, T14, and T29 (experiment II) and at T0, T14, and T21 (experiment III). The fluidization velocity was used for linear regression analyses.

$$FV = Q \div A \quad (2)$$

where

FV: fluidization velocity ($cm{\cdot}s^{-1}$)

Q: flow ($cm^3.s^{-1}$) measured in the flowmeter

A: cross-sectional area of the bottle (cm^2)

Seed specific weight ($kg{\cdot}m^{-3}$; Table 1) was calculated for each species using all data from each EU for experiment II (data from T0, T14, and T29; n = 28) and for experiment III (data from T0, T14, and T21; n = 36).

Seed number, survival, TV, IV, PSS, TWW, IWW, FV, shell height, and shell length data were tested for the basic assumptions for analysis of variance (ANOVA) using the Shapiro–Wilk test for the normality of residues and Levene's test (for data with two factors) and Bartlett's test (for data with one factor) for the homogeneity of variance. Pairwise comparisons of stocking density and sampling time means were carried out using Tukey's test ($p < 0.05$) for parametrical data, the Wilcoxon test, and the Permutation with t-test for nonparametric data. The fluidization velocity was calculated and linear regression analysis was performed to evaluate the relationship between FV (y-axis) and TWW (x-axis) using all data at T0, T7, T14, T21, and T29 (experiment II) and at T0, T7, T14, and T21 (experiment III) for each flow. All statistical tests were performed in RStudio. The data of specific weight were not used for statistical analysis.

3. Results

3.1. Survival and Total Number of Seeds

The tested stocking density (D15, D30, and D60) did not significantly ($p > 0.05$) affect oyster survival after 14 and 29 days, with the final survivals (T29) of 95.3 ± 2.1%, 91.3 ± 3.4%, and 90.2 ± 6.5%, respectively, for the treatments D15, D30, and D60. The seed number in D60 was significantly ($p < 0.05$; Tukey's test) higher than in D15 and D30 at T14 and T29 (Figure 2).

In experiment II, the diet did not significantly ($p > 0.05$) affect oyster survival and seed number after 14 and 29 days, with a final survival (T29) of 88.9 ± 2.8%, 94.2 ± 4.5%, and 89.6 ± 4.9%, respectively, for the treatments IC, IN, and RC. The seed numbers showed no differences between the treatments (Figure 3).

In experiment III, the diet showed a significant difference ($p < 0.05$; Tukey's test) between survival at T14 and T21 independent of the treatment with higher survival at T14 (96.0 ± 5.7%) than at T21 (89.7 ± 7.9%). However, no differences were observed in the survival between the diet treatments (I, IN, and N) independent of the time sampling. No interaction was observed between the diets tested and time samplings in the survival. The final survival (T21) was 90.2 ± 4.4%, 82.7 ± 8.9%, and 96.2 ± 1.0%, respectively, for the treatments I, IN, and N. The seed numbers showed a significant difference ($p < 0.05$; Tukey's test) between T14 and T21 independent of the treatment, with a higher number at T14 (22657 ± 1351 seeds) than at T21 (21169 ± 1874 seeds). However, no differences were observed in the seed numbers between the diet treatments (I, IN, and N) independent of the time sampling. No interaction was observed between the diets tested and time samplings in the seed numbers. The seed numbers at T14 and T21 showed no differences in all tested treatments (I, IN, and N; Figure 3).

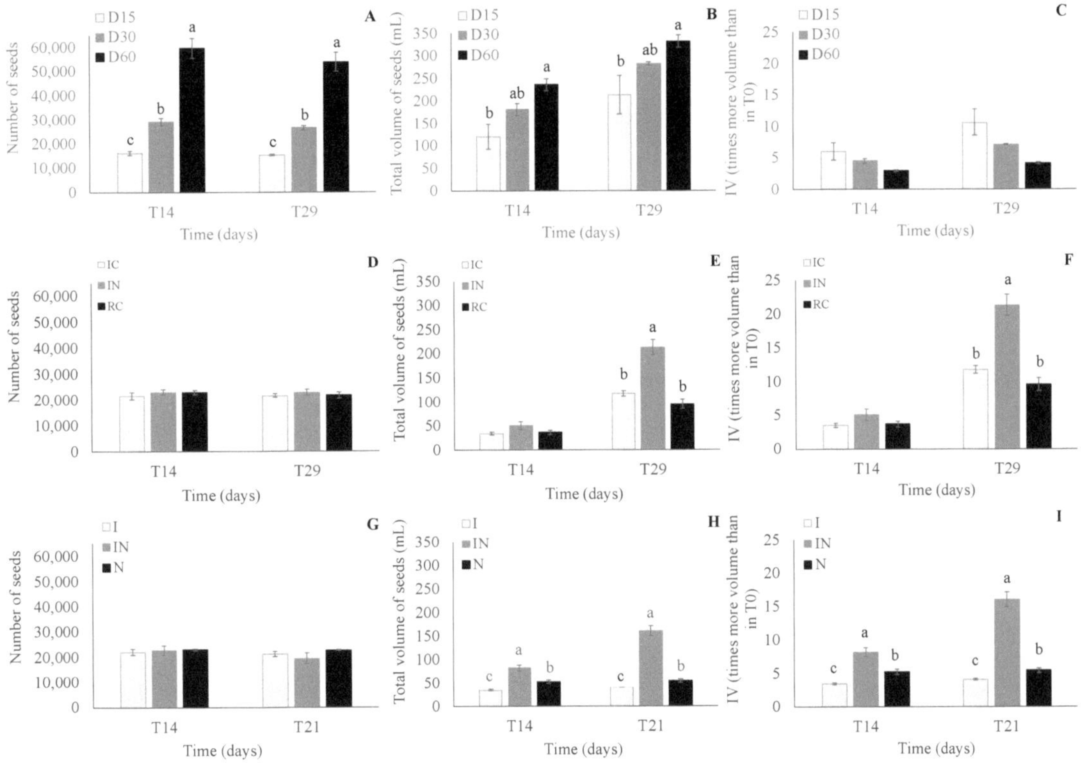

Figure 3. Seed number, total volume, and times more volume than in T0 (increase in volume; IV) in experiment I (*C. gasar*; (**A**), (**B**), and (**C**), respectively) in each stocking density treatment (D15, D30, and D60) at T14 and T29, in experiment II (*C. gasar*; (**D**), (**E**), and (**F**), respectively) in each diet treatment (IC, IN, and RC, respectively) at T14 and T29, and in experiment III (*C. gigas*; (**G**), (**H**), and (**I**), respectively) in each diet treatment (I, IN, and N) at T14 and T21, where, D15: 15 seeds·mL^{-1}; D30: 30 seeds·mL^{-1}; D60: 60 seeds·mL^{-1}; IC: 50% of *Isochrysis galbana* and 50% of *Chaetoceros muelleri*; IN: 50% of *I. galbana* and 50% of *Nannochloropsis oculata* (N); RC: 50% of *Rhodomonas salina* and of 50% of *C. muelleri*; I: 100% of *I. galbana*; N: 100% of *N. oculata*; and T = time. Different letters indicate statistical differences in each time sampling ((**A**, **B**, **D**, **E**, **F**, and **G**): Tukey's test; (**C**, **H**, and **I**): Permutation *t*-test). The absence of letters represents that there were no significant differences between the treatments.

3.2. Seed Growth

3.2.1. Experiment I

The total volume (TV) of the seeds increased significantly ($p < 0.05$; Tukey's test) from T14 (179.2 ± 50.9 mL) to T29 (275.8 ± 54.8 mL), independent of the density, and the density, independent of the time sampling, also affect TV, being higher ($p < 0.05$; Tukey's test) in D60 (283.7 ± 49.5 mL) than in D15 (167.2 ± 58.7 mL) and both not being different from D30 (231.7 ± 51.9 mL). No interaction was observed between the stocking density tested and time samplings in the TV. The seed TV in D60 was significantly ($p < 0.05$; Tukey's test) higher than in D15, and both (D15 and D60), were not different from D30 (Figure 3) at T14 and T29.

The increase in volume (IV) was significant ($p < 0.05$; Tukey's test) from T14 (4.5 ± 1.5) to T29 (7.3 ± 2.9) independent of the treatment. Also, the treatment affects IV independent of the sampling time in D15 (8.3 ± 2.9) and D30 (5.8 ± 1.3), with no differences between them, and both being higher ($p < 0.05$; Tukey's test) than in D60 (3.5 ± 0.6). No interaction was observed between the stocking density tested and sampling times in the IV. At T14 and T29, the seed IV showed no difference between the treatments (Permutation *t*-test) (Figure 3).

Stocking density affected significantly ($p < 0.05$; Tukey's test) the PSS in the size class <#1, #1–2, #2–3, and #3–4, independent of the sampling time. Analyzing sampling time independent of the stoking density, a significant effect ($p < 0.05$; Tukey's test) was observed in the size class <#1, #1–2, and #3–4. The PSS in the size class >#1 was higher ($p < 0.05$) in D30 compared to D15 and D60 at T14, and the PSS in D30 was higher ($p < 0.05$) than in D15 at T29. The PSS in the size class #1–2 was higher ($p < 0.05$) in D60 compared to D15, at T14 and T29, and compared to D30, at T29. The PSS in the size class #2–3 was higher ($p < 0.05$) in D15 and D30 compared to D60 at T14 and T29, and PSS in D30 was not different from in D60, at T14 and T29. The PSS in the size class #3–4 showed no differences between the tested stocking densities at T14 but was higher ($p < 0.05$) in D15 compared to D30 and D60 at T29. No seeds were observed in the size class ≥#4 at T14 and T29, and no differences were observed in the PSS between the tested stocking densities (Figure 4).

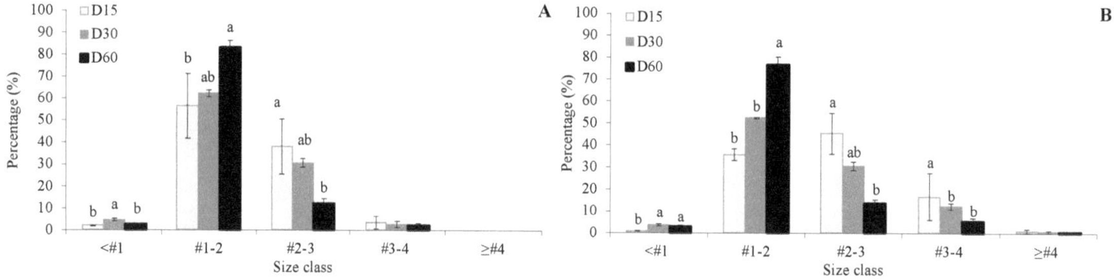

Figure 4. Percentage (%) of seeds (*C. gasar*) by size class in each treatment (D15, D30, and D60) at T14 (**A**) and T29 (**B**) in experiment I, where D15: 15 seeds·mL^{-1}; D30: 30 seeds·mL^{-1}; D60: 60 seeds·mL^{-1}. Different letters indicate statistical differences in each size class (**A** and **B**: Tukey's test). The absence of letters represents that there were no significant differences between the treatments.

3.2.2. Experiment II

In experiment II, the total volume (TV) of the seeds increased significantly ($p < 0.05$; Tukey's test) from T14 (41.9 ± 9.2 mL) to T29 (142.4 ± 52.1 mL), independent of the diet, and the diet treatment independent of the time sampling also affect TV, being higher ($p < 0.05$; Tukey's test) in IN (132.1 ± 82.2 mL) than in IC (76.5 ± 41.7 mL) and RC (66.4 ± 30.4 mL). Interaction ($p < 0.05$) between the tested diets and time samplings in the TV was observed. At T14, no differences in TV were observed between the diets, but at T29, the seed TV in the IN diet was higher ($p < 0.05$; Tukey's test) than in the IC and RC diets, and both (IC and RC) had no difference between them (Figure 3).

The increase in volume (IV) was significant ($p < 0.05$; Tukey's test) from T14 (4.1 ± 0.9) to T29 (14.2 ± 5.2) independent of the treatment. Also, the treatment affected IV independently of the sampling time, with IN (13.2 ± 8.2) IV being significantly higher ($p < 0.05$; Tukey's test) than in IC (7.6 ± 4.2) and RC (6.6 ± 3.0), both showing no difference between them. The interaction between the tested diets and time samplings in the IV was observed. At T14, no differences were observed between the treatments, but at T29, the IV of the diet treatment IN was higher ($p < 0.05$; Tukey's test) than in IC and RC (Figure 3). The differences ($p < 0.05$; Tukey's test) between time sampling (T14 and T29) in each treatment (IC, IN, and RC) were observed, and IV in each treatment at T29 was higher than at T14.

TWW was different ($p < 0.05$; Tukey's test) from T14 (35.7 ± 6.4 g) to T29 (115.7 ± 36.1 g) independent of the treatment. Also, the diet treatment affected TWW independent of the sampling time, with it being higher ($p < 0.05$; Tukey's test) in the IN diet (103.5 ± 61.5 g) than in IC (62.0 ± 31.5 g) and RC (61.4 ± 29.7 g). The interaction ($p < 0.05$) between the tested diets and the sampling times in the TWW was observed. At T14 and T29, the seed TWW in the diet IN were significantly ($p < 0.05$; Tukey's test) higher than in IC and RC, and both (IC and RC) showed no difference between them (Figure 5).

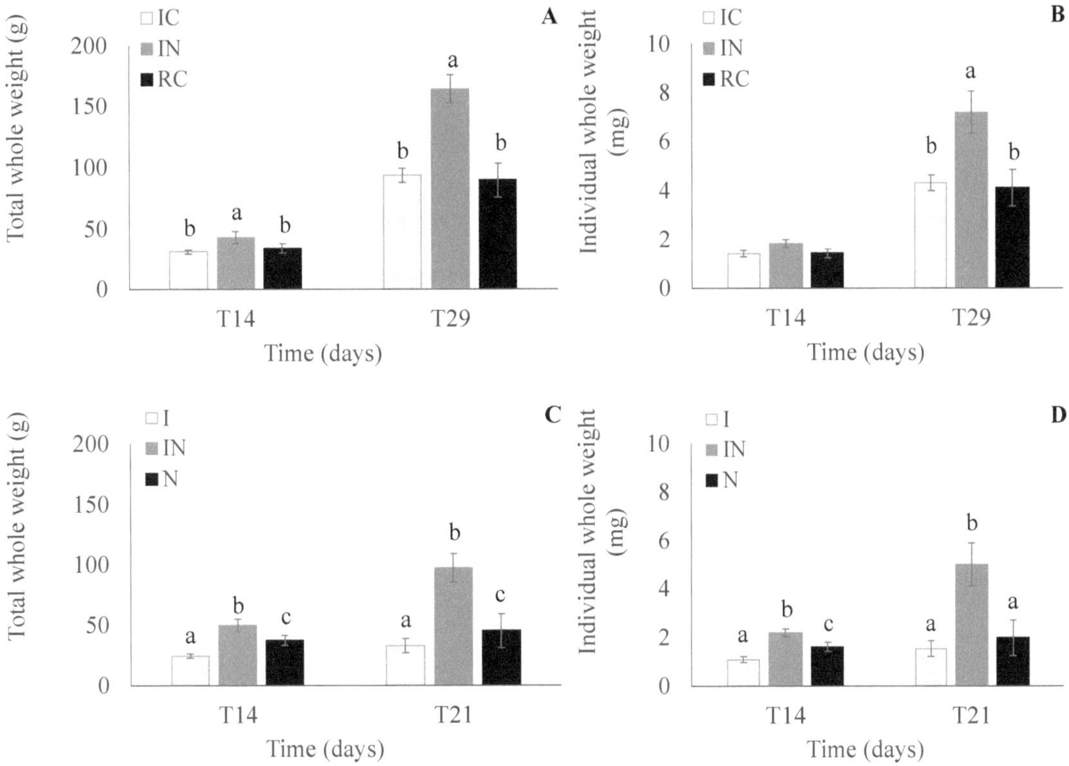

Figure 5. Seed total whole weight and individual whole weight in experiment II (*C. gasar*; (**A**) and (**B**), respectively) in each diet treatment (IC, IN, and RC, respectively) at T14 and T29 and in experiment III (*C. gigas*; (**C**) and (**D**), respectively) in each diet treatment (I, IN, and N) at T14 and T21, where IC: 50% of *Isochrysis galbana* and 50% of *Chaetoceros muelleri*; IN: 50% of *I. galbana* and 50% of *Nannochloropsis oculata* (N); RC: 50% of *Rhodomonas salina* and of 50% of *C. muelleri*; I: 100% of *I. galbana*; N: 100% of *N. oculate*; and T = time. Different letters indicate statistical differences in each time sampling ((**A**,**B**,**D**): Tukey's test; (**C**): Permutation *t*-test). The absence of letters represents that there were no significant differences between the treatments.

The IWW of the seeds increased significantly ($p < 0.05$; Tukey's test) from T14 (1.6 ± 0.2 g) to T29 (5.2 ± 1.6 g), independent of the diet. The diet treatment, independent of the time sampling, affected IWW, with it being higher ($p < 0.05$; Tukey's test) in IN (4.7 ± 2.7 mg) than in IC (2.9 ± 1.5 mg) and RC (2.8 ± 1.4 mg). The interaction ($p < 0.05$) between the tested diets and the sampling times in the IWW was observed. At T14, no differences in IWW were observed between the diets, but at T29, the seed IWW in the IN diet was higher ($p < 0.05$; Tukey's test) than in the IC and RC diets and both (IC and RC) showed no difference between them (Figure 5).

3.2.3. Experiment III

In experiment III, total volume (TV) showed no significant difference between T14 (56.4 ± 19.8 mL) and T21 (85.6 ± 53.7 mL), independent of the diet. However, the diet treatment, independent of the time sampling, affected TV, with it being higher ($p < 0.05$; Wilcoxon test) in IN (121.19 ± 40.42 mL) than in I (38.1 ± 3.4 mL) and in N (53.7 ± 3.8 mL) diets and the TV of seeds in the diet treatment N being higher ($p < 0.05$; Wilcoxon test) than in I diet. The interaction ($p < 0.05$; Tukey's test) between the tested diets and the time samplings in the TV was observed. The seed TVs at T14 and T21 were higher ($p < 0.05$; Permutation *t*-test) in the IN diet than in the I and N diets, and in the N diet were higher ($p < 0.05$; Permutation test) than in I diet (Figure 3).

The increase in volume (IV) showed no difference between time sampling T14 (5.6 ± 2.0) and T21 (8.5 ± 5.4) independent of the treatment. However, the treatment affected IV independent of the sampling time, with IV in IN (12.1 ± 4.0) being significantly higher ($p < 0.05$; Tukey's test) than in I (3.8 ± 0.3) and N (5.4 ± 0.4) and IV in N being higher ($p < 0.05$; Tukey's test) than in I. The interaction ($p < 0.05$; Tukey's test) between the tested diets and sampling times in the IV was observed. At T14 and T21, the IV of the diet treatment IN was higher ($p < 0.05$; Permutation *t*-test) than in I and in N and IV in N was higher ($p < 0.05$; Permutation *t*-test) than in I (Figure 3). Differences ($p < 0.05$; Tukey's test) between the time samplings (T14 and T21) in the diet treatments I and IN were observed, being at T21 higher ($p < 0.05$; Permutation *t*-test) than at T14 for both treatments, and for diet N, no differences in the IV between T14 and T21 were observed.

The seed TWWs showed no significant difference between T14 (37.3 ± 10.9 g) and T21 (58.5 ± 28.1 g), independent of the diet. However, the diet treatment, independent of the time sampling, affected TWW, being higher ($p < 0.05$; Wilcoxon test) in IN (73.6 ± 24.4 g) than in the I (28.8 ± 4.3 g) and in N (41.4 ± 4.6 g) diets and the TWW of seeds in the diet treatment N being higher ($p < 0.05$; Wilcoxon test) than in I. The interaction ($p < 0.05$) between the tested diets and time samplings in the TWW was observed. The seed TWWs at T14 and at T21 were higher ($p < 0.05$; Permutation test) in the IN diet than in the I and N diets, and in the N diet were higher ($p < 0.05$; Permutation test) than in the I diet (Figure 5).

The IWW of the seeds increased significantly ($p < 0.05$; Tukey's test) from T14 (1.6 ± 0.5 mg) to T21 (2.8 ± 1.5 mg), independent of the diet. The diet treatment, independent of the sampling time, also affected IWW, being higher ($p < 0.05$; Tukey's test) in IN (3.6 ± 1.4 mg) than in the I (1.3 ± 0.2 mg) and N (1.8 ± 0.2 mg) diets, and the IWW of the seeds in the diet treatment N being higher ($p < 0.05$; Tukey's test) than in the I diet. The interaction ($p < 0.05$) between the tested diets and sampling times in the IWW was observed. The seed IWW at T14 was higher ($p < 0.05$; Tukey's test) in the IN diet than in the I and N diets, and in the N diet, it was higher ($p < 0.05$; Tukey's test) than in the I diet (Figure 5). At T21, the IWW of the seeds in the diet IN was higher ($p < 0.05$; Tukey's test) than in the I and N diets, and in both (I and N) diets, IWW were not significantly different (Figure 5).

In the shell height and shell length analysis in experiment III, in the diet I, shell height showed significant ($p < 0.05$; Wilcoxon test) growth from T7 to T14 and no difference from T14 to T21, and shell length showed a significant ($p < 0.05$; Wilcoxon test) increase in shell length from T7 to T14 and decrease in shell length from T14 to T21. In the diet IN, the shell height and shell length showed significant ($p < 0.05$; Wilcoxon test) growth from T7 to T14 and from T14 to T21. In diet N, the shell height and shell length showed significant ($p < 0.05$; Wilcoxon test) growth from T7 to T14 and no difference from T14 to T21. Analyzing the shell height and shell length in each time sampling (T7, T14, and T21), high ($p < 0.05$; Tukey's test) values were observed in the diet IN (Figure 6).

3.3. *Fluidization Velocity, Specific Weight, and Linear Regression of FV and TWW*

The fluidization velocity (FV) varies from 0.7 to 2.5 $cm \cdot s^{-1}$ in experiment I, from 0.5 to 2.5 $cm \cdot s^{-1}$ in experiment II, and from 0.5 to 1.5 $cm \cdot s^{-1}$ in experiment III (Table 1). The general seed specific weight of *C. gasar* (experiment II) varies from 724.9 to 1004.2 $kg \cdot m^3$

with a mean of 860.3 ± 72.7 kg·m^3 and of *C. gigas* (experiment III) from 571.4 to 841.6 kg·m^3 with a mean of 706.5 ± 73.7 kg·m^3.

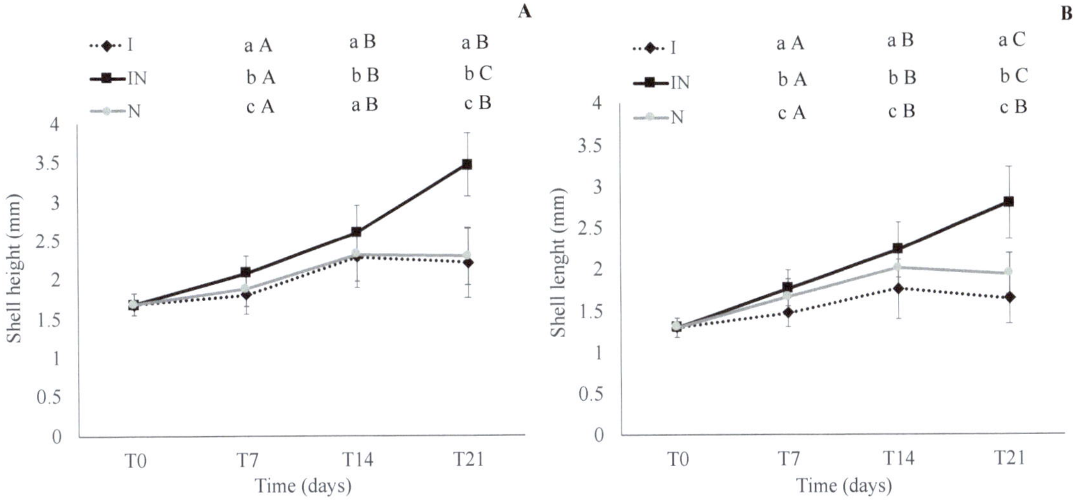

Figure 6. Shell height (**A**) and shell length (**B**) (*C. gigas*) at T0 (beginning of the experiment) and after 7, 14, and 21 days (T7, T14, and T21, respectively) in each treatment (I, IN, and N) in experiment III, where I = diet with 100% of *Isochrysis galbana*; IN = diet with 50% of *I. galbana* and 50% of *Nannochloropsis oculata*; N = diet with 100% of *N. oculate*; and T = time. Different lower letters indicate statistical differences in each time sampling (T7, T14, and T21) comparing diets (I, IN, and N) (shell height at T7: Tukey's test; shell height T14 and T21 and shell length at T7, T14, and T21: Wilcoxon test). Different capital letters indicate statistical differences in each diet treatment (I, IN, and N) comparing time sampling (T7, T14, and T21) (shell height in the diets I, IN, and N and shell length in the diet I and N: Wilcoxon test; shell length in the diet N: Tukey's test). The absence of letters represents that there were no significant differences between the treatments.

The linear regression analysis of TWW and FV was significant ($p < 0.05$) in both experiments (II and III), with an R-squared of 0.8286 (adjusted R-squared of 0.8256) for experiment II with *C. gasar* and of 0.8644 (adjusted R-squared of 0.8414) for experiment III with *C. gigas* (Figure 7).

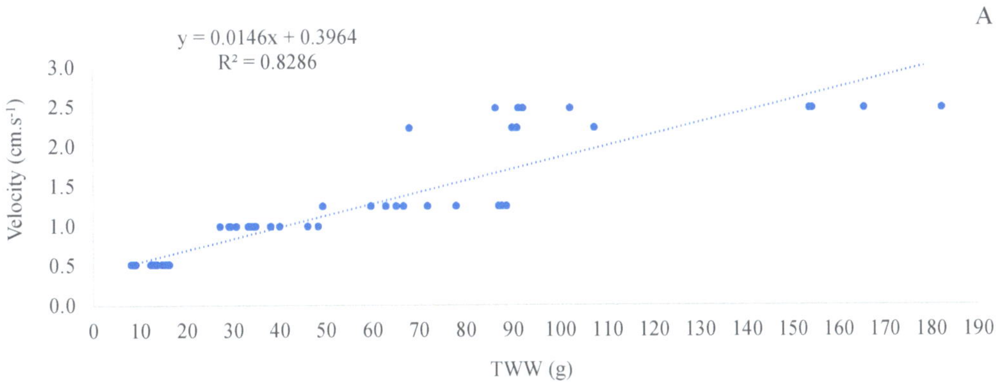

Figure 7. *Cont.*

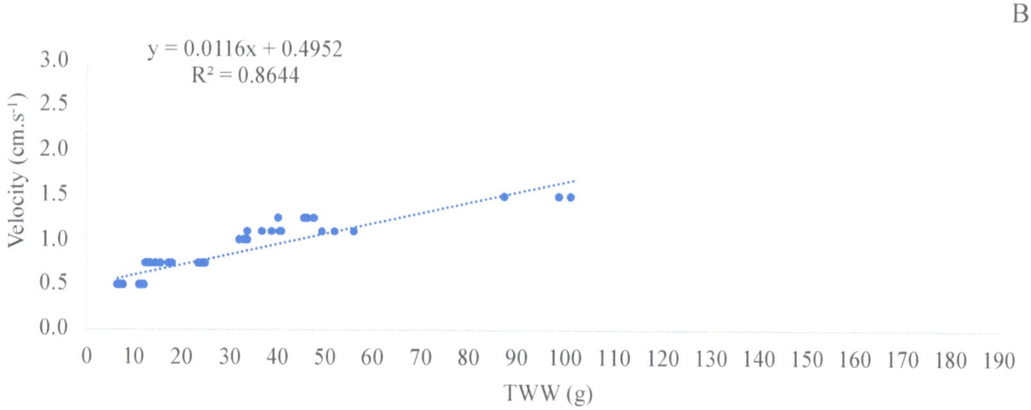

Figure 7. Regression of fluidization velocity (cm·s^{-1}) and total whole weight (TWW; g) in experiment II (*C. gasar*; **A**) and III (*C. gigas*; **B**) during the experiment period. Data of TWW and velocity used were obtained from T0, T7, T14, T21, and T29 (experiment II) and from T0, T7, T14, and T21 (experiment III) of all treatments and all experimental units.

4. Discussion

The seeds of oysters *C. gasar* and *C. gigas* showed high zootechnical performance in the fluidized bed bottle nursery in closed aquaculture system (FBBN-CAS) system used in the present study. The fluidized bed bottle system promotes faster oyster growth than the downwelling system [19]. The FBBN-CAS system designed for this study and flow used promote good conditions for seed development, where the seeds were constantly exposed to feed, observed by the microalgae concentration in the FBBN system before daily handling and by the constant biodeposits removed from the bottle observed in the outlet promoted by the seawater flow.

The fluidization velocity of the upward flow used in this study maintained the seeds fluidized in the bottle by visual analysis and demonstrated by the linear regression analysis was significant (with R-square values up to 80%). In the present study, the seeds of *C. gasar* with a shell height of 4.0 ± 1.7 mm (D60; experiment I) were fluidized with a velocity of 2.5 cm·s^{-1} and *C. gigas* with a shell height of 3.5 ± 0.4 mm (IN; experiment III) fluidized with a velocity of 1.5 cm·s^{-1}, respectively. In a study developed by Ver and Wang [10] with *C. virginica* (6.5 mm of shell height), similar velocity (2 and 2.5 cm·s^{-1}) was described as velocity at minimum fluidization maintained oyster seeds fluidized. The standardization of the FV by the seed TWWs in experiments II and III showed no variation between values, corroborating that the fluidization velocity used in the present study was adequate for the FBBN system tested.

The fluidization velocity of *C. gasar* and *C. gigas* in the FBBN system can be obtained from the linear regression equation presented in the present study with the TWW of seeds or by using specific weights to calculate TWW if the seed volume is known. The seeds of *C. gasar* showed a higher specific weight than *C. gigas*. More studies are suggested to evaluate the specific weight of this species in earlier seed stages and in different nursery culture systems.

For *C. gasar*, the initial bottle seed volumes with all tested stocking densities (D15, D30, and D60) can be feasible. The density of D60, which represents 8.8% of the total bottle volume, produces more seeds per bottle compared to D30 and D15, which represent 4.4% and 2.2% of the total bottle volume, respectively. Despite these higher seed numbers per bottle and consequently the higher total volume of seeds in the initial stocking density D60, oyster seeds in the density D15 showed more animals in the size class of #2–3 and #3–4 than in D60, suggesting that animals after 29 days grew more with the initial bottle area occupation of 2.2%. Both results are interesting for seed production, depending on the hatchery objective. If the aim is to produce a higher seed number or the hatchery has small infrastructures, higher initial stocking densities can be used. However, if the proposal is to produce a higher seed size, a low (2.2% of the bottle volume occupation) initial stocking density is recommended.

Higher growth in lower densities can be related to algae availability and feeding behavior in the bottle due to the relation between water flow and the number of seeds in each bottle. According to James [20], flow is more critical for oyster growth than population density. It could be possible that there was competition for food, even when residual algae was observed after each management. That flow in the high density promoted the fluidization of the seeds. Feeding rates and scope for growth analysis could help better understand seed growth dynamic in the different stocking densities tested.

Bialgae and monoalgae diets tested in the present study showed no effects on the seed's survival and number, though these diets affected the seed's growth. The bialgae diets with microalgae *N. oculata* improved the total volume of the seeds and increased the volume for both species (*C. gasar* and *C. gigas*). Seed growth of both oyster species with a bialgae diet of *I. galbana* and *N. oculata* was also observed by the high seed weight (total whole weight and individual whole weight) and shell height and shell length for both oyster species (*C. gasar* and *C. gigas*). The seed growth of both oyster species (*C. gasar* and *C. gigas*) observed in the bialgae diet of *I. galbana* with *N. oculata* can be related to the nutritional contribution of *N. oculata*. Analysis of fatty acid of *N. oculata* retorted that it is a microalgae rich in fatty acid, as in eicosatetraenoic acid (ETA 20:4n-3; [21]), eicosapentaenoic acid (20:5n-3; [22]) and palmitic acid (C16:0; [22]). Ohse et al. [23] observed that *N. oculata* has more total lipid content than *C. muelleri*, and Sühnel et al. [24] observed that *I. galbana* has more total lipid content than *C. muelleri*.

The bialgae diet, compared to the monoalgae diet, is preferred for better bivalve nutrition; however, even though feed was calculated by biomass as in experiment II, the appropriate selection of microalgae species for algae combination is important for better seed growth. Although Brown et al. [14] did not obtain good results using a monoalgae diet of *Nannochloropsis*-like sp. for *C. gigas* spat diet, in the present study, the monoalgae diet with the microalgae species *N. oculate* showed better zootechnical performance compared to the monoalgae diet with *I. galbana* for *C. gigas*.

The diet of *I. galbana* with *C. muelleri* did not provide the same nutritional value for the seeds to grow compared with *N. oculata*, as can be observed in seed weight growth with *C. gasar*. Lagreze et al. [25] observed that different combination of microalgae species in a bialgae diet promotes different survivals and shell growth of the clam *Anomalocardia brasiliana* (Gmelin 1791) larvae.

The combination of *I. galbana* and *N. oculata* was shown to be appropriate for both oyster species (*C. gasar* and *C. gigas*). Already for the clam *A. brasiliana*, Lagreze et al. [25] suggest a combination of *N. oculata* with *C. muelleri, Pavlova lutheri* (Green 1975; syn *Diacronema lutheri*) with *C. calcitrans, and P. lutheri* with *C. muelleri*. There is no one good microalgae combination for all bivalve species, but for each species and life cycle stage, this combination of two or more algae species in the diet needs to be evaluated. Diet of *I. galbana* with *C. muelleri* (IC) and *R. salina* with *C. muelleri* was unsuitable for *C. gasar*, with lower weight growth. However, if this microalgae species is the only species available, seed survival will not be affected by these diets (IC and RC) but will grow lower. For *C. gigas*, the monoalgae diet of *I. galbana* and *N. oculata* also promotes lower seed weight and shell growth than the bialgae diet.

Seed growth showed an increase in volume for the seeds fed with *N. oculata* compared to the seeds fed with *I. galbana*. However, the differences in seed volumes between both treatments are very low, as both monoalgae diets are not recommended if the hatchery objective is seed growth. However, if only one microalgae species is available, *I. galbana* and *N. oculata* can be used as monoalgae diets without affecting seed survival.

The reduction in seawater temperature, caused by cold weather (wintertime), observed from T14 to T21 in experiment III could explain the increasing residual algae in the treatments with *N. oculata* (IN and N). This fact can be related to the effect of temperature on seed clearance rate. Casas et al. [26] observed a lower clearance rate for *C. virginica* at 10 °C, compared to 20 °C and 30 °C.

The high survival rates observed in the present study for *C. gasar* (from 88 to 95%) and for *C. gigas* (from 82 to 96%) were also reported for other oyster species in the nursery phase (i.e., *C. gigas*, from 75 to 98%; and *O. edulis*, from 82 to 87%; [27]). The growth rate also demonstrated that the present configuration of the system is a good alternative for hatchery operation, reinforcing that FBBN is recommended to reduce hatchery space and optimize water use for oyster seed production. For *C. gasar*, the present study showed that the initial stocking density of 8.8% can produce high seed numbers, and the diet of *I. galbana* and *N. oculate* high seed growth for *C. gasar* and *C. gigas* in the nursery phase using the FBBN-CAS system.

5. Conclusions

In conclusion, the FBBN-CAS system developed for this study promotes a high zootechnical development for the seeds of *C. gasar* and *C. gigas*. For *C. gasar*, an initial stocking density of 8.8% of bottle occupation provides a higher seed number after 29 days of culture. Still, at 2.2% of the initial bottle occupation, bigger seeds can be achieved. All tested diets did not affect oyster seed survival, but the diet composition affects seed growth and the better seed growth of *C. gasar* and *C. gigas* can be achieved with a bialgae diet with *I. galbana* and *N. oculata*.

Author Contributions: Conceptualization, S.S., F.J.L.-S. and C.M.R.d.M.; methodology, S.S., F.J.L.-S. and C.M.R.d.M.; investigation, S.S., G.N.C., J.A., G.d.S., J.P.R.F., F.C.d.S. and C.H.A.d.M.G.; data curation, S.S., G.N.C. and G.d.S.; formal analysis, S.S., F.J.L.-S. and C.M.R.d.M.; writing—original draft preparation, S.S. and F.J.L.-S.; writing—review and editing, S.S., F.J.L.-S. and C.M.R.d.M.; supervision, F.J.L.-S. and C.M.R.d.M.; project administration, S.S. and C.M.R.d.M.; funding acquisition, S.S. and C.M.R.d.M. All authors have read and agreed to the published version of the manuscript.

Funding: This research was funded by the Ministry of Fisheries and Aquaculture grant number 003500063832013-82.

Institutional Review Board Statement: The study was bivalves, and the invertebrate animals do not need ethical approval.

Informed Consent Statement: Not applicable.

Data Availability Statement: The data that support this study are available from the corresponding author upon reasonable request.

Acknowledgments: We express out gratitude to the Ministry of Fisheries and Aquaculture (Process 003500063832013-82) for the post-doctoral scholarship for the first author (*S. Sühnel*) and project financial support. C.M.R. de Melo is a beneficiary of the "National Council for Scientific and Technological Development—Brazil (CNPq)" productivity fellowship (Process 305807/2022-6).

Conflicts of Interest: The authors declare no conflicts of interest.

References

1. Guo, X.; Li, C.; Wang, H.; Xu, Z. Diversity and evolution of living oysters. *J. Shellfish Res.* **2018**, *37*, 755–771. [CrossRef]
2. FAO. *The State of World Fisheries and Aquaculture 2020. Sustainability in Action*; FAO: Rome, Italy, 2020. [CrossRef]
3. FAO. Global Aquaculture Production Quantity (1950–2021). 2024. Available online: https://www.fao.org/fishery/statistics-query/en/aquaculture/aquaculture_quantity (accessed on 21 March 2024).
4. FAO. Global Aquaculture Production Value (1950–2021). 2024. Available online: https://www.fao.org/fishery/statistics-query/en/aquaculture/aquaculture_value (accessed on 21 March 2024).
5. Mao, Y.; Lin, F.; Fang, J.; Fang, J.; Li, J. Bivalve production in China. In *Goods and Services of Marine Bivalves*; Smaal, A.C., Ferreira, J.G., Grant, J., Petersen, J.K., Strand, Ø., Eds.; Springer Open: Cham, Switzerland, 2019; pp. 51–72. [CrossRef]
6. Sühnel, S.; Picanço, T.; Medeiros, S.C.; Magalhães, A.R.M.; Melo, C.M.R. Effects of seeding data and seed size on *Crassostrea gigas* (Thunberg, 1793) culture in a subtropical climate. *J. Shellfish Res.* **2017**, *36*, 303–313. [CrossRef]
7. Helm, M.M.; Bourne, N.; Lovatelli, A. *Hatchery Culture of Bivalves. A Practical Manual*; FAO Fisheries Technical Paper No. 471; FAO: Rome, Italy, 2004; pp. 5–166.

8. Appleyard, C.L.; Dealteris, J.T. Growth of the northern quahog, *Mercenaria mercenaria*, in an experimental-scale upweller. *J. Shellfish Res.* **2002**, *21*, 3–12.
9. Rodhouse, P.G.; O'Kelly, M. Flow requirements of the oysters *Ostrea edulis* L. and *Crassostrea gigas* Thunb. in an upwelling column system of culture. *Aquaculture* **1981**, *22*, 1–10. [CrossRef]
10. Ver, L.M.B.; Wang, J.W. Design criteria of a fluidized bed oyster nursery. *Aquacult. Eng.* **1995**, *44*, 229–249. [CrossRef]
11. Pfeiffer, T.J.; Rusch, K.A. An integrated system for microalgal and nursery seed clam culture. *Aquacult. Eng.* **2000**, *24*, 15–31. [CrossRef]
12. Whyte, J.N.C. Biochemical composition and energy content of six species of phytoplankton used in mariculture of bivalves. *Aquaculture* **1987**, *60*, 231–241. [CrossRef]
13. Langdon, C.J.; Waldock, M.J. The effect of algal and artificial diets on the growth and fatty acid composition of *Crassostrea gigas* spat. *J. Mar. Biol. Assoc.* **1981**, *61*, 431–448. [CrossRef]
14. Brown, M.R.; McCausland, M.A.; Kowalski, K. The nutritional value of four Australian microalgal strains fed to Pacific oyster *Crassostrea gigas* spat. *Aquaculture* **1998**, *165*, 281–293. [CrossRef]
15. Laing, I.; Millican, P.F. Relative growth and growth efficiency of *Ostrea edulis* L. spat fed various algal diets. *Aquaculture* **1986**, *54*, 245–262. [CrossRef]
16. Ronquillo, J.D.; Fraser, J.; Mcconkey, A.J. Effect of mixed microalgal diets on growth and polyunsaturated fatty acid profile of European oyster (*Ostrea edulis*) juveniles. *Aquaculture* **2012**, *360*, 64–68. [CrossRef]
17. Mazón-Suástegui, J.M.; Ruíz-Ruíz, K.M.; Parres-Haro, A.; Saucedo, P.E. Combined effects of diet and stocking density on growth and biochemical composition of spat of the Cortez oyster *Crassostrea corteziensis* at the hatchery. *Aquaculture* **2008**, *284*, 98–105. [CrossRef]
18. O'Connor, W.A.; Nell, J.A.; Diemar, J.A. The evaluation of twelve algal species as food for juvenile Sydney rock oysters *Saccostrea commercialis* (Iredale & Roughley). *Aquaculture* **1992**, *108*, 277–283. [CrossRef]
19. O'Connor, W.; Dove, M.; Finn, B.; O'Connor, S. *Manual for Hatchery Production of Sydney Rock Oyster (Saccostrea glomerata)*; Nsw Department of Primary Industries: Cronulla, NSW, Australia, 2008; pp. 41–42.
20. James, C. Soda Bottle Upweller System: Optimizing Production of Eastern Oyster Seed. 2022. Available online: https://projects.sare.org/wp-content/uploads/Coke-Bottle-Poster.pdf (accessed on 19 March 2024).
21. Cheng, P.; Zhou, C.; Chu, R.; Chang, T.; Xu, J.; Ruan, R.; Chen, P. Effect of microalgae diet and culture system on the rearing of bivalve mollusks: Nutritional properties and potential cost improvements. *Algal Res.* **2020**, *51*, 102076. [CrossRef]
22. Gu, N.; Lin, Q.; Li, G.; Tan, Y.; Huang, L.; Lin, J. Effect of salinity on growth, biochemical composition, and lipid production of *Nannochloropsis oculata* CS179. *Eng. Life Sci.* **2012**, *12–16*, 631–637. [CrossRef]
23. Ohse, S.; Derner, R.B.; Ozório, R.A.; Correa, E.B.F.; Cunha, P.C.R. Lipid content and fatty acid profiles in ten species of microalgae. *IDESIA* **2015**, *33*, 93–101. [CrossRef]
24. Sühnel, S.; Lagreze, F.; Pereira, A.; Da Silva, F.C.; Gurney-Smith, H.; Bercht, M.; Maraschin, M.; Magalhães, A.R.M.; Ferreira, J.F. Effects of astaxanthin on reproductive success in the tropical scallop *Nodipecten nodosus* (Linnaeus, 1758). *J. Shellfish Res.* **2014**, *33*, 89–98. [CrossRef]
25. Lagreze, F.J.S.; Albuquerque, M.C.P.; Araujo, J.; Sühnel, S.; Melo, C.M.R. Survival and growth of the native clam *Anomalocardia brasiliana* (Gmelin, 1791) larvae in laboratory. *Bol. Inst. Pesca SP* **2015**, *41*, 133–143.
26. Casas, S.M.; Figueira, R.; Lavaud, R.; Comeau, L.A.; LaPeyre, M.K.; Peyre, J.F. Combined effects of temperature and salinity on the physiology of two geographically-distant eastern oyster populations. *J. Exp. Mar. Bio. Ecol.* **2018**, *506*, 82–90. [CrossRef]
27. Spencer, B.E.; Akester, M.J.; Mayer, I. Growth and survival of seed oysters in outdoor pumped upwelling systems supplied with fertilized sea water. *Aquaculture* **1986**, *55*, 173–189. [CrossRef]

Disclaimer/Publisher's Note: The statements, opinions and data contained in all publications are solely those of the individual author(s) and contributor(s) and not of MDPI and/or the editor(s). MDPI and/or the editor(s) disclaim responsibility for any injury to people or property resulting from any ideas, methods, instructions or products referred to in the content.

fishes

MDPI

Article

Reproductive Conditioning of the Peruvian Scallop *Argopecten purpuratus* in Different Environments

Rafael Octavio Crisóstomo [1,2,3,4,*], Renzo Pepe-Victoriano [1], Sheda Méndez-Ancca [1,2,5], Abel Walter Zambrano-Cabanillas [6], Olegario Marín-Machuca [6], Hernan Mauricio Perez [4,7], Víctor Yana-Mamani [8] and Mario Ruiz-Choque [5]

[1] Universidad Arturo Prat, Facultad de Recursos Naturales Renovables, Área de Biología Marina y Acuicultura Avda. Arturo Prat 2120, Iquique 1100000, Chile; rpepev@unap.cl (R.P.-V.); smendeza@unam.edu.pe (S.M.-A.)
[2] Programa de Magíster en Acuicultura, Mención en Cultivos de Recursos Hidrobiológicos y Mención en Acuaponía, Universidad Arturo Prat, Avda. Santa María 2998, Arica 1000000, Chile
[3] Universidad de Atacama, Centro de Investigaciones Costeras (CIC-UDA), Avda. Copayapu 485, Copiapó 1530000, Chile
[4] Universidad Católica del Norte, Facultad de Ciencias del Mar, Centro de Innovación Acuícola Aqua Pacífico, Larrondo 1281, Coquimbo 1780000, Chile; hernan.perez@aquapacifico.cl
[5] Universidad Nacional de Moquegua (UNAM), Facultad de Ingeniería y Arquitectura, Departamento Académico de Ingeniería Pesquera, Ciudad Jardín s/n, Ilo 18601, Peru; mruizc@unam.edu.pe
[6] Universidad Nacional Federico Villarreal, Facultad de Oceanografía, Pesquería, Ciencias Alimentarias y Acuicultura, Departamentos Académicos de Acuicultura y Ciencias Alimentarias, Calle Roma 350, Miraflores, Lima 15001, Peru; azambrano@unfv.edu.pe (A.W.Z.-C.); omarin@unfv.edu.pe (O.M.-M.)
[7] Universidad Católica del Norte, Facultad de Ciencias del Mar, Laboratorio Central de Cultivos Marinos, Larrondo 1281, Coquimbo 1780000, Chile
[8] Universidad Nacional de Moquegua (UNAM), Facultad de Ingeniería y Arquitectura, Departamento Académico de Ingeniería de Sistemas e Informática, Ciudad Jardín s/n, Ilo 18601, Peru; vyanam@unam.edu.pe
* Correspondence: rafael.crisostomo@uda.cl

Citation: Crisóstomo, R.O.; Pepe-Victoriano, R.; Méndez-Ancca, S.; Zambrano-Cabanillas, A.W.; Marín-Machuca, O.; Perez, H.M.; Yana-Mamani, V.; Ruiz-Choque, M. Reproductive Conditioning of the Peruvian Scallop *Argopecten purpuratus* in Different Environments. *Fishes* **2024**, *9*, 9. https://doi.org/10.3390/fishes9010009

Academic Editors: Wagner C. Valenti, Andrew G. Jeffs and Eduardo Luis Ballester

Received: 5 September 2023
Revised: 10 November 2023
Accepted: 11 December 2023
Published: 24 December 2023

Copyright: © 2023 by the authors. Licensee MDPI, Basel, Switzerland. This article is an open access article distributed under the terms and conditions of the Creative Commons Attribution (CC BY) license (https://creativecommons.org/licenses/by/4.0/).

Abstract: Obtaining viable *Argopecten purpuratus* seeds faces challenges, especiallyci the unpredictability of the marine environment and high production costs in hatcheries. However, improving the method of "Broodstock Conditioning In Hatcheries" is key to ensure permanent seed supplies by minimizing the dependence on marine conditions and by maximizing economic viability in hatcheries. In an effort to overcome these barriers, broodstock were conditioned into two different environments: (a) Natural Environment: Natural marine conditions located in Bahía Inglesa, Atacama Region, Chile. (b) Hatchery: Laboratory conditions to achieve gonadal maturation, spawning induction, fertilization and larval development. The purpose of this research was to evaluate how the type of reproductive conditioning affects the reproductive potential and nutritional quality of the progeny. Both methods were successful at inducing the necessary maturity for reproduction, obtaining viable gametes and larvae. On the other hand, it was observed that in the natural environment, the oocytes and D larvae reached a greater size and nutritional value, being the most significant differences with ($p < 0.05$): the size of the D larvae reached figures of 95.8 ± 3.1 µm and 91.2 ± 2.7 µm in the environment and hatchery, respectively; the lipid content in dry mass was 25.2 ± 3.1 mg g^{-1} and 13.5 ± 1.9 mg g^{-1} for the natural environment and hatchery, respectively. Although quality indicators in hatcheries were slightly lower compared to the natural environment, the possibility of conditioning *A. purpuratus* broodstock independently of environmental variability highlights the importance of further optimizing broodstock conditioning aspects in hatcheries that would allow more predictable and sustainable production.

Keywords: reproductive conditioning; hatchery; broodstock; *A. purpuratus*

Key Contribution: This study makes a detailed contribution to the conditioning of *A. purpuratus* broodstock to optimize gonadal maturation and facilitate effective spawning, presenting a carefully designed technical strategy that aims to reach continuous improvement in reproduction and in the nutritional quality of oocytes and D-larvae, which is a key factor for the success and sustainability of seed supply for the aquaculture of this species.

1. Introduction

The cultivation of *A. purpuratus* has been an important economic activity in the regions of Coquimbo and Atacama for more than 30 years [1,2]. Nevertheless, the development of this scallop has been subject to considerable production and commercial fluctuations, which have prevented its consolidation [3,4]. An aspect that has been a constant problem for this culture has been the variability of naturally captured juveniles (seeds), which has always been the main source of seeds used for the farmingo of *A. purpuratus* [5]. An attempt has been made to address this by using larval and juvenile crop systems under laboratory (hatchery) conditions [6,7]. However, for these systems to be viable, it is necessary to have sexually mature broodstock that can provide gametes of adequate quality for a successful seed production [8].

To obtain mature broodstock of this species, two modalities were used: (a) Broodstock were maintained under natural or environmental conditions (natural environment) until favorable environmental conditions, for adequate gametogenesis development, existed. (b) Broodstock were bred under laboratory conditions (hatchery) with temperature controlled at 15 ± 1 °C. Both methods were used to produce seeds of *A. purpuratus* for large-scale cultivation of this species [9]. The broodstock was provided with microalgae as a food source [10]. Microalgae such as *Isochrysis galbana* var (t-iso) and *Nannochloropsis oculata* were used, which are necessary for the sexual maturation of *A. purpuratus* broodstock, They were supplied for the hatchery [9]; these microalgae thrive in a wide range of temperatures, from 15 °C to 30 °C [11], and improve the survival of the scallop larvae during in the early stages [12]. This scallop species inhabits semi-enclosed bays and is also cultivated there, close to upwelling areas, where the subsurface water rises and generates sudden changes in nutrients, temperature, and oxygen concentration levels [13–15]; thus, the variability and unpredictability of natural environmental conditions tend to be the greatest disadvantage of conditioning in the sea. In the case of laboratory conditioning, or the hatchery, the main difficulties are related to the high cost of microalgae production, temperature maintenance, and obtaining filtered and sterilized seawater for the process. Temperature and food availability are the most relevant environmental factors for the proper development of gametogenesis (cell proliferation and vitellogenesis) in bivalves [11,16].

Temperature is considered a crucial factor in the regulation of bivalve reproduction, and the seasonal changes it undergoes have often been correlated with gonadal growth [17]. In general, it has been argued that increased temperature, in the marine environment, accelerates the gonadal maturation process in bivalves, since observations of the reproductive cycles of several of these species indicate that the main spawning season occurs during the spring–summer period [15,17] and the reproduction of *A. purpuratus* is only viable at temperatures above 15 °C [18]. It has been found that under laboratory conditions, gonadal maturation is more appropriate under stable conditioning temperatures, where thermal stability would prevent the occurrence of partial spawning that delays the gonadal development process. In general, conditioning with a specific temperature regime must be analyzed in conjunction with other relevant factors of the process, as an individual analysis of isolated factors may lead to erroneous conclusions [17,19,20].

Regarding food availability [21], it was observed that the recovery of gonadal function was accelerated in *Ostrea edulis* broodstock and was also found that the larvae obtained from them were better developed when feeding levels were increased [22–24]. They also postulated that differences in gamete´s quality would depend on the environmental conditions to which the parents were exposed to. In bivalves, ingested food accumulates in the adductor muscle as energy reserve in the form of glycogen, a component that can be used as an energy source for the gametogenesis process [25,26]. In terms of food quality, it has been observed that the number of eggs and larvae produced is directly related to the nutritional value of the food provided during the parental reproductive conditioning process [12,27]. Even if after parental food reserves have been depleted and the larvae begin to consume exogenous food, growth and survival are positively affected if gonadal maturation conditions are adequate. In this regard, oocyte size is a factor that has been considered by several authors as a valid indicator of progeny quality [25,28,29]. While Working with *O. edulis* [21], it was found that maximum viability and survival in larval crops are directly related to the lipid content of the broodstock; for this reason, the diet is based on polyunsaturated fatty acids, such as docosahexaenoic acid (DHA), this component is found in concentrations of 31.31 ± 2.92% in the microalgae *N. oculata* [30] and eicosapentaenoic acid (EPA), which reaches 37.88 ± 0.66% in the microalgae *I. galbana* [31]. These microalgae complement each other and are used in aquaculture as an important source of nutrients to stimulate the gonadal maturation of broodstock.

Interestingly, in [32], a similar result was obtained from *Mercenaria mercenaria* and *Crassostrea virginica* larvae, where they found a high correlation between larval survival and the initial amount of lipids present in the oocytes. Consequently, lipids play a fundamental role both energetical and functionally, as they are not only the energy reserve source but serve as precursors of hormones and are also the main component of cell membranes [33,34].

Based on the information presented above, this study aimed to examine the quality of the offspring by using two methods of reproductive conditioning: natural (environment) and laboratory (hatchery) conditions. It is important to emphasize that the natural condition is characterized by its constant variability and it is unpredictability, although seed production does eventually occur. This contrasts sharply with the second method (laboratory), where factors such as feeding and temperature are more stable, providing a more regular environment for the continuous development and supply of seed. This sustainable approach contributes to the viability of *A. purpuratus* farming.

2. Materials and Methods

2.1. Obtaining Biological Material

A total of 300 *A. purpuratus* oysters were collected at the (CIC-UDA) Coastal Research Center at the University of Atacama marine concession, located in the "El Morro" sector, Bahía Inglesa, in the Atacama Region of Chile, at 27°8′13″ S latitude and 70°54′22″ W longitude, They were transferred to the (CIC-UDA) hatchery, located at 27°8′11″ S latitude and 70°54′18″ W longitude. The oysters were kept in tanks with circulating seawater to reduce the stress caused by transportation.

2.2. Experimental Design

To evaluate the influence of the two different environments on the reproductive capacity of adult *A. purpuratus* and the nutritional quality of the offspring, 180 immature broodstock with an average valve length of 8 ± 1 cm were selected from the biological material previously described. The broodstock were evenly distributed into two treatments: (environment) gonadal maturation in the natural marine environment, in suspended farms such as lantern nets; (hatchery) conditioning in farming tanks in a controlled environment, called as hatchery. Each treatment was replicated three times to estimate outcomes reliably.

2.2.1. Natural Maturation of *A. purpuratus* Broodstock in Marine Conditions

In this treatment (environment), 90 immature broodstock of *A. purpuratus* were selected for gonadal maturation. Coming from the crop in three lantern nets at a rate of 30 individuals per lantern net, each of which represented a replica of the trial. The nets were then submerged under a depth of 5 m below the ocean surface to facilitate optimal exposure of the broodstock to the marine environment. During this stage, the gonadal maturation process was observed every two weeks by SCUBA diving professionals at 5 m in depth (Figure 1).

In the last phase, the sexually mature broodstock of *A. purpuratus* were placed, temporarily, in ponds of 7.2 m^3 capacity until their spawning and fertilization were achieved; after this, embryonic development was carried out in tanks until larva D was obtained to estimate reproductive and production indicators.

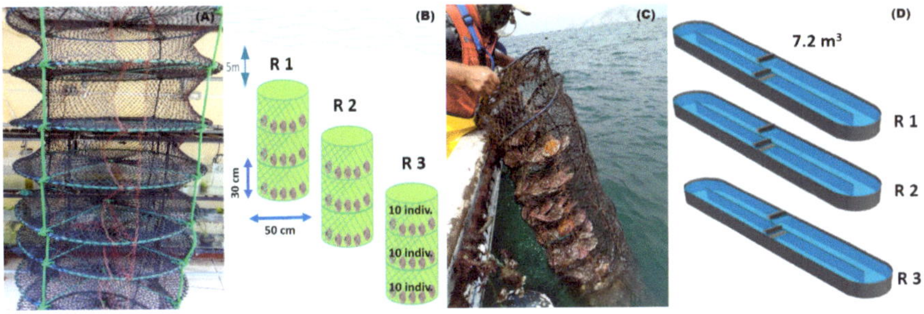

Figure 1. (**A**) Photograph of a lantern-type net where immature broodstock of *A. purpuratus* were placed for gonadal maturation, (**B**) diagram showing three replicates of the treatment (environment) organized in three divisions, in each division, a density of 10 broodstock was placed due to the reduced dimensions of the lantern net of 50 cm in diameter and 30 cm high, (**C**) photograph showing the collection of a lantern net containing mature broodstock, which were transferred to the hatchery of the (CIC-UDA), (**D**) figure showing three 7.2 m^3 farming tanks in which the reproduction of scallop was carried out. These tanks were operated by using the recirculatory aquaculture system (RAS).

2.2.2. Conditioning of Broodstock in the Hatchery

In a previous phase, to start the treatment (hatchery) and to feed the broodstock of *A. purpuratus* in the hatchery, the intensive growing of two varieties of marine microalgae was carried out: *I. galbana* (Tiso) and *N. oculata*. Two fiberglass tanks of 12 m in length, 2 m in width, and 0.5 m in useful height, and with a capacity of 9.6 m^3 were installed to ensure the proper circulation of seawater, There was also an agitation system installed, consisting of a vane driven by a three-phase motor was installed, generating a rotating flow of water at a speed of 1.57 rad/s. It is also important to note that the operating water level was 0.4 m. Keeping the water temperature constant at 15 ± 1 °C (Figure 2).

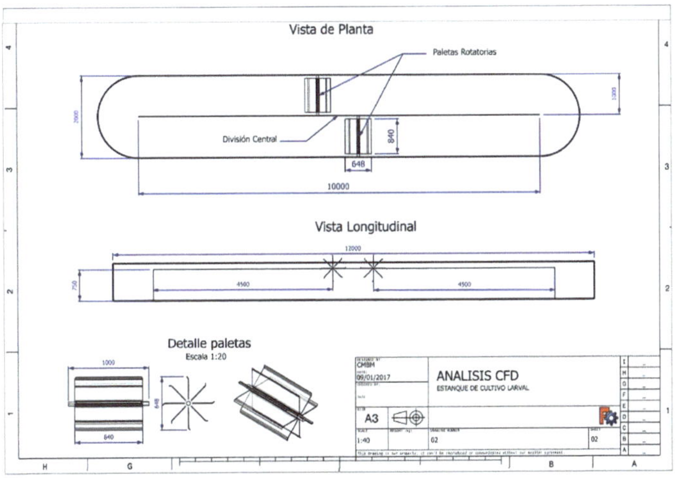

Figure 2. Details of the size of the tank used for the cultivation of microalgae, and the configuration of the propeller paddle used in the cultivation of live feed for the broodstock of *A. purpuratus*.

To carry out the treatment (hatchery), the remaining 90 immature broodstock of *A purpuratus* from the cultivation lines of the marine concession (CIC-UDA) were conditioned in three fiberglass tanks of 0.3 m^3 capacity. These tanks were installed in the hatchery of the (CIC-UDA), whose dimensions were 1.2 m long, 0.6 m wide and 0.6 m high. Each contained thirty scallops and represented a replica. In these tanks, to achieve gonadal maturation continuous aeration was ensured, and a complete water change was carried out, including cleaning every 48 h, so it was a static water management system, maintaining a constant water temperature at 15 ± 1 °C thanks to a 24,000 BTU air conditioning system (Figure 3)

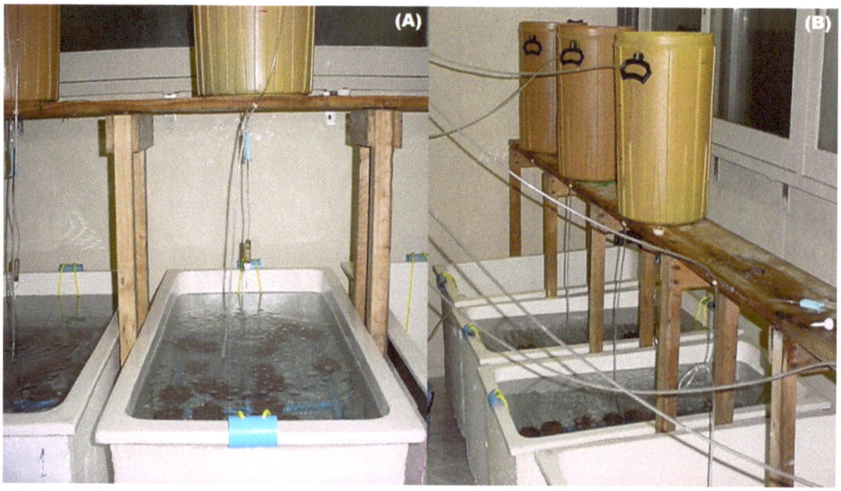

Figure 3. (**A**) Photo of tanks used to condition immature broodstock of *A. purpuratus* for gonadal maturation. They were rectangular to facilitate feed distribution and improve oxygen exchange, (**B**) Photo of the distribution of the culture tanks of the static system.

In a later phase, the gonadally mature broodstock were migrated to three fiberglass crops tanks of 7.2 m^3 capacity, 12 m long, 2 m wide, 0.75 m high, each tank representing a replica that served for broodstock spawning, oocyte fertilization, embryogenesis, and obtaining larva D of *A. purpuratus*. The pond was equipped with 2 mechanical water propulsion systems (propeller vanes); the rotation speed of the vanes was set at 1.57 rad/s, and the fluid level in the pond was 0.3 m high. In addition, a heater was installed to regulate the temperature in a range of 18 ± 2 °C; this equipment had a power of 6 kW, voltage of 220 V, current of 16 A, three-phase power, seawater-proof housing, electronic panel for temperature range, temperature sensors for upper and lower limit to prevent overheating or cooling of the water. The culture system model was the Recirculatory Aquaculture System (RAS), maintaining average velocities from 0.02 m/s near the walls to 0.18 m/s near the pallet rotation´s sector. The average flow within the pond was 307 m^3/h (Figure 4).

The three phases of hatchery conditioning are summarized in the following diagram, where the first part of the procedure of microalgae farming is represented, the second phase consists of the treatment (hatchery) through conditioning of immature broodstock of *A. purpuratus* for sexual maturation. And the third phase consists of induction to spawn of mature broodstock, fertilization of oocytes, embryonic and larval development, and to measure the reproductive capacity and nutritional quality of the progeny (Figure 5).

2.2.3. Spawning, Fertilization, Early Larval Development and Sampling of Treatments

After 50 days of treatment, 5 mature broodstock (n = 5) of each replicate were sampled to determine gonadal maturity by using the following reproductive parameters:

- The gonadosomatic index (GSI) was calculated by using the following formula: GSI = (Gonad Weight/Total Body Weight) × 100;

- Percentage of spawning assessed using the fertility calculation method: Spawning% = (N° of fertilized eggs/N° of spawning broodstock)
- The number of oocytes per individual was counted in a Sedgewick-Rafter counting chamber: N° of oocytes = (N° of oocyte cells counted/Area of the counting chamber) × gram of gonadal tissue. According to the protocol used by [35]

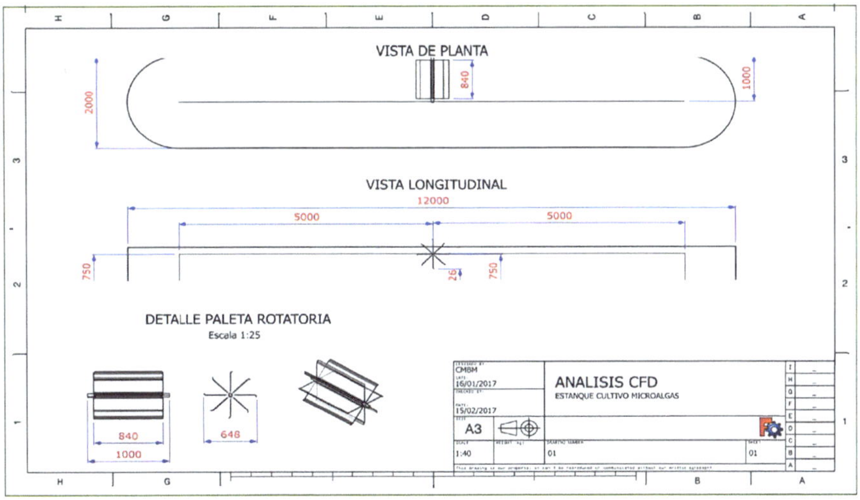

Figure 4. Design of the culture tanks adapted for *A. purpuratus* mature broodstock. In this controlled environment, the process of spawning, fertilization of oocytes, embryonic development, and obtaining D. larvae was successfully accomplished.

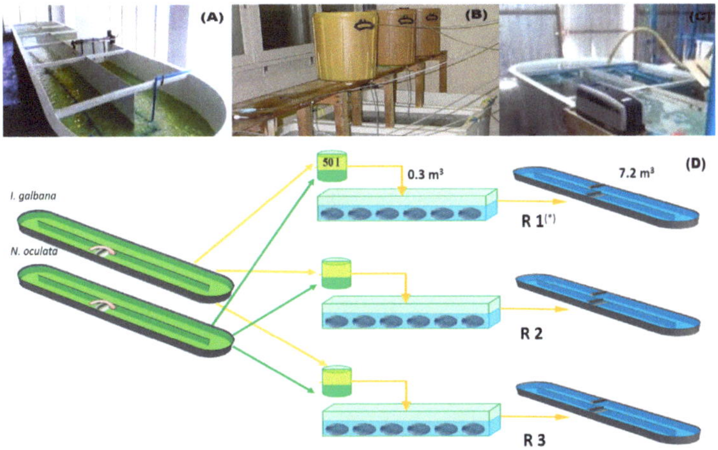

Figure 5. (**A**) Photo of the production of microalgae for broodstock feed, (**B**) photo conditioning of immature broodstock in hatchery, (**C**) spawning photo tank, oocyte/ embryo fertilization and larval development D, (**D**) diagram of the phases of the conditioning treatment of the broodstock in the hatchery (hatchery),where (*) is the number of treatment replicates.

Once the sexual maturation was verified, the broodstock of the two treatments were placed into six cultive tanks with a capacity of 7.2 m³ (Figure 4), where controlled spawning was induced in the hatchery, increasing the water temperature gradually until it reached 18 °C, triggering the spawning and fertilization reflex the eggs. After the reproduction process was completed, the broodstock of *A. purpuratus* were removed from the tanks. The oocytes were maintained with light aeration and were not fed for 48 h, until they reached the straight hinge stage of larval development D larvae. Subsequently, the water was homogenized, and 20 mL were collected from each of the tanks to determine the following production parameters:

- Percentage of fertilized oocytes, which were counted with the help of a 1 mL (50 × 20 × 1 mm) Sedgewick-Rafter counting camera using a Leica MZ10 stereoscope. The percentage of fertilized eggs was calculated with the following formula: Percentage of Fertilized Eggs = (N° of Fertilized Eggs/(N° of Fertilized Eggs + N° of Unfertilized Eggs) × 100
- Larval survival (a). The survival of the D larvae of *A. purpuratus* was determined 48 h after fertilization by taking samples of 20 mL from each tank and using the volumetric method: Larval survival = (N° of surviving larvae/N° of initial larvae) × 100

2.3. Feeding of the Broodstock

The microalgae diet was fed to the broodstock of *A. purpuratus* in the form of living food, consisting of a 50:50 mixed diet by the number of *I. galbana* (T-iso) cells and *N. oculata* cells, using the Neubauer chamber [36,37] through an automatic drip system. The amount of food supplied was equivalent to 6% of the scallop's body weight (shell plus flesh) [17]. The Maintenance of the broodstock at the hatchery ended when more than 80% of the scallops had spawned.

2.4. Size Measurements and Survival

The fertilized eggs and 48 h old D larvae from the two reproductive conditioning treatments were measured (n = 3)/They replicate using the Leica ICC50 W microscope with LAS EZ 3.1.0 imaging software. Oocyte diameter and anteroposterior length of D larvae were measured. The D larvae were counted and their survival was determined according to the number of fertilized eggs from each treatment distributed in each 7.2 m^3 incubation tank.

2.5. Proximal Biochemical Composition

The oocytes and straight-hinge larvae ("D" larvae) were analyzed to determine the lipid [38], protein [39], and carbohydrate [40] contents.

2.6. Statistical Analyses

For the statistical analysis, homoscedasticity and normality were established, and later a completely randomized design was applied, establishing two treatments, with three replicates, consisting of the cultivation of *A. purpuratus* broodstock in two conditions: natural (environment) and laboratory-controlled (hatchery). The variables studied were the sexual maturation of *A. purpuratus*; the following were defined as response variables: spawning, number of oocytes/embryo, fertilized eggs, larval size, survival, and progeny quality (biochemical composition of larvae). All evaluated parameters were compared between conditioning treatments by using Students' t-test.

3. Results

3.1. Development of Gonadal Maturation

After 50 days of conditioning, both experimental groups indicated an increase in their GSI during the conditioning period. The values obtained in both treatments reflected no significant difference at the end of the process $p < 0.05$, (Table 1).

Table 1. Gonadosomatic index (GSI), percentage of spawning, and fecundity index (expressed as the average number of released oocytes per scallop) of *Argopecten purpuratus* held in two different conditions.

Treatments	GSI	Spawning (%)	Number of Oocytes/ Ind.
Environment	27 ± 2.9	88	1,654,343
Hatchery	25 ± 2.5	82	1,534,576

Each value is a mean ± SD (n = 5).

3.2. Oocyte Evaluation

The size of the oocytes obtained from the spawning of broodstock conditioned in the natural environment (Environment treatment) was significantly greater than those obtained from the broodstock conditioned in the laboratory (Hatchery treatment) ($p < 0.05$, Table 2). Despite this difference in size, the fertilization percentage did not present significant differences ($p > 0.05$, Table 2).

Table 2. Size and fertilization success of oocytes from *Argopecten purpuratus* held in two different conditions.

Treatments	Fertilized Eggs (%)	Oocyte Size (µm)
Environment	93.2 ± 3.3	65.9 ± 1.7 *
Hatchery	91.7 ± 2.8	61.4 ± 1.2

Each value is a mean ± SD (n = 3). * indicates significantly different values (t-test, $p < 0.05$).

Regarding the biochemical composition of the oocytes, the lipid content was significantly higher in those obtained from broodstock conditioned in the natural environment, and there were no significant differences with ($p < 0.05$) in the protein and carbohydrate contents (Table 3).

Table 3. Biochemical components of oocytes from *Argopecten purpuratus* held in two different conditions.

Treatments	Proteins (mg g^{-1} Dry Mass)	Lipids (mg g^{-1} Dry Mass)	Carbohydrates (mg g^{-1} Dry Mass)
Environment	356.7 ± 93.8	98.9 ± 8.3 *	39.4 ± 6.3
Hatchery	342.5 ± 85.7	41.6 ± 6.7	36.7 ± 7.1

Each value is a mean ± SD (n = 3). * indicates significantly different values (t-test, $p < 0.05$).

3.3. Straight-Hinge Larvae ("D" Larvae) Evaluation

The size of D larvae from the spawning of broodstock conditioned in the natural environment (environment treatment) was significantly larger than those obtained from broodstock conditioned in the laboratory (hatchery treatment) ($p < 0.05$, Table 4). Survival showed no significant differences between both experimental groups $p > 0.05$ (Table 4).

Table 4. Size and survival of straight-hinge larvae from *Argopecten purpuratus* held in two different conditions.

Treatments	Larval Size (µm)	Larval Survival (%)
Environment	95.8 ± 3.1 *	52.1 ± 2.5
Hatchery	91.2 ± 2.7	49.5 ± 3.3

Each value is a mean ± SD (n= 3). * indicates significantly different values (t-test, $p < 0.05$)

In relation to the biochemical composition of D larvae, the lipid and protein contents were significantly higher in those from broodstock conditioned in the natural environment ($p < 0.05$, Table 5). There were no significant differences in the carbohydrate content $p > 0.05$, (Table 5).

Table 5. Biochemical components of oocytes from *Argopecten purpuratus* held in two different conditions

Treatments	Proteins (mg g^{-1} Dry Mass)	Lipids (mg g^{-1} Dry Mass)	Carbohydrates (mg g^{-1} Dry Mass)
Environment	256.5 ± 21.1 *	25.2 ± 3.1 *	13.1 ± 3.3
Hatchery	198.8 ± 17.4	13.5 ± 1.9	11.3 ± 3.7

Each value is a mean ± SD (n = 3). * indicates significantly different values (t-test, $p < 0.05$).

3.4. Breeding Stock Conditioning in Hatchery

In the hatchery (hatchery), three tanks of 300 L each were installed to create an environment under controlled conditions. Inside these tanks, the broodstock were provided with food using microalgae obtained from 50 L plastic containers. These containers contained cultures of *I. galbana* and *N. oculata*, and constant aeration was maintained (Figure 1).

4. Discussion

Both treatments used for reproductive conditioning of *A. purpuratus* proved to be viable. However, in regions where the temperature drops below 15 °C, it is essential to carry out reproduction under laboratory conditions (hatchery). This helps to reduce the uncertainties associated with natural conditions and ensures a steady supply of seed. It is important to note that the second condition (hatchery) must be carefully calibrated to replicate key aspects of the natural habitat conducive to reproduction. This includes the provision of appropriate microalgal diets rich in essential fatty acids such as polyunsaturated fatty acids (HUFA): C20:5n3 (EPA), C22:6n3 (DHA) and C22:5n6 (DPA). These fatty acids are abundant in different microalgae species, such as *I. galbana*, with concentrations of (EPA) at 37.38 ± 0.66% and 26.47 ± 0.55%, according to [31] and [12], respectively. Similarly, *N. oculata* presents (EPA) levels of 30.77 ± 1.19%, according to [12], and (DHA) levels that reach up to 31.31 ± 2.92%, according to [30]. It is important to highlight that (DHA) is crucial, serving as a precursor for the synthesis of prostaglandins, which plays a fundamental role in processes such as gametogenesis, vitellogenesis, and spawning [41]

In this study, by feeding the broodstock with the microalgae *I. galbana* and *N. oculata*, spawning rates of 88% and 82% were achieved in the natural environment and in hatchery conditions, respectively. Similar results were obtained with [42]: on a high-protein diet, scallops achieved 83.9% reproductive efficiency; on a standard diet, this dropped to 62.6%, and a low-protein diet resulted in suboptimal growth and reproductive performance. In contrast to the lower spawning percentage, [17] achieved 70% under similar laboratory hatchery conditions, maintaining a stable temperature of 15 °C and providing a mixed diet of microalgae, including *I. galbana Chaetoceros gracilis*. Other researchers, such as [43,44], fed broodstock of *A. purpuratus* by using three different treatments. The first diet consisted of a mixture of microalgae species, including *Isochrysis galbana (T-Iso clone), Tetraselmis suecica, Pavlova lutheri and Chaetoceros gracilis*. The second was microalgae supplemented with an emulsion rich in 22:6n-3 (DHA), which resulted in spawning occurring 5 to 7 h earlier than those fed only by microalgae or microalgae supplemented with EPA. The third treatment consisted of microalgae supplemented with 20:5n-3 (EPA), which resulted in a 2% higher lipid content found in the oocytes than in the previous treatments, and *Nannochloropsis* sp. [45]. It appears that gonadal maturation responses are linked to the lipid and protein content of the diet of the purple scallop *A. purpuratus* [46].

Similarly, for oocytes, a larger diameter was observed in the natural condition (environment), indicating a positive correlation between oocyte diameter and subsequent larval viability [12]. Even an increase in oocyte diameter can imply an increase in yolk content [19]. Therefore, the larger oocyte size observed in the natural state is associated with a high accumulation of energy reserves, especially lipids [38,43,46,47], such as polyunsaturated fatty acids (HUFA), which are part of the matrix and cell membranes [10,48]. In this study, significant differences were found in the lipid content of the oocytes with values of 98.9 ± 8.3 mg g^{-1} dry mass and 41.6 ± 6.7 mg g^{-1} dry mass in the natural (environment) and (hatchery) conditions, respectively. In contrast, proteins and carbohydrates showed no significant differences, suggesting that lipids have a significant effect on fertilization and oocyte survival. This conclusion is supported by the findings of [49], who indicated that lipids contributed the greatest percentage (46.7%) of the energy required for embryogenesis, followed by proteins and carbohydrates, which contributed 43.5% and 9.8%, respectively. A diet rich in a variety of microalgae, particularly diatoms and green algae, is essential to produce high-quality oocytes and to improve the reproductive performance of scallops in hatcheries [50]. Regarding the size of "D" larvae, there were significant differences between the two treatments —natural environment and hatchery conditions— with sizes reaching 95.8 ± 3.1 µm and 91.2 ± 2.7 µm, respectively. In another similar work [51], in which the bivalve *Gari solida* was cultivated in a hatchery, the D-type larvae of the bivalve reached a size of 78 ± 4.7 µm, depending exclusively on microalgae. The ability of phytoplankton to accumulate endogenous reserves in "D" larvae depends on the quality and quantity of proteins, which are essential for larval development [19]. Other authors indicate that lipids, proteins, and carbohydrates contribute up to 47.6%, 44.9%, and 7.5%, respectively, of the energy expended during larval shell formation in the prodissoconcha I phase of the species *Patinopecten yessoensis* [49,52].

As a result, early embryonic and larval development requires a significant amount of lipids and proteins that are rapidly consumed during the formation of new larval structures such as the larval shell and velum [10,29]. In addition to being the primary source of energy during the early stages of development, lipids provide essential polyunsaturated fatty acids that are crucial for the formation of cell membranes [29,45,47]. Therefore, a higher content of lipids stored in the oocytes would increase the developmental rates of the early larvae [10,53], explaining the larger size achieved by the "D" larvae from the naturally conditioned broodstock [45]. Demonstrating that the diet consumed by oysters in the natural environment during reproductive conditioning provides high levels of polyunsaturated fatty acids, suggesting that the resulting larvae have cell membranes with elevated levels of EPA (22:6 n-3), positively influencing their development and survival. However, these contributions would depend on the quality and availability of certain phytoplankton species containing high concentrations of polyunsaturated fatty acids [54], whose presence depends on the prevailing environmental conditions, which are highly variable in upwelling areas [13,14,31].

During the study period, the environmental conditions in the Bahía Inglesa were favorable for the reproductive process of *A. purpuratus*, with recorded phytoplankton blooms that enhanced the conditioning process. However, these conditions are not always present, which limits the availability of high-quality broodstock. Therefore, it is necessary to resort to conditioning of broodstock under laboratory conditions (hatchery) because of the success of the spawning events and the high survival rate of the larvae at 48 h of development, and like that demonstrating the feasibility of conditioning under controlled conditions. Nevertheless, significant advances could be made in this process by identifying native microalgae present in marine phytoplankton with high levels of lipids (polyunsaturated fatty acids) and proteins that can be cultivated in the hatchery to replace or supplement commercial microalgae diets, such as (*I. galbana* var T-iso, *Chaetoceros calcitrans, N. oculata, Pavlova lutheri*) diet. Ultimately, these improvements will increase production and economic benefits for mariculturists [8]. Furthermore, this practice will promote sustainability by reducing dependence on natural resources and mitigating the environmental impacts associated with overexploitation of marine resources [14].

5. Conclusions

This study demonstrates the effectiveness of broodstock conditioning for *A. purpuratus* seed production, both under natural marine (environment) and (hatchery) conditions. Nevertheless, certain indicators of oocyte (size and lipid content) and larval (size, lipid, and protein content) quality were found to be lower in laboratory-conditioned broodstock in the hatchery than in the natural conditions. However, the successful spawning and robust survival of 'D' larvae at 48 h of development underlines the feasibility of conditioning under controlled conditions. This opens up the possibility of becoming less dependent on the vagaries of the marine conditions (environment) and represents a significant step towards sustainable *A. purpuratus* seed production.

Author Contributions: Conceptualization: R.O.C., S.M.-A., R.P.-V. and O.M.-M.; data curation: S.M.-A., H.M.P., A.W.Z.-C., O.M.-M. and M.R.-C.; formal analysis: S.M.-A., R.P.-V., R.O.C., A.W.Z.-C., O.M.-M. and V.Y.-M.; funding acquisition and investigation: R.O.C., H.M.P., R.P.-V.; methodology: S.M.-A., R.O.C., H.M.P., A.W.Z.-C., O.M.-M. and V.Y.-M.; supervision R.O.C., H.M.P. and O.M.-M., validation: R.O.C., H.M.P., S.M.-A., A.W.Z.-C. and R.P.-V.; writing—original draft: R.O.C., H.M.P., S.M.-A. and M.R.-C.; writing—review: and editing S.M.-A., A.W.Z.-C., O.M.-M., V.Y.-M. and M.R.-C. All authors have read and agreed to a published version of the manuscript.

Funding: This research was funded by the National University of Centro de Investigaciones Costeras, Universidad de Atacama, Chile, and Centro de Innovación Acuícola Aqua Pacífico, Facultad de Ciencias del Mar, Universidad Católica del Norte, Larrondo 1281, Coquimbo (Directoral Resolution N°. 953-2022). This work was also financed by the internal research project fund, DIUDA inserción, of the Universidad de Atacama.

Institutional Review Board Statement: Ethical review and approval were waived for this study due to the following reasons: The main contribution of this research is to develop a controlled environment to facilitate the successful reproduction of the mollusk *A. purpuratus*. This breakthrough boosts the development of sustainable invertebrate aquaculture and, in perspective, promotes future food security. The research also focuses on marine invertebrates belonging to one mollusk species. These organisms, lacking a central nervous system and exhibiting rudimentary behavioral responses, have limited cognitive abilities. Importantly, the methodology employed in our research has been carefully designed to minimize any potential impact on the welfare of the organisms studied. Our approach focuses on practices that ensure maximum respect for animals while maintaining the integrity and health of the individuals involved. Given the context of our study and the nature of the organisms under investigation, we strongly believe that the exemption from ethical approval is relevant and by accepted ethical guidelines in the scientific community.

Data Availability Statement: The data presented in this study are available upon request to the corresponding author.

Acknowledgments: The author is extremely grateful to Carlos Gonzalez Allende for his technical assistance.

Conflicts of Interest: The authors declare no conflicts of interest. The authors declare that the research was conducted in the absence of any commercial or financial relationship that could be construed as a potential conflict of interest.

References

1. Molina, R.; Cerda, R.; González, E.; Hurtado, F. Simulation model of the scallop (*Argopecten purpuratus*) farming in northern Chile: Some applications in the decision making process. *Lat. Am. J. Aquat. Res.* **2012**, *40*, 679–693. [CrossRef]
2. Stotz, W. When aquaculture restores and replaces an overfished stock: Is the conservation of the species assured? The case of the scallop Argopecten purpuratus in northern Chile. *Aquac. Int.* **2000**, *8*, 237–247. [CrossRef]
3. Bakit, J.; Álvarez, G.; Díaz, P.A.; Uribe, E.; Sfeir, R.; Villasante, S.; Bas, T.G.; Lira, G.; Pérez, H.; Hurtado, A.; et al. Disentangling Environmental, Economic, and Technological Factors Driving Scallop (*Argopecten purpuratus*) Aquaculture in Chile. *Fishes* **2022**, *7*, 380. [CrossRef]
4. Bakit, J.; Hurtado, A.; Márquez Porras, R.; Villasante, S. Navigating transformations from artisanal fishers to entrepreneurial scallop farmers in Chile. *Front. Mar. Sci.* **2023**, *10* . [CrossRef]
5. Avendaño, M.; Cantillánez, M.; Riascos, J.M. The decreasing availability of settlement surfaces affects the transition from larvae to early recruitment of the scallop Argopecten purpuratus through El Niño and La Niña episodes. *Front. Mar. Sci.* **2019**, *6*, 630. [CrossRef]
6. Merino, G.; Uribe, E.; Soria, G.; Von-Brand, E. A comparison of larval production of the northern scallop, Argopecten purpuratus, in closed and recirculating culture systems. *Aquac. Eng.* **2009**, *40*, 95–103. [CrossRef]
7. Soria, G.; Merino, G.; Uribe, E.; Von Brand, E. Effect of increasing salinity on weight-specific filtration rate of juvenile scallop *Argopecten purpuratus* reared at two temperatures: Is any effect related to ammonia buildup? *J. Shellfish. Res.* **2011**, *30*, 279–286. [CrossRef]
8. Velasco, L.; Barros, J. Potential for hatchery broodstock conditioning of the Caribbean scallops *Argopecten nucleus* and *Nodipecten nodosus*. *Aquaculture* **2007**, *272*, 767–773. [CrossRef]

9. Utting, S.; Millican, P. Techniques for the hatchery conditioning of bivalve broodstocks and the subsequent effect on egg quality and larval viability. *Aquaculture* **1997**, *155*, 45–54. [CrossRef]
10. Berntsson, K.P.J.; Wangberg, S.; Carlsson, A. Effects of broodstock diets on fatty acid composition, survival and growth rates in larvae of the European flat oyster, Ostrea edulis. *Aquaculture* **1997**, *154*, 139–153. [CrossRef]
11. Benavides, C.K.; Borrell, J.M.; García, C.R.; Gómez, K.K.; Gonzales, A.K. *Design and Management of a Bivalve Cultivation System in Sechura Bay*; University of Piura: Piura, Perú, 2018; pp. 34–35. Available online: https://hdl.handle.net/11042/3618 (accessed on 10 December 2023).
12. Fonseca, P. Improving Larval Survival in Mollusk Cultures by Improving Microalgae Cultivation Techniques and Introducing New Microalgae Species. Ph.D. Thesis, University of Santiago de Compostela, Santiago de Compostela, España, 2016. Available online: https://dialnet.unirioja.es/servlet/tesis?codigo=125679 (accessed on 10 December 2023).
13. Olivares, J. Temporal variation of oceanographic conditions in Herradura de Guayacá Bay, Chile. *Biota* **1988**, *4*, 89–106.
14. Uribe, E.; Blanco, J. Capacity of aquatic systems to sustain the culture of pectinids: The case of *Argopecten purpuratus* in Tongoy Bay, Chile. *Los Moluscos Pectínidos Iberoam. Cienc. Acuic. Cap* **2001**, *12*, 233–248.
15. Pernet, F.; Tremblay, R.; Comeau, L.; Guderley, H. Temperature adaptation in two bivalve species from different thermal habitats: Energetics and remodelling of membrane lipids. *J. Exp. Biol.* **2007**, *210*, 2999–3014. [CrossRef] [PubMed]
16. Ascencio, L.A.; Enriquez, M.; Martínez, I.; Aldana, D. Effect of temperature and salinity on the reproductive cycle of female and male *Crassostrea virginica* (Bivalvia Ostreidae). *Trop. Biol. Mag.* **2016**, *64*, 449–459. [CrossRef] [PubMed]
17. Martinez, G.; Perez, H. Effect of different temperature regimes on reproductive conditioning in the scallop *Argopecten purpuratus*. *Aquaculture* **2003**, *228*, 153–167. [CrossRef]
18. Lardies, M.A.; Benitez, S.; Osores, S.; Vargas, C.A.; Duarte, C.; Lohrmann, K.B.; Lagos, N.A. Physiological and histopathological impacts of increased carbon dioxide and temperature on the scallops *Argopecten purpuratus* cultured under upwelling influences in northern Chile. *Aquaculture* **2017**, *479*, 455–466. [CrossRef]
19. Martinez, G.; Brokordt, K.; Aguilera, C.; Soto, V.; Guderley, H. Effect of diet and temperature upon muscle metabolic capacities and biochemical composition of gonad and muscle in *Argopecten purpuratus* Lamarck 1819. *J. Exp. Mar. Biol. Ecol.* **2000**, *247*, 29–49. [CrossRef]
20. Marshall, R.; McKinley, S.; Pearce, C.M. Effects of nutrition on larval growth and survival in bivalves. *Rev. Aquac.* **2010**, *2*, 33–55. [CrossRef]
21. Helm, M.; Holland, D.; Stephenson, R. The effect of supplementary algal feeding of a hatchery breeding stock of *Ostrea edulis* L. on larval vigour. *J. Mar. Biol. Assoc. UK* **1973**, *53*, 673–684. [CrossRef]
22. Caers, M.; Utting, S.; Coutteau, P.; Millican, P.; Sorgeloos, P. Impact of the supplementation of a docosahexaenoic acid rich emulsion on the reproductive output of oyster broodstock, Crassostrea gigas. *Mar. Biol.* **2002**, *140*, 1157–1166.
23. Courtens, N.N.V.; Gajardo, M.H.G.; Sorgeloos, P. Effect of lipid emulsions on production and fatty acid composition. *Mar. Biol.* **2003**, *143*, 327–338.
24. Hendriks, I.E.; van Duren, L.A.; Herman, P.M. Effect of dietary polyunsaturated fatty acids on reproductive output and larval growth of bivalves. *J. Exp. Mar. Biol. Ecol.* **2003**, *296*, 199–213. [CrossRef]
25. Martinez, G.; Mettifogo, L. Mobilization of energy from adductor muscle for gametogenesis of the scallop, *Argopecten purpuratus* Lamarck. *J. Shellfish. Res.* **1998**, *17*, 113–116.
26. Brokordt, K.B.; Guderley, H.E. Energetic requirements during gonad maturation and spawning in scallops: Sex differences in *Chlamys islandica* (Muller 1776). *J. Shellfish. Res.* **2004**, *23*, 25–33.
27. Brown, M.R.; Jeffrey, S.W.; Volkman, J.K.; Dunstan, G.A. Nutritional properties of microalgae for mariculture. *Aquaculture* **1997**, *151*, 315–331. [CrossRef]
28. Martinez, G.; Torres, M.; Uribe, E.; Diaz, M. Biochemical composition of broodstock and early juvenile Chilean scallops, *Argopecten purpuratus* Lamarck, held in two different environments. *J. Shellfish. Res.* **1992**, *11*, 307.
29. Uriarte, I.; Farías, A.; Hernandez, J.; Schäfer, C.; Sorgeloos, P. Reproductive conditioning of *Chilean scallop* (*Argopecten purpuratus*) and the *Pacific oyster* (*Crassostrea gigas*): Effects of enriched diets. *Aquaculture* **2004**, *230*, 349–357. [CrossRef]
30. Andriopoulos, V.; Gkioni, M.D.; Koutra, E.; Mastropetros, S.G.; Lamari, F.N.; Hatziantoniou, S.; Kornaros, M. Total phenolic content, biomass composition and antioxidant activity of selected marine microalgae species with potential as aquaculture feed. *Antioxidantes* **2022**, *11*, 1320. [CrossRef]
31. Blanco-Llamero, C.; Señoráns, F.J. Biobased solvents for pressurized liquid extraction of nannochloropsis gaditana Omega-3 lipids. *Mar. Drugs* **2021**, *19*, 107. [CrossRef]
32. Gallager, S.M.; Mann, R. Growth and survival of larvae of *Mercenaria mercenaria* (L.) and *Crassostrea virginica* (Gmelin) relative to broodstock conditioning and lipid content of eggs. *Aquaculture* **1986**, *56*, 105–121. [CrossRef]
33. Sargent, J.R.; Tocher, D.R.; Bell, J.G. The lipids. In *Fish Nutrition*; Springer: Berlin/Heidelberg, Germany, 2003; pp. 181–257.
34. Pérez, H.; Brokordt, K.; Martínez, G.; Guderley, H. Effect of spawning on force production during escape responses of the scallop *Argopecten purpuratus*. *Mar. Biol.* **2009**, *156*, 1585–1593. [CrossRef]
35. Dionicio-Acedo, J.; Cabrera-Simon, A.; Rosado-Salazar, M.; Aguirre-Velarde, A. Optimización del tratamiento térmico para la inducción al desove de *Argopecten purpuratus* (Mollusca: Bivalvia). *Rev. Biol. Mar. Oceanogr.* **2021**, *56*, 145–150. [CrossRef]

36. Castañeda, V.; Zevallos, S.; Ayerbe, R.; Castillo, C.R. Experiences in Controlled Systems for Obtaining Fan Scallop *Argopecten purpuratus* Seeds (Lamarck, 1819). Ilo, Moquegua. Technical Report, Instituto del Mar del Perú. 2008. Available online: https://repositorio.imarpe.gob.pe/bitstream/20.500.12958/2176/1/INF%2038%284%29-6.pdf (accessed on 10 December 2023).
37. Méndez, S.; Alvarez, Y.; Sosa, L.E.; Vizcarra, Y.G. Cellular concentration and dry biomass in three species of marine microalgae: Chlorella vulgaris, Nannochloropsis oculata, and Tetraselmis striata. *Rev. Investig. Altoandinas* **2020**, *22*, 155–160. [CrossRef]
38. Mann, R.; Gallager, S.M. Physiological and biochemical energetics of larvae of *Teredo navalis* L. and *Bankia gouldi* (Bartsch) (Bivalvia: Teredinidae). *J. Exp. Mar. Biol. Ecol.* **1985**, *85*, 211–228. [CrossRef]
39. Dorsey, T.E.; McDonald, P.W.; Roels, O.A. A heated biuret-Folin protein assay which gives equal absorbance with different proteins. *Anal. Biochem.* **1977**, *78*, 156–164. [CrossRef] [PubMed]
40. DuBois, M.; Gilles, K.A.; Hamilton, J.K.; Rebers, P.T.; Smith, F. Colorimetric method for determination of sugars and related substances. *Anal. Chem.* **1956**, *28*, 350–356. [CrossRef]
41. Reza, M.A. Implementation of Prostaglandin Analysis in Oyster Tissues and Comparison of Prostaglandin e2 after Sampling Stress in Pleasure Oyster (*Crassostrea corteziensis*) and Japanese Oyster (*C. Gigas*) and during Spawning of *C. Gigas*. Master's Thesis, Master of Science in the Use, Management and Preservation of Natural Resources, Northwest Biological Research Centre, La Paz, México, 2009. Available online: https://cibnor.repositorioinstitucional.mx/jspui/bitstream/1001/272/1/reza_m.pdf (accessed on 10 December 2023).
42. Farías, A.; Uriarte, I.; Varas, P. Study of the nutritional requirements of the northern scallop *Argopecten purpuratus* (Lamarck, 1819) during the reproductive conditioning. *Oceanogr. Lit. Rev.* **1998**, *8*, 1447.
43. Caers, M.; Coutteau, P.; Cure, K.; Morales, V.; Gajardo, G.; Sorgeloos, P. The Chilean scallop *Argopecten purpuratus* (Lamarck, 1819): II. Manipulation of the fatty acid composition and lipid content of the eggs via lipid supplementation of the broodstock diet. *Comp. Biochem. Physiol. Part B Biochem. Mol. Biol.* **1999**, *123*, 97–103. [CrossRef]
44. Caers, M.; Coutteau, P.; Sorgeloos, P.; Gajardo, G. Impact of algal diets and emulsions on the fatty acid composition and content of selected tissues of adult broodstock of the Chilean scallop *Argopecten pupuratus* (Lamarck, 1819). *Aquaculture* **2003**, *217*, 437–452. [CrossRef]
45. Pérez, H.M.; Brokordt, K.; Gallardo, A.; Vidal, I.; Guderley, H. A diet rich in polyunsaturated fatty acids improves the capacity for HSP70 synthesis in adult scallop *Argopecten purpuratus* and their offspring. *Mar. Biol.* **2016**, *163*, 193. [CrossRef]
46. Palma-Fleming, H.; Navarro, J.M.; Peña, E.; Martínez, G. Effect of three conditioning diets on the fatty acid composition of gonads and muscle of *Argopecten purpuratus*. *N. Z. J. Mar. Freshw. Res.* **2002**, *36*, 605–620. [CrossRef]
47. Farıas, A.; Bell, J.; Uriarte, I.; Sargent, J. Polyunsaturated fatty acids in total lipid and phospholipids of Chilean scallop *Argopecten purpuratus* (L.) larvae: Effects of diet and temperature. *Aquaculture* **2003**, *228*, 289–305. [CrossRef]
48. Rojas, I.; Cárcamo, C.B.; Defranchi, Y.; Jeno, K.; Rengel, J.; Araya, M.; Tarnok, M.E.; Aguilar, L.; Álvarez, G.; Schmitt, P.; et al. A Diet Rich in HUFAs Enhances the Energetic and Immune Response Capacities of Larvae of the Scallop *Argopecten purpuratus*. *Animals* **2023**, *13*, 1416. [CrossRef] [PubMed]
49. Whyte, J.; Bourne, N.; Ginther, N. Biochemical and energy changes during embryogenesis in the rock scallop Crassadoma gigantea. *Mar. Biol.* **1990**, *106*, 239–244. [CrossRef]
50. Cerpa, J.; González, F.; Becerra, J.; Lépez, I. Lipid and polypeptide profiles in the female portion of gonads from diet-conditioned broodstock of North Chilean scallops *Argopecten purpuratus*. *North Am. J. Aquac.* **2003**, *65*, 1–7. [CrossRef]
51. Contreras, R.; Pacheco, E.; Puebla, C. Desarrollo embrionario y larval temprano de *Gari solida* (Gray, 1828) (Bivalvia Psammobiidae). *Lat. Am. J. Aquat. Res.* **2014**, *42*, 283–288. [CrossRef]
52. Whyte, J.; Bourne, N.; Ginther, N. Depletion of nutrient reserves during embryogenesis in the scallop *Patinopecten yessoensis* (Jay). *J. Exp. Mar. Biol. Ecol.* **1991**, *149*, 67–79. [CrossRef]
53. Nevejan, N.; Saez, I.; Gajardo, G.; Sorgeloos, P. Supplementation of EPA and DHA emulsions to a Dunaliella tertiolecta diet: Effect on growth and lipid composition of scallop larvae, *Argopecten purpuratus* (Lamarck, 1819). *Aquaculture* **2003**, *217*, 613–632. [CrossRef]
54. Volkman, J.K.; Jeffrey, S.W.; Nichols, P.D.; Rogers, G.I.; Garland, C.D. Fatty acid and lipid composition of 10 species of microalgae used in mariculture *Exp. Mar. Biol. Ecol.* **1989**, *128*, 219–240. [CrossRef]

Disclaimer/Publisher's Note: The statements, opinions and data contained in all publications are solely those of the individual author(s) and contributor(s) and not of MDPI and/or the editor(s). MDPI and/or the editor(s) disclaim responsibility for any injury to people or property resulting from any ideas, methods, instructions or products referred to in the content.

fishes

MDPI

Article

Effect of Organic or Inorganic Fertilization on Microbial Flocs Production in Integrated Cultivation of *Ulva lactuca* with *Oreochromis niloticus* and *Penaeus vannamei*

Andrezza Carvalho [1,*], Hellyjúnyor Brandão [1], Julio C. Zemor [1], Alessandro Pereira Cardozo [1], Felipe N. Vieira [2], Marcelo H. Okamoto [1], Gamze Turan [3] and Luís H. Poersch [1]

[1] Marine Aquaculture Station, Institute of Oceanography, Federal University of Rio Grande—FURG, Rua do Hotel, no.2, Cassino, Rio Grande 96210-030, RS, Brazil; hellyjunyor@gmail.com (H.B.); juliozemor@hotmail.com (J.C.Z.); ocalessandro@hotmail.com (A.P.C.); mar.okamoto@gmail.com (M.H.O.); lpoersch@gmail.com (L.H.P.)

[2] Laboratório de Camarões Marinhos, Departamento de Aquicultura, Centro de Ciências Agrárias, Univesidade Federal de Santa Catarina, Rua dos Coroas 503, Barra da Lagoa, Florianópolis 88061-600, SC, Brazil; felipe.vieira@ufsc.br

[3] Aquaculture Department, Fisheries Faculty, Ege University, Izmir 35100, Turkey; gamzeturan2000@gmail.com

* Correspondence: andrezzachagas@hotmail.com

Abstract: Different fertilization regimes in biofloc systems influence the predominance of distinct bacterial populations, impacting water quality and organism performance. This study evaluates the growth and nutrient absorption of the macroalgae *Ulva lactuca* when cultivated in an integrated system with *Penaeus vannamei* and *Oreochromis niloticus* in chemoautotrophic and heterotrophic systems. The experiment lasted 45 days and comprised two treatments, each with three replicates: chemoautotrophic—utilizing chemical fertilizers; heterotrophic—employing inoculum from mature biofloc shrimp cultivation, supplemented with organic fertilizers. Each treatment consisted of three systems, each containing a 4 m^3 tank for shrimp, 0.7 m^3 for tilapia, and 0.35 m^3 for macroalgae, with continuous water circulation between tanks and constant aeration. Water quality analyses were carried out during the experiment, as were the performances of the macroalgae and animals. The data were subjected to a statistical analysis. Results revealed an increase in macroalgae biomass and the removal of nitrate (57%) and phosphate (47%) during cultivation, with a higher specific growth rate observed in the chemoautotrophic treatment. Nonetheless, the heterotrophic treatment exhibited higher levels of protein in the macroalgae (18% dry matter) and phosphate removal rates (56%), along with superior maintenance of water quality parameters. Tilapia performance varied across treatments, with a higher final weight and weight gain recorded in the heterotrophic treatment. The recycling of water from an ongoing biofloc cultivation with organic fertilization demonstrated viability for macroalgae cultivation within an integrated system involving shrimp and fish.

Keywords: nitrate; phosphate; growth; biocompounds; water renewal

Key Contribution: The use of macroalgae in an integrated system with different fertilizations showed a high rate of nitrate and phosphate removal. The use of partial harvests promoted high biomass production with a high protein content.

Citation: Carvalho, A.; Brandão, H.; Zemor, J.C.; Cardozo, A.P.; Vieira, F.N.; Okamoto, M.H.; Turan, G.; Poersch, L.H. Effect of Organic or Inorganic Fertilization on Microbial Flocs Production in Integrated Cultivation of *Ulva lactuca* with *Oreochromis niloticus* and *Penaeus vannamei*. *Fishes* **2024**, *9*, 191. https://doi.org/10.3390/fishes9060191

Academic Editor: Chunsheng Liu

Received: 25 March 2024
Revised: 16 May 2024
Accepted: 21 May 2024
Published: 23 May 2024

Copyright: © 2024 by the authors. Licensee MDPI, Basel, Switzerland. This article is an open access article distributed under the terms and conditions of the Creative Commons Attribution (CC BY) license (https://creativecommons.org/licenses/by/4.0/).

1. Introduction

Numerous recent studies have explored marine macroalgae as a source of human food, bioactive compounds, and supplements for marine organisms [1,2]. According to Chopin [3], macroalgae cultivation can be deemed sustainable due to their minimal need for feed, reduced land footprint, and bioremediation capabilities. The genus Ulva exhibits a global distribution, with the morphology and composition adapting to the environmental variables of the production site, rendering it a feasible and intriguing option for cultivation [4]. The cultivation of macroalgae integrated with other aquatic organisms has gained traction in aquaculture, referred to as Integrated Multitrophic Aquaculture (IMTA) [5]. Resende et al. [6] demonstrated the feasibility of incorporating macroalgae for nutrient absorption and biomass production in open cultivation systems with Sea bream *Sparus aurata* and European sea bass *Dicentrarchus labrax*. In addition to macroalgae serving as inorganic consumers, the IMTA system also includes organic consumers to consume the solids produced in the system. The tilapia *Oreochromis niloticus*, known for its ease of handling and feeding habits, has been utilized in integrated systems, showing positive results in solid filtration [7,8]. Due to its robustness, tilapia can also be produced in brackish water without negative effects on its zootechnical performance [9]. Both species could bring benefits when integrated into the cultivation of a main species, as in the farming of the Pacific white shrimp *Penaeus vannamei*, the most cultivated shrimp in the world, especially because it is an euryhaline [10], easy to manage and adapt [11], and holds significant economic interest [12], which has been employed as a primary species in integrated systems.

In conjunction with integrated systems, the utilization of biofloc technology has intensified production by improving the utilization of accumulated waste in cultivation. Studies by Brito et al. [13], Legarda et al. [14], and Morais et al. [15] have reported positive results from the production of shrimp, tilapia, and macroalgae in an integrated biofloc system. In general, biofloc technology offers enhanced biosecurity with reduced water exchange, facilitates water quality control through microbial activity, and serves as a supplementary food source for cultivated species [16]. Various fertilization approaches promote the growth and dominance of distinct bacterial groups within the system, including heterotrophic bacteria, chemoautotrophic bacteria, or a combination of both in mixed or mature systems [17]. The bacterial groups play a crucial role in nitrogen consumption and oxidation within the system, contributing to the maintenance of water quality for the organisms. Heterotrophic bacteria are favored by daily carbon source fertilization, typically at a ratio of 15 g of carbohydrate per gram of available nitrogen in the system [18]. Ammonia consumption by heterotrophic bacteria leads to bacterial biomass production, increasing the total suspended solids concentration in the water, which should ideally be maintained within the range of 100 to 350 mg L^{-1}, as suggested by Gaona et al. [19] to avoid adverse effects on animal performance.

Another significant bacterial group comprises chemoautotrophs, which, according to Ebeling et al. [18], oxidize nitrogen within the system, converting ammonia to nitrite and eventually to nitrate, a less toxic final product for organisms. Chemical fertilization is utilized for system establishment, requiring approximately 30 to 45 days of chemical fertilization prior to cultivation initiation to maintain low concentrations of ammonia and nitrite [17]. Unlike heterotrophic bacteria, chemoautotrophic bacteria generate fewer solids in the system and consume less oxygen. However, they utilize more inorganic carbon, requiring alkalinity adjustments to maintain levels above 150 mg of $CaCO_3$ L^{-1} [20]. Nitrate accumulates as the final nitrogen product in this system's oxidation process, posing toxicity risks to cultivated organisms at high concentrations [21], and, when discharged untreated, can lead to diseases such as methemoglobinemia in humans [22]. Another viable option is to utilize a biofloc inoculum from an ongoing cultivation, providing enhanced stability in nitrogen control and greater sustainability through water reuse [17]. This approach results in a mixed system containing both heterotrophic and chemoautotrophic bacteria, aimed at regulating water quality by promoting bacterial biomass production and nitrification [23]. Organic fertilization is typically employed at the onset of cultivation to expedite the stabilization of ammonia until nitrifying bacteria become established [24].

The biofloc system is complex and subject to variations based on the fertilization strategy utilized, which can impact water quality, production costs, and animal performance. Brandão et al. [23] reported greater shrimp growth in mixed systems compared to heterotrophic systems. Conversely, tilapia performance was negatively impacted in chemoautotrophic systems due to low organic matter loads [8]. However, limited information exists regarding macroalgae performance within these systems. High concentrations of solids produced by heterotrophic bacteria may accumulate on macroalgae, hindering photosynthesis and affecting their performance [25]. Additionally, elevated nutrient concentrations present in the chemoautotrophic system can induce stress in macroalgae and trigger reproductive events [26]. Choosing the right cultivation system can provide better growing conditions and biomass production for the macroalgae. In addition to growth and nutrient absorption, the specific physical and chemical variables inherent to each cultivation system also influence the nutritional composition of macroalgae [27]. The production of biomass with enhanced nutritional value can offer economic advantages for the system through the generation of valuable by-products. For instance, utilizing the biomass of macroalgae cultivated in the integrated system as a food source for shrimp and fish could prove beneficial for aquaculture [2]. Therefore, the objective of this study was to evaluate the growth performance, nutrient absorption, and bioactive compounds of the macroalga *Ulva lactuca* when cultivated in an integrated system with Pacific white shrimp *Penaeus vannamei* and tilapia *Oreochromis niloticus* using two biofloc fertilization strategies: a chemoautotrophic system and a heterotrophic system.

2. Materials and Methods

2.1. Location and Origin of the Animals

The experiment was conducted in an agricultural greenhouse situated at the Marine Station of Aquaculture (EMA), Institute of Oceanography, Federal University of Rio Grande (IO-FURG), located on Cassino Beach, Rio Grande, Rio Grande do Sul. The greenhouse was devoid of shading, and aeration within the tanks was provided by a blower through continuous air injection via micro-perforated hoses (aerotubes).

2.2. Animal Materials

The shrimp originated from a biofloc cultivation system within a grow-out greenhouse at EMA, with an initial weight of 7.13 ± 0.18 g. Tilapia were sourced from a recirculation system grow-out cultivation, starting with an initial weight of 412.33 ± 72.58 g. The macroalgae were cultivated in a greenhouse in a 1 m^3 tank containing water with 35.1 ± 2.74 mg L^{-1} of nitrate and 2.24 ± 1.2 mg L^{-1} of phosphate.

2.3. Experimental Design

The experiment, which spanned 45 days, was conducted on six experimental production systems. Each system comprised a 4 m^3 tank for shrimp (350 shrimp m^{-2}), a 0.7 m^3 tank for tilapia (10 fish per m^3), and a 0.35 m^3 tank for macroalgae cultivation (0.1 g m^3 of the useful volume of the entire system). A submerged pump circulated the system, transferring water into the macroalgae tank, which then flowed by gravity into the shrimp tank before returning to the tilapia tank (Figure 1). The macroalgae were contained within the tank using a circular structure with a diameter of 0.60 m positioned near the surface, constructed from polyethylene netting with 5 mm mesh openings.

2.4. Treatments

Two treatments were employed, each with three replicates: chemoautotrophic—a system utilizing chemical inorganic fertilization; heterotrophic—a system supplemented with organic fertilizer. Inoculum preparation for the chemoautotrophic system involved maintaining water with a salinity of 20 ppt in an 8 m^3 tank. Over 35 days, daily fertilization with sodium nitrite (Neon Comercial, São Paulo, SP, Brazil) and ammonium chloride (Neon Comercial, São Paulo, SP, Brazil) was conducted to achieve a concentration of 1 mg L^{-1} for each compound in the water. To establish bacterial populations in the system, six pillow-like structures containing biological media were placed in the main tank and then distributed among the replicates. The tank was continuously aerated and devoid of light, and no heaters were utilized to simulate greenhouse cultivation conditions.

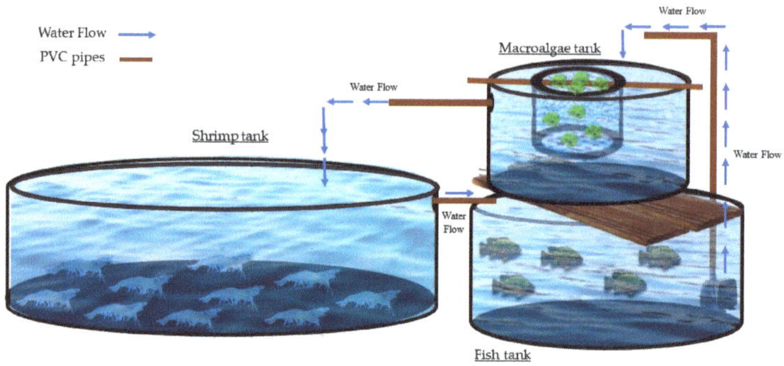

Figure 1. Design of the experimental system, consisting of a shrimp tank, a fish tank, and a macroalgae tank, with water recirculating between them.

Once the ammonia and nitrite concentrations stabilized and were converted into nitrate, the experiment started. Chemoautotrophic treatment replicates were prepared by blending 40% inoculum with water of salinity 20, up to a useful volume of 5 m^3. At the onset of the experiment, the water parameters were as follows: a temperature of 26.0 ± 0.4 °C, dissolved oxygen of 7.2 ± 0.5 mg L^{-1}, pH of 8.18 ± 0.3, alkalinity of 170.0 ± 2. 0 mg $CaCO_3$ L^{-1}, total ammonia nitrogen of 0.02 ± 0.02 mg L^{-1}, nitrite of 1.5 ± 0.2 mg L^{-1}, nitrate of 64.0 ± 1.7 mg L^{-1}, phosphate of 1.2 ± 0.4 mg L^{-1}, and total suspended solids of 160.0 ± 5.8 mg L^{-1}.

The tanks designated for the heterotrophic treatment were prepared with 40% mature biofloc inoculum and seawater at a salinity of 20 ppt, up to a useful volume of 5 m^3. The biofloc inoculum was sourced from a shrimp cultivation system with a useful volume of 237 m^3, a density of 184 shrimp m^{-2}, and an average weight of 7.1 ± 1.2 g, cultivated for 68 days. The initial water quality parameters in the shrimp production tank before the experiment were as follows: a temperature of 25.6 °C, dissolved oxygen of 5.4 mg L^{-1}, pH of 7.47, alkalinity of 215.0 mg $CaCO_3$ L^{-1}, total ammoniacal nitrogen of 0.20 mg L^{-1}, nitrite of 0.20 mg L^{-1}, nitrate of 147.0 mg L^{-1}, phosphate of 4.0 mg L^{-1}, and total suspended solids of 700.00 mg L^{-1}.

2.5. Chemical and Physical Water Parameters

Water quality analyses were conducted on samples collected from the shrimp tanks, considering water homogenization due to the circulation within the systems. Temperature and dissolved oxygen were measured twice daily using a Pro-20 model (YSI Inc., Yellow Springs, OH, USA), and pH was measured daily with a bench pH meter (Seven2Go S7 Basic, Mettler Toledo, São Paulo, SP, Brazil). Salinity was assessed twice a week using a Pro-20 model (YSI Inc., OH, USA), and, if necessary, fresh water was added to maintain salinity at 20. Alkalinity (mg $CaCO_3$ L^{-1}) was monitored twice a week following the APHA methodology [28], with calcium hydroxide added to both treatments when alkalinity fell below 150 mg $CaCO_3$ L^{-1}, as per Furtado et al. [20] recommendation.

Total ammoniacal nitrogen (mg L^{-1}) and nitrite (mg L^{-1}) were initially measured daily and then twice a week after nutrient stabilization, according to UNESCO methodology [29]. Nitrate (mg L^{-1}) and phosphate (mg L^{-1}) were measured twice a week, according to the method proposed by Aminot and Chaussepied [30]. Total suspended solids (mg L^{-1}−TSS) and settleable solids (ml L^{-1}−SS) were quantified twice a week, using the methodology described by Baumgarten et al. [31] and APHA [28], respectively. For the heterotrophic system, organic carbon (molasses) was added when the total ammoniacal nitrogen exceeded 1 mg L^{-1} to promote nitrogen uptake through heterotrophic bacteria growth, as proposed by Wasielesky et al. [32]. In the chemoautotrophic system, inorganic carbon (calcium hydroxide) was added when ammonia and nitrite concentrations exceeded 1 mg L^{-1} and

5 mg L^{-1}, respectively. In this treatment, alkalinity was maintained at 300 mg $CaCO_3$ L^{-1} for optimal nitrifying bacteria performance, as recommended by Furtado [20].

2.6. Macroalgae Growth and Biochemical Analysis

Macroalgae biomass was weighed every 15 days. Before the weighing process, the macroalgae were gently shaken inside the holding structure to eliminate any solids adhering to the surface. Subsequently, the circular holding structure was removed from the tank and set aside to air-dry for 10 min to remove excess water before weighing. The initial weight of macroalgae in each replicate was 502.7 ± 0.5 g. After each weighing, the extra macroalgae biomass was removed, ensuring that the initial weight of the macroalgae was maintained. The following formula was used to calculate the macroalgae specific growth rate (SGR) [33]:

$$\text{SGR } (\%\ d^{-1})\text{: } 100 \times [\ln\ (\text{final weight (g)}/\text{initial weight (g)})/(\text{final time} - \text{initial time})] \quad (1)$$

The nutrient absorption efficiency (NRR) of the macroalgae was calculated using the following formula [33]:

$$\text{NRR } (\%)\text{: } 100 \times [(\text{nutrient concentration at initial time (mg } L^{-1}) - \text{nutrient concentration at final time (mg } L^{-1}))/\text{nutrient concentration at initial time (mg } L^{-1})] \quad (2)$$

At the conclusion of the experiment, random samples of macroalgae were collected from each replicate. Wet samples were weighed and then subjected to drying in an oven at 60 °C for 24 h after obtaining the dry weight. To determine the concentration of chlorophyll-a, chlorophyll-b, and carotenoids, 500 mg of the dry sample was macerated and then incubated in 5 mL of methanol in the dark for 60 min at 4 °C. After that, the solution was centrifuged (12,000× *g*, 10 min), and the supernatant was used to quantify the pigments. The wavelengths of 664 and 647 ηm were used to calculate chlorophyll a (Chla = 11.75 × A664 − 2.35 × A647), chlorophyll b (Chlb = 18.61 × A647 − 3.91 × A664), and carotenoids (Car = (1000 × A470 − 2.27 × Chla − 81.4 Chlb)/227), according to the methodology of Lichtenthaler & Wellburn [34].

Protein quantification was conducted using the Bradford method. An extract was obtained from the dried macroalgae, following the protocol of Barbarino & Lourenço [35], with the addition of 1 mL of sodium hydroxide and centrifugation. The extract and TCA (25%) were added in a ratio of 2.5:1 (*v*/*v*) to precipitate the protein and kept in an ice-cold bath for 30 min. The solution was then centrifuged and washed with dilutions of TCA (10 and 5%), removing the supernatant, until the protein pellet was formed. To the pellet suspension, 0.5 mL of sodium hydroxide (0.1 N) was added, and 20 μL of the solution was combined with 1 mL of the total protein kit for the final analysis procedure.

2.7. Feed Management and Performance of the Animals

Shrimp were fed twice daily with 1.6 mm feed (Guabi aqua QS 1–2 mm, Guabi Nutrition and Animal Health S.A., Campinas, São Paulo, SP, Brazil), and weekly biometrics were conducted to adjust feed quantities following the method proposed by Jory et al. [36]. The tilapia were fed twice a day with commercial feed containing 40% protein (Guabi Tech, Guabi Nutrition and Animal Health S.A., Campinas, São Paulo, SP, Brazil) at a rate of 1% of the biomass to encourage biofloc consumption. To evaluate shrimp performance, measurements were taken at the beginning, middle, and end of the experiment. Fish biometrics were conducted at the beginning and end of the experiment. The animals' performance was assessed using the following formulas:

- Final average weight (g): final biomass of live animals (g)/total number of animals;
- Weekly weight gain (g week^{-1}): weight gain (g)/number of weeks;
- Final biomass (g): ∑ final weight of all live animals (g);
- Feed conversion rate (FCR) = ∑ feed offered (g)/(biomass gain (g));
- Survival (%) = (final number of animals/initial number of animals) × 100;

– Yield (kg m^{-3}): (final biomass (kg)/tank volume (m^3);
– Weight gain rate (%) = 100 × [(final mean weight − initial mean weight)/initial mean weight]

2.8. Statistical Analysis

The data mean (±standard deviation) values are presented in Tables 1–3. Data normality and homoscedasticity were assessed using the Shapiro–Wilk and Levene tests, respectively. Upon meeting these assumptions, a Student's *t*-test was employed to compare treatment differences. In cases where the assumptions of the Student's *t*-test were not met, the non-parametric Kruskal–Wallis test was utilized. Additionally, a one-way ANOVA followed by a Tukey post-hoc test was conducted to evaluate nitrate and phosphate concentrations over time in each treatment. A significance level of 5% ($p \leq 0.05$) was applied to all analyses. The tests were carried out using the PAST 4.03 2020 software [37].

3. Results

3.1. Physical and Chemical Parameters

During the 45-day trial period, there were significant differences ($p < 0.05$) observed in pH, alkalinity, and calcium hydroxide consumption between the treatments. The chemoautotrophic system exhibited the highest values, along with higher consumption of calcium hydroxide (Table 1).

Regarding nutrient levels, the chemoautotrophic system demonstrated higher concentrations of ammonia and nitrite, reaching maximums of 3.1 and 20.0 mg L^{-1}, respectively. Conversely, the heterotrophic system exhibited higher concentrations of total suspended solids and settleable solids (Table 1). Significant nitrate and phosphate removal ($p < 0.05$) was observed in both treatments, although the heterotrophic treatment displayed a higher phosphate removal rate compared to the chemoautotrophic system.

Table 1. Water quality parameters (mean ± standard deviation) (maximum–minimum) of chemoautotrophic and heterotrophic biofloc systems during the 45 days of integrated cultivation of *Ulva lactuca* with *Oreochromis niloticus* and *Penaeus vannamei*.

	Treatments	
Parameters	**Chemoautotrophic**	**Heterotrophic**
Temperature (°C)	22.82 ± 0.30 (27.4–15.5)	22.67 ± 0.38 (27.3–14.9)
DO (mg L^{-1})	7.06 ± 0.04 (9.9–5.3)	6.96 ± 0.06 (9.7–5.2)
pH	8.15 ± 0.02 ^{a} (8.9–7.7)	7.89 ± 0.04 ^{b} (8.1–7.5)
Salinity (g L^{-1})	20.20 ± 0.26 (22.1–19.1)	21.83 ± 0.76 (23.3–20.3)
Alkalinity (mg $CaCO_3$ L^{-1})	280.00 ± 8.19 ^{a} (365.0–155.0)	189.58 ± 7.02 ^{b} (230.0–150.0)
TAN (mg L^{-1})	0.95 ± 0.07 ^{b} (3.1–0.0)	0.18 ± 0.08 ^{a} (1.9–0.0)
N—Nitrite (mg L^{-1})	7.96 ± 1.16 ^{b} (20.0–0.0)	0.86 ± 0.60 ^{a} (5.2–0.0)
N—Nitrate (mg L^{-1})	26.41 ± 2.72 (68.0–10.0)	26.04 ± 4.03 (75.0–15.0)
P—Phosphate (mg L^{-1})	1.04 ± 0.25 (2.0–0.3)	1.19 ± 0.21 (2.2–0.4)
SS (ml L^{-1})	0.39 ± 0.21 ^{a} (3.0–0.0)	8.44 ± 2.91 ^{b} (15.0–3.0)
TSS (mg L^{-1})	189.22 ± 26.02 ^{a} (270.0–70.0)	335.38 ± 47.92 ^{b} (452.5–175.0)
Calcium hydroxide (g L^{-1}) #	0.29 ± 0.02 ^{b} (0.32–0.28)	0.08 ± 0.03 ^{a} (0.10–0.04)
Water exchange (m^{-3}) &	2.0 ± 2.0 ^{a}	0.0 ± 0.0 ^{a}
	Removal rate	
Nitrate (%)	56.47 ± 4.93	57.00 ± 7.00
Phosphate (%)	47.75 ± 4.75 ^{b}	56.14 ± 1.14 ^{a}

DO (dissolved oxygen); TAN (total ammonium nitrogen); SS (settleable solids); TSS (total suspended solids). # Use of calcium hydroxide during cultivation. & Volume of water used for renovations. Different letters in the same line represent significant differences ($p \leq 0.05$) between treatments after Student's *t*-test.

Over the weeks of cultivation, there was a reduction in the concentration of nitrate and phosphate (Figures 2 and 3). The highest nitrate concentrations were observed at the beginning of cultivation, with a decrease from day 14 onward in both treatments (Figure 2). Similarly, phosphate concentrations were higher during the initial week, after which they stabilized in the heterotrophic treatment. In contrast, the chemoautotrophic treatment exhibited a decrease in phosphate concentration until the first week, followed by stabilization and a subsequent increase in the final week (Figure 3).

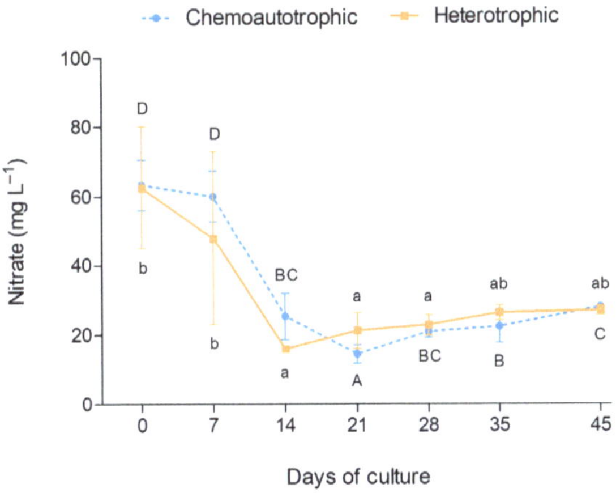

Figure 2. Average weekly nitrate concentrations (mg L^{-1}) during the experimental period in the chemoautotrophic (chemical fertilization prior to stocking) and heterotrophic (use of an inoculum from an ongoing biofloc shrimp cultivation) treatments in an integrated cultivation of *Ulva lactuca* with *Oreochromis niloticus* and *Penaeus vannamei*. Capital letters show differences between the chemoautotrophic treatments over time. Lowercase letters show statistical differences over time in the mature treatment.

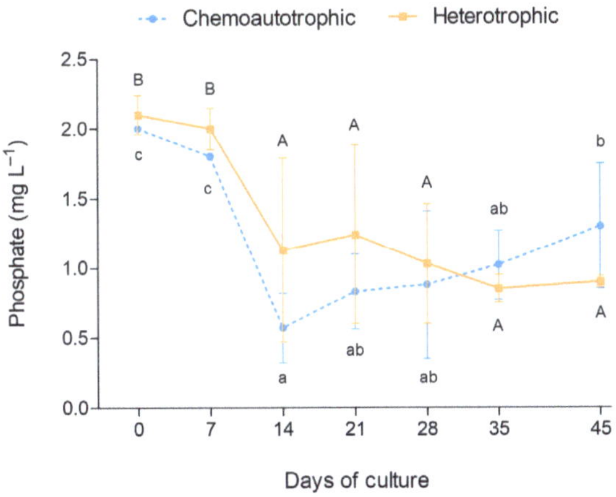

Figure 3. Weekly average phosphate concentrations (mg L^{-1}) during the experimental period in the chemoautotrophic (chemical fertilization prior to stocking) and heterotrophic (use of an inoculum from an ongoing biofloc shrimp cultivation) treatments in an integrated cultivation of *Ulva lactuca* with *Oreochromis niloticus* and *Penaeus vannamei*. Capital letters show differences between the chemoautotrophic treatments over time. Lowercase letters show statistical differences over time in the mature treatment.

Total suspended solids exhibited significant differences ($p < 0.05$) between treatments during most of the experimental weeks. High concentrations of solids reaching 452 mg L^{-1} were observed in the heterotrophic treatment, in contrast to maximum concentrations of 270 mg L^{-1} in the chemoautotrophic treatment (Figure 4).

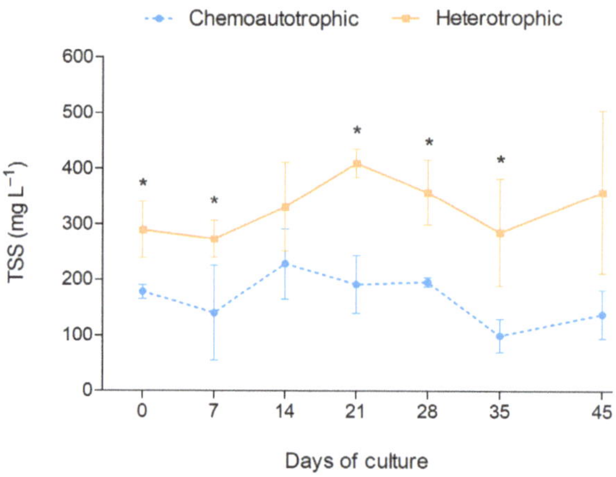

Figure 4. Average weekly concentrations of total suspended solids (mg L^{-1}) during the experimental period in the chemoautotrophic (chemical fertilization prior to stocking) and heterotrophic (use of an inoculum from an ongoing biofloc shrimp cultivation) treatments in an integrated cultivation of *Ulva lactuca* with *Oreochromis niloticus* and *Penaeus vannamei*. An asterisk (*) means a statistical difference on the same day between treatments.

3.2. *Macroalgae Growth and Biochemical Analysis*

There was an increase in macroalgae biomass in both treatments, with higher concentrations of protein in the macroalgae tissue in the heterotrophic treatment ($p < 0.05$). However, no significant differences ($p \geq 0.05$) were found in biomass gain, chlorophyll-a, chlorophyll-b, or carotenoids between the treatments (Table 2).

Table 2. Performance and biochemistry of the macroalgae (mean ± standard deviation) in the chemoautotrophic and heterotrophic treatments at the end of 45 days of an integrated culture of *Ulva lactuca* with *Oreochromis niloticus* and *Penaeus vannamei*.

	Treatments	
	Chemoautotrophic	**Heterotrophic**
Initial mean weight (g–FW)	502.76 ± 0.65	502.64 ± 0.44
Biomass gain (g–FW)	964.63 ± 290.57	708.11 ± 141.70
Protein (%)	15.13 ± 0.56 [b]	18.49 ± 0.56 [a]
Chlorophyll-*a* (mg g^{-1})	2.20 ± 0.02	2.18 ± 0.06
Chlorophyll-*b* (mg g^{-1})	3.28 ± 0.02	3.26 ± 0.08
Carotenoids (mg g^{-1})	0.06 ± 0.00	0.04 ± 0.02

Different letters in the same line represent significant differences ($p \leq 0.05$) between treatments after Student's *t*-test.

The specific growth rate indicates that the growth of macroalgae in the chemoautotrophic treatment remained consistent throughout the entire experiment, with no significant difference ($p \geq 0.05$) observed between weighings. In contrast, for the heterotrophic treatment, the highest growth rate was recorded on day 21, followed by a subsequent decrease in growth rate. Notably, a significant difference in growth rate was only detected in the last weighing among the treatments (Figure 5).

3.3. *Performance of the Animals*

Shrimp performance remained unaffected by the different biofloc strategies, with no discernible differences observed between treatments. However, the performance of the fish exhibited a significant difference ($p < 0.05$) between treatments, with a higher final weight, weight gain, and weight gain rate recorded in the heterotrophic treatment compared to the chemoautotrophic treatment (Table 3).

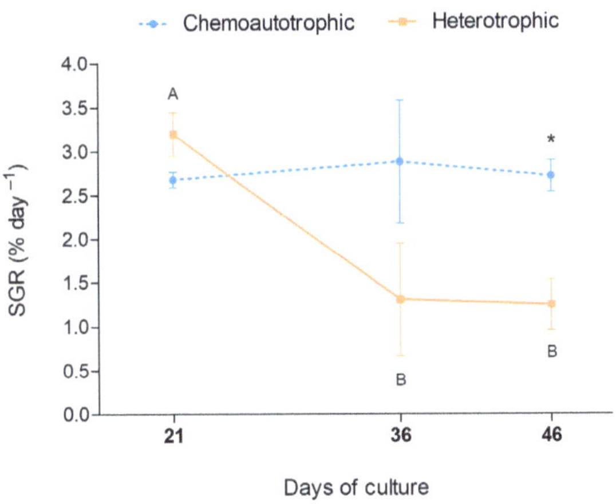

Figure 5. Macroalgae specific growth rate (% day^{-1}) in the chemoautotrophic (chemical fertilization prior to stocking) and heterotrophic (use of an inoculum from an ongoing biofloc shrimp cultivation) treatments in an integrated cultivation of *Ulva lactuca* with *Oreochromis niloticus* and *Penaeus vannamei*. An asterisk (*) means a statistical difference on the same day between treatments. Capital letters mean differences in the same treatment between sampling days.

Table 3. Animal performance (mean ± standard deviation) in the chemoautotrophic (chemical fertilization prior to stocking) and heterotrophic (use of an inoculum from an ongoing biofloc shrimp cultivation) treatments at the end of 45 days of an integrated cultivation of *Ulva lactuca* with *Oreochromis niloticus* and *Penaeus vannamei*.

	Treatments	
	Chemoautotrophic	**Heterotrophic**
Shrimp		
Final mean weight (g)	9.60 ± 0.60	10.75 ± 0.49
WWG (g week^{-1}) #	0.41 ± 0.10	0.60 ± 0.08
Final biomass (kg)	10.02 ± 0.43	10.79 ± 0.60
FCR ##	2.16 ± 0.38	2.36 ± 0.24
Survival (%)	84.71 ± 4.50	84.08 ± 5.22
Fish		
Final mean weight (g)	447.55 ± 1.05 ^{b}	456.25 ± 3.25 ^{a}
WWG (g week^{-1}) #	5.93 ± 0.18 ^{b}	7.38 ± 0.55 ^{a}
Final biomass (kg)	2.89 ± 0.54	2.98 ± 0.25
FCR ##	1.03 ± 0.10	1.18 ± 0.02
Survival (%)	90.48 ± 16.50	95.24 ± 8.25
WGR (%) ###	8.63 ± 0.25 ^{b}	10.74 ± 0.79 ^{a}

WGW (weekly weight gain); ## FCR (food conversion rate); ### WGR (weight gain rate). Lowercase letters mean differences between treatments.

4. Discussion

It is widely acknowledged that the utilization of biofloc technology, compared to conventional cultivation methods, leads to reduced water consumption and enhanced control over water quality parameters [16]. The various fertilization strategies employed in this experiment resulted in differences in water quality maintenance, organism performance, water usage, and inputs. The chemoautotrophic system utilizes more inputs, such as sodium hydroxide, compared to the heterotrophic system. This variance is necessary to maintain optimal alkalinity values. In the chemoautotrophic system, the optimal functioning of nitrifying bacteria occurs at values maintained at around 300 mg $CaCO_3$ L^{-1}, consequently resulting in higher pH values [20]. As a result, there is a more frequent application of sodium hydroxide to correct these values and stimulate the growth of nitrifying bacteria with higher alkalinity. Furthermore, differences in nutrient concentrations were observed between the adopted systems, which play a crucial role in macroalgae development. According to Messyasz et al. [38], most marine *Ulva* species thrive in environments with high concentrations of ammonia or nitrate. Ammonia, originating from waste feed and animal excretion, is the primary nitrogenous compound formed in the system and can be lethal to cultivated organisms at low levels [39]. In both systems in this study, an initial increase in ammonia concentration was observed due to animal stocking. Wasielesky et al. [32] suggest that the use of organic carbon fertilization could promote the growth of heterotrophic bacteria in the system, which consume produced ammonia and generate bacterial biomass. For the chemoautotrophic system, only inorganic fertilization with calcium hydroxide was employed to encourage the growth of nitrifying bacteria with higher alkalinity [20]. However, the slow establishment of nitrifying bacteria resulted in maximum values of 3.1 mg L^{-1} of ammonia in the system, which were higher than those found in the heterotrophic system. Nevertheless, according to Lin & Chen [39], the values obtained in our study were not toxic to the organisms.

The produced ammonia is oxidized into nitrite by ammonium-oxidizing bacteria and subsequently into nitrate by nitrite-oxidizing bacteria. However, the observed increase in nitrite levels in the chemoautotrophic treatment suggests that the nitrite-oxidizing bacteria were not fully established in the system to facilitate this transformation. Despite the use of artificial substrate in this experiment, it is likely that the bacterial population was insufficient to oxidize the nitrite produced following the stocking of shrimp. The use of artificial substrate in the system is necessary for bacterial adherence and to increase their numbers [40]. According to Lin & Chen [41], the safe level for nitrite at a salinity of 25 is 15.2 mg L^{-1}, and concentrations exceeding this limit can be lethal to shrimp. Consequently, in our experiment, we carried out partial water exchange, and a reduction in shrimp and fish feeding was necessary to control nitrite levels in the system, resulting in higher water usage than in the heterotrophic system and the dilution of nutrients.

In biofloc systems, elevated concentrations of nitrate and phosphate are common in long-term production due to low water exchange rates and high animal densities, providing an advantageous environment for macroalgae development. Carneiro [42] noted that when macroalgae inhabit eutrophicated environments, they tend to absorb significant nutrient concentrations initially for storage, serving as a precautionary measure in case of sudden nutrient depletion. Additionally, Hanisak et al. [43] suggested that a constant high nitrogen availability in the environment does not necessarily result in increased removal, as macroalgae nitrogen absorption capacity saturates quickly at high concentrations. This phenomenon may have occurred in both treatments in our study, resulting in a substantial reduction in nitrate and phosphate concentrations at the onset of cultivation. Following the second week, nutrient stabilization occurred. It is documented that 57% of nitrogen is lost from the water daily, with an increase over time [44], suggesting that the stabilization of these nutrients in the experiment may be attributed to the continuous absorption carried out by the macroalgae. Studies such as Massocato et al. [45] have demonstrated that 85% of the nitrate from a fish cultivation was absorbed within the first five days of algae cultivation.

Phosphorus is also another compound accumulated in the system and produced daily through waste feed [44]. It is an important element in photosynthesis and the transfer of energy from macroalgae [46], which shows the advantage of integrating macroalgae into closed systems. Phosphorus absorption is connected with nitrogen absorption, with an ideal ratio of 30:1 (nitrogen:phosphorus), so that phosphorus or nitrogen are not limiting [47]. The higher removal rate found in the heterotrophic treatment may be linked to the pH values. According to Rathod et al. [48], higher phosphate absorption occurs at pH levels below neutrality. The maintenance of high alkalinity and pH in the chemoautotrophic treatment may have negatively impacted phosphate absorption.

The utilization of macroalgae as a biofilter has advanced due to their excellent performance in nutrient absorption, ease of management, and high biomass production [49].

Alencar et al. [50] demonstrated that the macroalgae *Ulva lactuca* absorbed 94% of the ammonia concentration in an integrated cultivation with shrimp. Conversely, the impact of the organic load generated in macroalgae cultivation remains poorly understood. Due to the intensive production of bacterial biomass, the heterotrophic system in this study exhibited higher concentrations of total suspended solids and settleable solids. In contrast, the chemoautotrophic system, with its use of inorganic fertilizers and water exchange, exhibited a lower organic load, with a maximum of 270 mg L^{-1}. Similar outcomes were reported by Ferreira et al. [17] in their study of the two biofloc systems. Despite the absence of a significant difference in macroalgae biomass gain between the treatments, a higher growth rate was observed toward the end of cultivation in the chemoautotrophic treatment, potentially attributable to the lower solids content in the system compared to the heterotrophic system. The accumulation of microbial biomass and waste in the heterotrophic system intensified toward the end of cultivation, likely directly affecting macroalgae growth. Carvalho et al. [51] demonstrated that the presence of macroalgae in the heterotrophic system led to solid deposition due to the formation of a physical barrier, reducing light exposure for the macroalgae and consequently impacting their performance.

Despite the lower concentration of solids in the chemoautotrophic system, they still accumulated on the surface of the macroalgae, representing one of the challenges of biofloc systems. Studies like Resende et al. [6] reported significantly higher growth rates, with a maximum growth rate of $15.33 \pm 2.87\%$ day^{-1} when macroalgae were cultivated freely in tanks with fish farm effluent, characterized by minimal solids concentrations. The results found in our experiment are in agreement with studies by Martins et al. [52], who observed a growth rate of $3.0 \pm 0.6\%$ day^{-1} with the macroalga *Ulva ohnoi* in a biofloc system. Studies with red algae in biofloc have also been carried out, showing a maximum growth rate of $1.19 \pm 0.04\%$ day^{-1} [53], similar to those observed in our heterotrophic treatment results in the last weeks of cultivation. However, unlike studies such as Carvalho et al. [25] and Legarda et al. [14], which did not observe macroalgae growth in biofloc systems, our use of partial harvests might have reduced macroalgae density in the cultivation structure and minimized shading, resulting in improved biomass production. Biancacci et al. [54] showed that the use of partial harvests in the cultivation of the macroalga *Macrocystis pyrifera* promoted greater biomass gain, a lower incidence of epiphytes, and a change in the macroalgae biochemical composition.

In addition to serving as a bioremediator, macroalgae possess economic value, as the biomass they produce can be utilized in the pharmaceutical and food industries [55], thereby fostering sustainability and profitability in production. Macroalgae serve as vital sources of nutrients and vitamins and possess antioxidant and immunostimulant properties [56]. The higher protein values observed in macroalgae from the heterotrophic system may be attributed to reduced luminosity in the system due to the gradual accumulation of solids over the cultivation period. Ganesan et al. [57] showed a correlation between high pigment concentrations in low-light and salinity environments in their study on the macroalga *Ulva fasciata*, indicating potential adaptations to environmental conditions. The observed high values of chlorophyll-a and chlorophyll-b in our study compared to those reported by Silva et al. [58] may be linked to the necessity of increasing pigment concentrations in macroalgae to maximize photosynthesis, likely due to reduced light penetration caused by suspended particles in a biofloc system. Similar trends were noted by Fillit et al. [59], who reported increased pigment concentrations during periods of low light availability.

In the integrated system, all species must have productivity in cultivation and economic potential [60]. Despite the elevated nitrite concentrations in the chemoautotrophic system, shrimp and fish performance was not affected. However, growth outcomes and survival in both treatments were lower than those reported by Ferreira et al. [17] in their study on shrimp cultivation in chemoautotrophic, heterotrophic, and mature systems. This can be attributed to temperature differences between the studies. The minimum temperature recorded in our experiment was 14.9 °C, directly impacting the survival of the organisms. Furthermore, the overall average temperature in our study (22.0 °C) was lower compared to studies conducted with shrimp and fish [61], which also influenced the growth of the animals due to their decreased metabolism. Fish performance in terms of weight gain was superior in the heterotrophic system compared to the chemoautotrophic system, possibly due to the higher availability of suspended organic matter. The reduced feed supply aimed to induce floc consumption in the system, as demonstrated by Holanda et al. [7], with floc serving as a supplementary food source for organisms [62]. Hence, the higher concentration of total suspended solids in the heterotrophic system might have positively influenced fish weight gain. Similar results were reported by Poli et al. [8], who observed lower fish growth in an integrated system with chemoautotrophic floc.

The use of integrated multi-trophic systems aims to balance system productivity with sustainability, ensuring that all organisms adapt to the cultivation conditions. According to Khanjani et al. [63], the utilization of integrated systems has been consistently increasing highlighting potential species for inclusion in the system, with crustaceans being among the most commonly produced target species. Zimmermann et al. [64] discuss the future of tilapia production, emphasizing multitrophic cultivation and biofloc technology as promising systems for maximizing production, considering greater sustainability, biosecurity, and increased density. However, the integration of macroalgae into biofloc systems has not yet been fully stabilized. The inclusion of macroalgae in biofloc systems has presented challenges due to their low productivity [14,15,25], but their role as bioremediators in nutrient absorption has shown promise, as demonstrated by the data presented in this study. Furthermore, the production of macroalgae biomass with an increase in nitrogen content in tissues, as reported by Legarda et al. [14] and Carvalho et al. [25], adds value to the product and enhances its applicability. The incorporation of macroalgae produced in integrated systems into fish and shrimp feed has yielded significant results, as evidenced by Marinho et al. [65] and Valente et al. [66]. Improved methods for managing the incorporation of macroalgae into biofloc systems are needed to enhance production and sustainability in intensive production systems.

5. Conclusions

The use of macroalgae in an integrated system with organic fertilization proved to be viable for increasing biomass production and nitrate and phosphate absorption, improving the system's sustainability. The use of a system with a low concentration of solids, as in the chemoautotrophic system, promoted better growth rates for the macroalgae. However, the use of an inoculum from a heterotrophic system intensified the removal of phosphate and nitrate and increased the protein content of the macroalgae. A better maintenance of water quality was found in the heterotrophic system with the use of organic fertilization, without the need for water renewal. Finally, the heterotrophic system contributed to the better performance of the tilapia, with an increase in weight gain and a higher average final weight.

Author Contributions: Conceptualization, A.C., J.C.Z., A.P.C., F.N.V., M.H.O., G.T. and L.H.P.; data curation, A.C.; formal analysis, A.C. and H.B.; funding acquisition, L.H.P.; investigation, H.B., J.C.Z., A.P.C., F.N.V., M.H.O. and L.H.P.; methodology, A.C., H.B., A.P.C., F.N.V. and L.H.P.; project administration, G.T. and L.H.P.; supervision, F.N.V., G.T. and L.H.P.; validation, A.C.; visualization, J.C.Z., A.P.C., M.H.O., G.T. and L.H.P.; writing—original draft, A.C.; writing—review and editing, H.B., J.C.Z., A.P.C., F.N.V., M.H.O., G.T. and L.H.P. All authors have read and agreed to the published version of the manuscript.

Funding: This research was funded by the ASTRAL Project—H2020 grant agreement 863034.

Institutional Review Board Statement: All applicable international, national, and/or institutional guidelines for the care and use of animals were followed by the authors. The experiment was approved by the Ethics and Animal Welfare Committee of FURG (Case number 23116.005895/2016-42).

Data Availability Statement: Data are contained within the article.

Acknowledgments: Special thanks to the Brazilian Council of Research (CNPq), the Coordination for the Improvement of Higher Level or Education Personnel (CAPES), and the Rio Grande do Sul State Government. Luís H. Poersch and Felipe N. Vieira received a productivity research fellowship from CNPq.

Conflicts of Interest: The authors declare no conflicts of interest.

References

1. Gordalina, M.; Pinheiro, H.M.; Mateus, M.; da Fonseca, M.M.R.; Cesário, M.T. Macroalgae as Protein Sources—A Review on Protein Bioactivity, Extraction, Purification and Characterization. *Appl. Sci.* **2021**, *11*, 7969. [CrossRef]
2. Wan, A.H.L.; Davies, S.J.; Soler-Vila, A.; Fitzgerald, R.; Johnson, M.P. Macroalgae as a Sustainable Aquafeed Ingredient. *Rev. Aquac.* **2019**, *11*, 458–492. [CrossRef]
3. Chopin, T. Marine Aquaculture in Canada: Well-Established Monocultures of Finfish and Shellfish and an Emerging Integrated Multi-Trophic Aquaculture (IMTA) Approach Including Seaweeds, Other Invertebrates, and Microbial Communities. *Fisheries* **2015**, *40*, 28–31. [CrossRef]
4. Queirós, A.S.; Circuncisão, A.R.; Pereira, E.; Válega, M.; Abreu, M.H.; Silva, A.M.S.; Cardoso, S.M. Valuable Nutrients from *Ulva rigida*: Modulation by Seasonal and Cultivation Factors. *Appl. Sci.* **2021**, *11*, 6137. [CrossRef]
5. Troell, M.; Joyce, A.; Chopin, T.; Neori, A.; Buschmann, A.H.; Fang, J.G. Ecological Engineering in Aquaculture—Potential for Integrated Multi-Trophic Aquaculture (IMTA) in Marine Offshore Systems. *Aquaculture* **2009**, *297*, 1–9. [CrossRef]
6. Resende, L.; Flores, J.; Moreira, C.; Pacheco, D.; Baeta, A.; Garcia, A.C.; Rocha, A.C.S. Effective and Low-Maintenance IMTA System as Effluent Treatment Unit for Promoting Sustainability in Coastal Aquaculture. *Appl. Sci.* **2022**, *12*, 398. [CrossRef]
7. Holanda, M.; Wasielesky, W.; de Lara, G.R.; Poersch, L.H. Production of Marine Shrimp Integrated with Tilapia at High Densities and in a Biofloc System: Choosing the Best Spatial Configuration. *Fishes* **2022**, *7*, 283. [CrossRef]
8. Poli, M.A.; Legarda, E.C.; de Lorenzo, M.A.; Pinheiro, I.; Martins, M.A.; Seiffert, W.Q.; do Nascimento Vieira, F. Integrated Multitrophic Aquaculture Applied to Shrimp Rearing in a Biofloc System. *Aquaculture* **2019**, *511*, 734274. [CrossRef]
9. de Souza, R.L.; de Lima, E.C.R.; de Melo, F.P.; Ferreira, M.G.P.; de Souza Correia, E. The Culture of Nile Tilapia at Different Salinities Using a Biofloc System. *Rev. Cienc. Agron.* **2019**, *50*, 267–275. [CrossRef]
10. Li, E.; Chen, L.; Zeng, C.; Chen, X.; Yu, N.; Lai, Q.; Qin, J.G. Growth, Body Composition, Respiration and Ambient Ammonia Nitrogen Tolerance of the Juvenile White Shrimp, *Litopenaeus vannamei*, at Different Salinities. *Aquaculture* **2007**, *265*, 385–390. [CrossRef]
11. Natori, M.M.; Sussel, F.R.; Santos, E.D.; Previero, T.D.C.; Viegas, E.M.M.; Gameiro, A.H. Desenvolvimento Da Carcinicultura Marinha No Brasil e No Mundo: Avanços Tecnológicos e Desafios. *Informações econômicas* **2011**, *41*, 61–73.
12. FAO. *State of the World Fisheries and Aquaculture—2022 (SOFIA)*; FAO: Rome, Italy, 2022; ISBN 9789251072257.
13. Brito, L.O.; Arantes, R.; Magnotti, C.; Derner, R.; Pchara, F.; Olivera, A.; Vinatea, L. Water Quality and Growth of Pacific White Shrimp *Litopenaeus vannamei* (Boone) in Co-Culture with Green Seaweed *Ulva lactuca* (Linaeus) in Intensive System. *Aquac. Int.* **2014**, *22*, 497–508. [CrossRef]
14. Legarda, E.C.; da Silva, D.; Miranda, C.S.; Pereira, P.K.M.; Martins, M.A.; Machado, C.; de Lorenzo, M.A.; Hayashi, L.; do Nascimento Vieira, F. Sea Lettuce Integrated with Pacific White Shrimp and Mullet Cultivation in Biofloc Impact System Performance and the Sea Lettuce Nutritional Composition. *Aquaculture* **2021**, *534*, 736265. [CrossRef]
15. de Morais, A.P.M.; Santos, I.L.; Carneiro, R.F.S.; Routledge, E.A.B.; Hayashi, L.; de Lorenzo, M.A.; do Nascimento Vieira, F. Integrated Multitrophic Aquaculture System Applied to Shrimp, Tilapia, and Seaweed (*Ulva ohnoi*) Using Biofloc Technology. *Aquaculture* **2023**, *572*, 739492. [CrossRef]
16. Crab, R.; Avnimelech, Y.; Defoirdt, T.; Bossier, P.; Verstraete, W. Nitrogen Removal Techniques in Aquaculture for a Sustainable Production. *Aquaculture* **2007**, *270*, 1–14. [CrossRef]
17. Ferreira, G.S.; Santos, D.; Schmachtl, F.; Machado, C.; Fernandes, V.; Bögner, M.; Schleder, D.D.; Seiffert, W.Q.; Vieira, F.N. Heterotrophic, Chemoautotrophic and Mature Approaches in Biofloc System for Pacific White Shrimp. *Aquaculture* **2021**, *533*, 736099. [CrossRef]
18. Ebeling, J.M.; Timmons, M.B.; Bisogni, J.J. Engineering Analysis of the Stoichiometry of Photoautotrophic, Autotrophic, and Heterotrophic Removal of Ammonia-Nitrogen in Aquaculture Systems. *Aquaculture* **2006**, *257*, 346–358. [CrossRef]
19. Gaona, C.A.P.; de Almeida, M.S.; Viau, V.; Poersch, L.H.; Wasielesky, W. Effect of Different Total Suspended Solids Levels on a *Litopenaeus vannamei* (Boone, 1931) BFT Culture System during Biofloc Formation. *Aquac. Res.* **2017**, *48*, 1070–1079. [CrossRef]
20. Furtado, P.S.; Poersch, L.H.; Wasielesky, W. Effect of Calcium Hydroxide, Carbonate and Sodium Bicarbonate on Water Quality and Zootechnical Performance of Shrimp *Litopenaeus vannamei* Reared in Biofloc Technology (BFT) Systems. *Aquaculture* **2011**, *321*, 130–135. [CrossRef]
21. Furtado, P.S.; Campos, B.R.; Serra, F.P.; Klosterhoff, M.; Romano, L.A.; Wasielesky, W. Effects of Nitrate Toxicity in the Pacific White Shrimp, *Litopenaeus vannamei*, Reared with Biofloc Technology (BFT). *Aquac. Int.* **2015**, *23*, 315–327. [CrossRef]
22. Macedo, C.F.; Sipaúba-Tavares, L. Eutrophication and Water Quality in Pisciculture: Consequences and Recommendations. *Bol. Do Inst. De Pesca* **2010**, *36*, 149–163.

23. Brandão, H.; Xavier, Í.V.; Santana, G.K.K.; Santana, H.J.K.; Krummenauer, D.; Wasielesky, W. Heterotrophic versus Mixed BFT System: Impacts on Water Use, Suspended Solids Production and Growth Performance of *Litopenaeus vannamei*. *Aquac. Eng.* **2021**, *95*, 102194. [CrossRef]
24. Krummenauer, D.; Samocha, T.; Poersch, L.; Lara, G.; Wasielesky, W. The Reuse of Water on the Culture of Pacific White Shrimp, *Litopenaeus vannamei*, in BFT System. *J. World Aquac. Soc.* **2014**, *45*, 3–14. [CrossRef]
25. Carvalho, A.; de Oliveira Costa, L.C.; Holanda, M.; Poersch, L.H.; Turan, G. Influence of Total Suspended Solids on the Growth of the Sea Lettuce *Ulva lactuca* Integrated with the Pacific White Shrimp *Litopenaeus vannamei* in a Biofloc System. *Fishes* **2023**, *8*, 163. [CrossRef]
26. Copertino, M.D.S.; Tormena, T.; Seeliger, U. Biofiltering Efficiency, Uptake and Assimilation Rates of *Ulva clathrata* (Roth) J. Agardh (Clorophyceae) Cultivated in Shrimp Aquaculture Waste Water. *J. Appl. Phycol.* **2009**, *21*, 31–45. [CrossRef]
27. Duke, C.S.; Litaker, W.; Ramus, J. Effects of the Temperature, Nitrogen Supply and Tissue Nitrogen on Ammonium Uptake Rates of the Chlorophyte Seaweeds *Ulva Curvata* and *Codium Decorticatum*. *J. Phycol.* **1989**, *25*, 113–120. [CrossRef]
28. American Public Health Association (APHA). *Standard Methods for the Examination of Water and Waste Water*, 16th ed.; American Public Health Association: Washington, DC, USA, 1989.
29. UNESCO. *Chemical Methods for Use in Marine Environmental Monitoring*; Intergovernmental Oceanographic Commission: Paris, France, 1983.
30. Aminot, A.; Chaussepied, M. *Manuel Des Analyses Chimiques En Milieu Marin*; Centre National pour L'exploitation des Océans: Paris, France, 1983.
31. Baumgarten, M.D.G.Z.; De Barros Rocha, J.M.; Niencheski, L.F.H. *Manual de Análises Em Oceanografia Química*; FURG: Rio Grande, Brazil, 1996.
32. Wasielesky, W.; Krummenauer, D.; Lara, G.; Fóes, G. Cultivo de Camarões Em Sistema de Bioflocos: Realidades e Perspectivas. **2013**, *15*, 30–36. *Revista ABCC*.
33. Loureiro, R.R.; Reis, R.P.; Critchley, A.T. In Vitro Cultivation of Three *Kappaphycus alvarezii* (Rhodophyta, Areschougiaceae) Variants (Green, Red and Brown) Exposed to a Commercial Extract of the Brown Alga *Ascophyllum nodosum* (Fucaceae, Ochrophyta). *J. Appl. Phycol.* **2010**, *22*, 101–104. [CrossRef]
34. Lichtenthaler, H.K.; Wellburn, A.R. Determinations of Total Carotenoids and Chlorophylls a and b of Leaf Extracts in Different Solvents. *Biochem. Soc. Trans.* **1983**, *11*, 591–592. [CrossRef]
35. Barbarino, E.; Lourenço, S.O. An Evaluation of Methods for Extraction and Quantification of Protein from Marine Macro- and Microalgae. *J. Appl. Phycol.* **2005**, *17*, 447–460. [CrossRef]
36. Jory, D.E.; Cabrera, T.R.; Dugger, D.M.; Fegan, D.; Lee, P.G.; Lawrence, L.; Jackson, C.J.; Mcintosh, R.P.; Castañeda, J.I.; Al, B.; et al. A Global Review of Shrimp Feed Management: Status and Perspectives. *Aquaculture* **2011**, *318*, 104–152.
37. Hammer, Ø.; David, A.; Harper, T.; Paul, D. Ryan National Knowledge Resource Consortium -a National Gateway of S&T on-Line Resources for CSIR and DST Laboratories. *Curr. Sci.* **2013**, *105*, 1352–1357.
38. Messyasz, B.; Rybak, A. Abiotic Factors Affecting the Development of *Ulva* Sp. (Ulvophyceae; Chlorophyta) in Freshwater Ecosystems. *Aquat. Ecol.* **2011**, *45*, 75–87. [CrossRef]
39. Lin, Y.C.; Chen, J.C. Acute Toxicity of Ammonia on *Litopenaeus vannamei* Boone Juveniles at Different Salinity Levels. *J. Exp. Mar. Bio. Ecol.* **2001**, *259*, 109–119. [CrossRef]
40. De Lara, G.R.; Poersch, L.H.; Wasielesky, W., Jr. The Quantity of Artificial Substrates Influences the Nitrogen Cycle in the Biofloc Culture System of *Litopenaeus vannamei*. *Aquac. Eng.* **2021**, *94*, 102171. [CrossRef]
41. Lin, Y.C.; Chen, J.C. Acute Toxicity of Nitrite on *Litopenaeus vannamei* (Boone) Juveniles at Different Salinity Levels. *Aquaculture* **2003**, *224*, 193–201. [CrossRef]
42. Carneiro, M.A.D.A. Estudo Do Crescimento, Eficiência de Biofiltração e Cinética de Absorção de Nutrientes (N-NH, N-NO e P-PO4^3) Da Macroalga *Gracilaria cervicornis* (Turner). Master's Thesis, Universidade Federal do Rio Grande do Norte, Rio Grande, Brazil, 2007.
43. Hanisak, M.D. The Use of *Gracilaria tikvahiae* (Gracilariales, Rhodophyta) as a Model System to Understand the Nitrogen Nutrition of Cultured Seaweeds. *Hydrobiologia* **1990**, *204–205*, 79–87. [CrossRef]
44. Da Silva, K.R.; Wasielesky, W.; Abreu, P.C. Nitrogen and Phosphorus Dynamics in the Biofloc Production of the Pacific White Shrimp, *Litopenaeus vannamei*. *J. World Aquac. Soc.* **2013**, *44*, 30–41. [CrossRef]
45. Massocato, T.F.; Robles-Carnero, V.; Moreira, B.R.; Castro-Varela, P.; Pinheiro-Silva, L.; da Silva Oliveira, W.; Vega, J.; Avilés, A.; Bonomi-Barufi, J.; Rörig, L.R.; et al. Growth, Biofiltration and Photosynthetic Performance of *Ulva* Spp. Cultivated in Fishpond Effluents: An Outdoor Study. *Front. Mar. Sci.* **2022**, *9*, 981468. [CrossRef]
46. Tsagkamilis, P.; Danielidis, D.; Dring, M.J.; Katsaros, C. Removal of Phosphate by the Green Seaweed *Ulva lactuca* in a Small-Scale Sewage Treatment Plant (Ios Island, Aegean Sea, Greece). *J. Appl. Phycol.* **2010**, *22*, 331–339. [CrossRef]
47. Zirino, A.; Elwany, H.; Facca, C.; Maicu', F.; Neira, C.; Mendoza, G. Nitrogen to Phosphorus Ratio in the Venice (Italy) Lagoon (2001–2010) and Its Relation to Macroalgae. *Mar. Chem.* **2016**, *180*, 33–41. [CrossRef]
48. Rathod, M.; Mody, K.; Basha, S. Efficient Removal of Phosphate from Aqueous Solutions by Red Seaweed, *Kappaphycus alverezii*. *J. Clean. Prod.* **2014**, *84*, 484–493. [CrossRef]

49. Chopin, T.; Buschmann, A.H.; Halling, C.; Troell, M.; Kautsky, N.; Neori, A.; Kraemer, G.P.; Zertuche-González, J.A.; Yarish, C.; Neefus, C. Integrating Seaweeds into Marine Aquaculture Systems: A Key toward Sustainability. *J. Phycol.* **2001**, *37*, 975–986. [CrossRef]
50. de Alencar, J.R.; Junior, P.A.H.; Celino, J.J. Cultivo de Camarão Branco *Litopenaeus vannamei* (Boone, 1931) Com a Macroalga *Ulva lactuca* Linneaus (Chlorophyta) No Tratamento de Efluentes Em Sistema Fechado de Recirculação. *Rev. Biol. e Ciências da Terra* **2010**, *10*, 117–137.
51. Carvalho, A.; de Oliveira Costa, L.C.; Holanda, M.; Gonçalves, M.; Santos, J.; Costa, C.S.B.; Turan, G.; Poersch, L.H. Growth of the Macroalgae *Ulva lactuca* Cultivated at Different Depths in a Biofloc Integrated System with Shrimp and Fish. *Phycology* **2023**, *3*, 280–293. [CrossRef]
52. Martins, M.A.; da SILVA, V.F.; Tarapuez, P.R.; Hayashi, L.; Vieira, F.D.N. Cultivation of the Seaweed *Ulva* Spp. With Effluent from a Shrimp Biofloc Rearing System: Different Species and Stocking Density. *Bol. do Inst. Pesca* **2020**, *46*, 1–6. [CrossRef]
53. Pires, C.M.; Bazzo, G.C.; Barreto, P.L.M.; do Espírito Santo, C.M.; Ventura, T.F.B.; Pedra, A.G.L.M.; Rover, T.; McGovern, M.; Hayashi, L. Cultivation of the Red Seaweed *Kappaphycus alvarezii* Using Biofloc Effluent. *J. Appl. Phycol.* **2021**, *33*, 1047–1058. [CrossRef]
54. Biancacci, C.; Visch, W.; Callahan, D.L.; Farrington, G.; Francis, D.S.; Lamb, P.; McVilly, A.; Nardelli, A.; Sanderson, J.C.; Schwoerbel, J.; et al. Optimisation of At-Sea Culture and Harvest Conditions for Cultivated *Macrocystis pyrifera*: Yield, Biofouling and Biochemical Composition of Cultured Biomass. *Front. Mar. Sci.* **2022**, *9*, 951538. [CrossRef]
55. Mabeau, S.; Fleurence, J. Seaweed in Food Products: Biochemical and Nutritional Aspects. *Trends Food Sci. Technol.* **1993**, *4*, 103–107. [CrossRef]
56. Peñalver, R.; Lorenzo, J.M.; Ros, G.; Amarowicz, R.; Pateiro, M.; Nieto, G. Seaweeds as a Functional Ingredient for a Healthy Diet. *Mar. Drugs* **2020**, *18*, 301. [CrossRef]
57. Ganesan, M.; Mairh, O.P.; Eswaran, K.; Subba Rao, P.V. Effect of Salinity, Light Intensity and Nitrogen Source on Growth and Composition of *Ulva fasciata* Delile (Chlorophyta, Ulvales). *Indian J. Mar. Sci.* **1999**, *28*, 70–73.
58. Silva, A.F.R.; Abreu, H.; Silva, A.M.S.; Cardoso, S.M. Effect of Oven-Drying on the Recovery of Valuable Compounds from *Ulva rigida, Gracilaria* Sp. and *Fucus vesiculosus*. *Mar. Drugs* **2019**, *17*, 90. [CrossRef] [PubMed]
59. Fillit, M. Seasonal Changes in the Photosynthetic Capacities and Pigment Content of *Ulva rigida* in a Mediterranean Coastal Lagoon. *Bot. Mar.* **1995**, *38*, 271–280. [CrossRef]
60. Buck, B.H.; Troell, M.F.; Krause, G.; Angel, D.L.; Grote, B.; Chopin, T. State of the Art and Challenges for Offshore Integrated Multi-Trophic Aquaculture (IMTA). *Front. Mar. Sci.* **2018**, *5*, 165. [CrossRef]
61. Wyban, J.; Walsh, W.A.; Godin, D.M. Temperature Effects on Growth, Feeding Rate and Feed Conversion of the Pacific White Shrimp (*Penaeus vannamei*). *Aquaculture* **1995**, *138*, 267–279. [CrossRef]
62. Ahmad, I.; Babitha Rani, A.M.; Verma, A.K.; Maqsood, M. Biofloc Technology: An Emerging Avenue in Aquatic Animal Healthcare and Nutrition. *Aquac. Int.* **2017**, *25*, 1215–1226. [CrossRef]
63. Khanjani, M.H.; Zahedi, S.; Mohammadi, A. Integrated Multitrophic Aquaculture (IMTA) as an Environmentally Friendly System for Sustainable Aquaculture: Functionality, Species, and Application of Biofloc Technology (BFT). *Environ. Sci. Pollut. Res.* **2022**, *29*, 67513–67531. [CrossRef] [PubMed]
64. Zimmermann, S.; Kiessling, A.; Zhang, J. The Future of Intensive Tilapia Production and the Circular Bioeconomy without Effluents: Biofloc Technology, Recirculation Aquaculture Systems, Bio-RAS, Partitioned Aquaculture Systems and Integrated Multitrophic Aquaculture. *Rev. Aquac.* **2023**, *15*, 22–31. [CrossRef]
65. Marinho, G.; Nunes, C.; Sousa-Pinto, I.; Pereira, R.; Rema, P.; Valente, L.M.P. The IMTA-Cultivated Chlorophyta *Ulva* Spp. as a Sustainable Ingredient in Nile Tilapia (*Oreochromis niloticus*) Diets. *J. Appl. Phycol.* **2013**, *25*, 1359–1367. [CrossRef]
66. Valente, L.M.P.; Araújo, M.; Batista, S.; Peixoto, M.J.; Sousa-Pinto, I.; Brotas, V.; Cunha, L.M.; Rema, P. Carotenoid Deposition, Flesh Quality and Immunological Response of Nile Tilapia Fed Increasing Levels of IMTA-Cultivated *Ulva* spp. *J. Appl. Phycol.* **2016**, *28*, 691–701. [CrossRef]

Disclaimer/Publisher's Note: The statements, opinions and data contained in all publications are solely those of the individual author(s) and contributor(s) and not of MDPI and/or the editor(s). MDPI and/or the editor(s) disclaim responsibility for any injury to people or property resulting from any ideas, methods, instructions or products referred to in the content.

fishes

Article

Influence of Total Suspended Solids on the Growth of the Sea Lettuce *Ulva lactuca* Integrated with the Pacific White Shrimp *Litopenaeus vannamei* in a Biofloc System

Andrezza Carvalho [1], Léa Carolina de Oliveira Costa [2], Mariana Holanda [3], Luís H. Poersch [1,*] and Gamze Turan [4]

1 Marine Aquaculture Station, Federal University of Rio Grande—FURG, Rua do Hotel, n° 02, Querência, Rio Grande 96210-030, RS, Brazil; andrezzachagas@hotmail.com
2 Instituto Federal de Educação Ciência e Tecnologia do Pará, Rodovia BR 316, KM 61, Castanhal 68740970, PA, Brazil leacarolinacosta@yahoo.com.br
3 Escuela de Ciencias del Mar, Pontificia Universidad Católica de Valparaíso, Valparaíso 2373223, Chile; mariana.holandalima@gmail.com
4 Aquaculture Department, Fisheries Faculty, Ege University, Izmir 35100, Turkey; gamze.turan@ege.edu.tr
* Correspondence: lpoersch@gmail.com

Abstract: A biofloc system is rich in nutrients, which favors the cultivation of macroalgae, but the influence of the system on the performance of macroalgae is unknown. The objective of this study was to analyze the feasibility of introducing the macroalgae *Ulva lactuca* into the culture of *Litopenaeus vannamei* in a biofloc system. The first experiment evaluated the influence of 400 mg L^{-1} and 30 mg L^{-1} solids concentration of the system in biofloc and von Stosch culture medium on macroalgae growth. In the second experiment, the densities of 1, 2, and 3 g L^{-1} of *U. lactuca* were cultivated in an integrated system with shrimp and monoculture treatment. Both experiments had 35 days of cultivation. There was no significant difference in macroalgae growth between the treatments with biofloc and von Stosch culture medium. In the integrated culture, the density of 1 g L^{-1} showed better nutrient absorption. Shrimp performance was not affected by macroalgae cultivation. In conclusion, the solids did not affect the growth of the macroalgae, and it could be cultivated in a biofloc system for nitrate uptake in integrated culture with shrimp.

Keywords: nutrients; *Ulva lactuca*; shrimp; density; bioremediation

Citation: Carvalho, A.; Costa, L.C.d.O.; Holanda, M.; Poersch, L.H.; Turan, G. Influence of Total Suspended Solids on the Growth of the Sea Lettuce *Ulva lactuca* Integrated with the Pacific White Shrimp *Litopenaeus vannamei* in a Biofloc System. *Fishes* **2023**, *8*, 163. https://doi.org/10.3390/fishes8030163

Academic Editors: Wagner C. Valenti and Eduardo Luis Ballester

Received: 31 December 2022
Revised: 27 February 2023
Accepted: 6 March 2023
Published: 15 March 2023

Copyright: © 2023 by the authors. Licensee MDPI, Basel, Switzerland. This article is an open access article distributed under the terms and conditions of the Creative Commons Attribution (CC BY) license (https://creativecommons.org/licenses/by/4.0/).

1. Introduction

Shrimp production using biofloc technology (BFT), normally carried out without water renewal, results in the accumulation of nutrients in the system [1]. This occurs due to the action of microorganisms that transform shrimp excreta and food remains into protein and inorganic nutrients [2]. Chemoautotrophic and heterotrophic bacteria grow during the culture. The first group of bacteria works in the oxidation of ammonia to nitrite and later to nitrate, which is accumulated in the system, along with phosphorus from the feed [3]. Heterotrophic bacteria participate in the conversion of ammonia into bacterial biomass and, together with the accumulation of feces and feed remains, increase the concentration of total suspended solids in the system. The production of waste is constant during cultivation, and when it is not properly disposed of or treated, it can generate problems in the water quality in the production system and environmental problems in the release of effluents without treatment [4].

The total suspended solids are important in the water quality of the system and must be maintained between 100 to 300 mg L^{-1} [5]. In addition to water quality, microbial flocs function as a complementary source of natural food within the crop [6]. Azim and Little [7] nutritionally described biofloc containing 38% protein and 3% lipid in dry matter, showing a high nutritional value, which may depend on the carbon source used in the system. Wasielesky et al. [8] showed that it is possible to reduce the feeding frequency of shrimp *L. vannamei* when cultivated in a biofloc system.

Despite the benefits of microbial flocs in the shrimp culture system, its effect on macroalgae growth is unknown. As macroalgae are photosynthetic organisms, the concentration of solids can interfere with the light capture that is essential for their growth. Brito et al. [9] showed the deposition of solids on macroalgae, which may have a negative effect on its growth. However, the nutritional value of macroalgae can also change when grown in biofloc. Legarda et al. [10] showed an increase in nitrogen, phosphorus, chlorophyll *a*, and carotenoids when macroalgae were cultivated in an integrated system in biofloc.

One way to take advantage of the nitrogen and phosphorus accumulated in the system is integration with other species of different trophic levels in production, as proposed by Chopin [11] in an integrated multi-trophic aquaculture (IMTA). In the system, residues are reused by different species, increasing the final productivity of the cultivation and sustainability. The IMTA system is composed of a species fed with commercial feed, such as shrimp or fish, and then a species capable of absorbing inorganic compounds dissolved in the water, such as macroalgae, and a species that consumes organic compounds, such as oyster, is inserted into the system, which will feed on suspended particles in the culture [12].

Considering the precepts in the IMTA and associating them with the biofloc, the presence of macroalgae in integrated cultures with shrimp can promote the absorption of nutrients from the culture, such as ammonia, nitrate and phosphate. Thus, the use of cultivation effluents for the cultivation of macroalgae or integrated cultivations can be an alternative with lower financial and environmental costs, bearing in mind that enrichment culture media, such as von Stosch, have a high cost due to expensive chemical compounds including essential vitamins that make their use unfeasible for large-scale production [13].

The choice of species for the composition of the systems can be a limiting factor for the success of the production. Macroalgae have rapid growth due to the efficiency on converting solar energy into biomass, due to their simple cellular structures compared to terrestrial plants [14], generating large biomasses in a short time. Alencar et al. [15] showed that the use of effluents from a shrimp culture provided a relative growth rate of 8.8% day^{-1} of the macroalgae *U. lactuca* and an absorption efficiency with an average of 90% for ammonium (NH_4^+) and orthophosphate (PO_4^{-3}). Ramos et al. [16], analyzing integration of another macroalgae, *U. fasciata*, with the cultivation of Pacific white shrimp (*L. vannamei*) with sedimentation and filtration systems by oysters, showed that the combination of systems enabled improvements in several aspects of water quality, using macroalgae in the removal of dissolved nutrients in the system.

Another factor to be considered is the commercial importance of the macroalgae cultivated in the system. Seaweed cultivation is economically viable due to the presence of high-value compounds in algae cells, which are extracted and used for the manufacture of cosmetics, pharmaceuticals, and chemical compounds. El-baz et al. [17] showed that among species of red and green algae, *U. lactuca* had a higher lipid concentration and inhibitory actions on viral and bacterial activities. In industry, macroalgae have wide applicability, with them being a good nutritional alternative and having a good acceptability, as shown by Turan and Tekogul [18].

The use of macroalgae as bioremediators has been widely used and has been shown to be efficient. Copertino et al. [19], carrying out a study with *U. chlatrata* recirculating water from a *L. vannamei* culture, showed a maximum relative growth rate of 20% day^{-1} and an uptake of 90% of total ammonia nitrogen (TAN) of the system, demonstrating the feasibility of this integration. However, depending on the area and dispersion of the effluent, large macroalgae biomasses are necessary and may be unfeasible [20]. There has been an attempt to adjust the proportion of macroalgae in the crop so that the absorption of nutrients is still effective. Alencar et al. [15] found better growth and nutrient absorption results with a density of 3 g L^{-1}; this high density is convenient in small volumes of water. Del Río et al. [21], using *U. rigida* as a biofilter for fish tanks, found that the results are good at densities between 1.5 and 2.5 g L^{-1}. However, in shrimp farming in a biofloc system with high organic load and nutrients, little is known about the ideal density for the maintenance of the system and on the performance of macroalgae in cultivation in a biofloc system and alternatives to optimize nutrient absorption. Therefore, the objective of this work is to evaluate the influence of the seaweed *Ulva lactuca* in the cultivation of the shrimp *Litopenaeus vannamei* in a biofloc system.

2. Materials and Methods

2.1. Study Location and Origin of the Animals and Macroalgae

The experiments were carried out at the Marine Aquaculture Station (EMA), Institute of Oceanography of the Federal University of Rio Grande (IO-FURG), located at Cassino Beach, Rio Grande, Rio Grande do Sul, Brazil. Two experiments were carried out with the macroalgae *U. lactuca*. Macroalgae were collected at Cassino Beach (32°17′52.30″ S–52°15′59.80″ W), Rio Grande, RS, Brazil. After collection, the algae samples were taken to the laboratory for removal of epiphytes and acclimatized until the beginning of the experiments. The shrimp used in the second experiment came from cultivation in a greenhouse at the Shrimp culture Laboratory, Marine Aquaculture Station (EMA).

2.2. Lab-Scale Experiments

This experiment aimed to evaluate the growth of macroalgae *U. lactuca* with different concentrations of solids from a cultivation in a biofloc system compared with a specific culture medium.

2.2.1. Experimental Design and Facilities

The experimental design was carried out with three treatments in triplicate, namely: (1) BFT: cultivation of *U. lactuca* in effluent from shrimp cultivation in a biofloc system, with 400 mg L^{-1} of TSS; (2) DEC: cultivation of *U. lactuca* in effluent from shrimp culture in a biofloc system, after a period of decantation of solids, with 30 mg L^{-1} of TSS; (3) VS: cultivation of *U. lactuca* in standard von Stosch enrichment solution [22] at a concentration of 10 mL L^{-1} (Table 1) and sea water, without the presence of TSS.

Table 1. Composition of von Stosch enrichment medium modified and adapted by Guiry and Cunningham [22].

Chemical Compounds	Medium Solution (g·L^{-1} dH_2O)	Concentration Used	Concentration in the Final Medium (M)
Na_2 β-glycerophosphate	5.36	10 mL	2.48×10^{-4}
$NaNO_3$	42.52	10 mL	5.00×10^{-3}
$FeSO_4 \cdot 7H_2O$	0.28	10 mL	1.00×10^{-5}
$MnCl_2 \cdot 4H_2O$	1.96	10 mL	1.00×10^{-4}
Na_2 EDTA·$2H_2O$	3.72	10 mL	1.00×10^{-4}
Vitamins	-	10 mL	
Vitamins			
Thiamine HCl	-	200 mg	5.93×10^{-6}
Biotin	0.1	1 mL	4.09×10^{-9}
Cyanocobalamin	0.2	1 mL	1.48×10^{-9}

Five litter transparent plastic containers (or carboys) with 3 L useful volume and an area of 0.13 m^{-2} exposed to light were used for lab cultivation of the macroalgae. After algae were transferred into the culture units, the top of the containers was covered with a transparent PVC film to prevent water evaporation and contamination.

The experiment was carried out for 35 days under controlled conditions of a temperature of 26.58 ± 0.05, with 12:12 h light/dark photoperiod, 3013.89 ± 107.64 LUX light intensity, and total light per day of 2.39 ± 0.09 micromole day^{-1}. Constant aeration was provided using a blower (3900 L $hour^{-1}$) that directed air through a porous airstone of

15 cm in length in each experimental unit. To carry out the experiment in the laboratory, a density of 2 g L^{-1} [21] of macroalgae was used.

2.2.2. Biofloc Effluent and Culture Medium

For the treatments with biofloc, an effluent from a shrimp culture in a BFT system in a greenhouse was used, which lasted 43 days, at a density of 400 shrimp m^3. The water quality of the system was measured with 66 mg L^{-1} of nitrate, 5.6 mg L^{-1} of phosphate and 400 mg L^{-1} of total suspended solids (TSS), indicating that the system was mature [3], and with an acceptable solids concentration for the cultivation of shrimp in biofloc [5].

This effluent was placed in natura in the BFT treatment for the macroalgae culture, with a concentration of 400 mg L^{-1} of total solids in suspension. For the DEC treatment, the effluent underwent a decantation process for 30 min, so that the denser solids could settle and be removed from the water. This resulted in a concentration of solids lower (30 mg L^{-1}) than that found in the shrimp tanks (400 mg L^{-1}).

For the control treatment, a 10 mL L^{-1} von Stosch enrichment solution modified by Guiry and Cunningham [22] was used. None of the treatments were renewed in order to maintain a standard. Therefore, the von Stosch solution was only inoculated into the experimental units at the beginning of the experiment. To prepare the von Stosch enrichment solution, 940 mL of filtered seawater was used and 10 mL of each solution made was added (Table 1). To prepare the vitamins, 950 mL of filtered and sterilized water was used, the Thiamine HCl was dissolved, and 1 mL of the other solutions produced was added (Table 1). The initial concentrations of nutrients in the treatment von Stosch solution were 0.0 ± 0.0, 0.0 ± 0.0, 5.0 ± 0.0, and 1.2 ± 0.0 mg·L^{-1} of total ammonium nitrogen, nitrite, nitrate, and phosphate, respectively.

2.3. Experiments in Greenhouse Conditions

The experiment was carried out in a greenhouse structure unit with the objective of evaluating the optimum density of the macroalgae, *U. lactuca*, in the integrated culture with the shrimp *L. vannamei*, to optimize the absorption of nutrients from the system during the 35 days of the experimental period.

2.3.1. Experimental Design and Facilities

Four treatments with three replications were applied as follows: IMTA 1 treatment: integrated shrimp culture with *U. lactuca* (1 g L^{-1}); IMTA 2 treatment: integrated shrimp culture with *U. lactuca* (2 g L^{-1}); IMTA 3 treatment: integrated shrimp culture with *U. lactuca* (3 g L^{-1}); and MONO C treatment: monoculture of shrimp without *U. lactuca*.

A total of 12 shrimp tanks with 300 L of total capacity with a base diameter of 0.81 m and a height of 0.53 m and 150 L of useful volume for each tank were used for this experiment. Six rectangular floating structures (40 × 30 × 5 cm) were used as *U. lactuca* cultivation units. Each shrimp tank contained a single floating structure for the macroalgae; they were placed close to the surface, by using a rectangular structure (40 × 30 × 5 cm) made of sponge and a 5 mm polyethylene mesh, covering an area of 0.12 m^{-2} of the tank surface, and were subject to a daily light intensity of 28.68 ± 8.53 moles day^{-1} (Figure 1).

The water exchange was not applied, and the aeration was maintained by a blower (4 CV) that injected air into three 20 cm-long micro-perforated hoses per tank. The density used in all the experimental treatments was 300 shrimp m^3 according to Krummenauer et al. [23].

2.3.2. Biofloc Effluent

To carry out the experiment in a greenhouse, water from shrimp cultivation in a biofloc system in progress (inoculum) was used. The experimental units were filled with 120 L of seawater (80%) and 30 L (20%) of inoculum from a shrimp biofloc culture.

The inoculum had 45 days of culture, with a well-established bacterial community according to Ferreira et al. [3], showing 70 mg L^{-1} of nitrate, 4 mg L^{-1} of phosphate, and

650 mg L^{-1} of total suspend solids (TSS). The concentrations of total ammonia nitrogen, nitrite, nitrate, phosphate, and total suspended solids were measured as 0.8 ± 0.1, 0.1 ± 0.0, 14.0 ± 0.0, 0.7 ± 0.1 and 127.5 ± 23.8 mg L^{-1}, respectively.

Figure 1. The rectangular floating structure used for *U. lactuca* cultivation in the shrimp tank.

2.4. Physical and Chemical Parameters

For water quality, parameters such as temperature (°C), salinity (‰), dissolved oxygen (DO, mg L^{-1}), and pH were measured daily in all of the experimental tanks, with the aid of a multiparameter probe (YSI, model Pro-20, USA) and a benchtop pH meter (Mettler Toledo, FEP20, Brazil). Salinity was measured weekly using a multiparameter (YSI, model Pro-20, USA). Water samples were collected from near the surface and near the aeration points from each experimental tank and the water samples were kept in plastic containers and taken for analysis immediately. The daily total ammonia nitrogen (or TAN, mg L^{-1}) and nitrite (NO_2, mg L^{-1}) were analyzed according to the methodology of UNESCO [24] and Bendschneider and Robinson [25]. When the concentration of total ammonia nitrogen (TAN) was greater than 1 mg L^{-1}, molasses was used in ratio 6:1 carbon: nitrogen to control water quality [26]. Nitrate (NO_3, mg L^{-1}) and phosphate (PO_4, mg L^{-1}) were analyzed using the methodology described by Aminot and Chaussepied [27] and they were monitored three times a week. Turbidity (NTU) was measured by a portable turbidimeter (Hach®, 2100P, Portugal) and total suspended solids (or TSS, mg L^{-1}) were quantified by filtration and gravimetry according to the methodology described by Baumgarten et al. [28]. Turbidity and TSS were determined weekly. The total alkalinity (mg $CaCO_3$ L^{-1}) was monitored according to the methodology presented by APHA [29] and it was measured weekly in the lab-scale experiments and twice a week in the pilot-scale experiments in greenhouse conditions. To maintain $CaCO_3$ above 150 mg L^{-1}, calcium hydroxide was used [30]. Weekly sedimentable solids (or SS, ml L^{-1}) were measured by using the Imhoff cone method [29] in the pilot-scale experiments conducted under greenhouse conditions.

2.5. Growth and Nutrient Absorption by U. lactuca

Initial and final algal biomass yields were measured. In order to determine the wet (or fresh) biomass, Ulva thallus was first collected by hand from the rectangular floating structures. The excess water was removed by using a hand centrifuge, followed by using paper towels. The samples were weighed in a Digital Balance machine (MARTE® BL3200H,

SP Labor, São Paulo, Brazil) to determine their wet weight. The following equations were used to calculate the relative growth rate (RGR) [31]:

$$\text{RGR}\ (\%\ \text{day}^{-1})\text{: [in (final weight (g)/initial weight (g))/(final time/initial time)} \times 100]. \quad (1)$$

The nutrient absorption efficiency [32] of *U. lactuca* was calculated as follows:

$$\text{Nutrient removal rate—NRR (\%): [(concentration of nutrient in the initial time (mg L}^{-1}) - \text{concentration of nutrients in the final time (mg L}^{-1}))/(\text{concentration of nutrient in the initial time (mg L}^{-1}))^{-1}] \times 100. \quad (2)$$

2.6. Protein Analysis in *U. lactuca*

Random algal samples with three replicates for protein analysis were hand-collected from each experimental unit, washed under running tap water, and rinsed with distilled water at the beginning and at the end of the experiments. The excess water was removed by using a hand centrifuge, followed by drying with paper towels. The samples were weighed to determine the wet weight and placed in an oven at 60 °C for 24 h and weighed again to obtain the dry weight. Subsequently, the samples were ground to powder by using a coffee bean grinder machine (Cadance®, Belo Horizonte, Brazil).

The nitrogen content of the algae was determined by using the Kjeldahl titration method according to AOAC [33] at the Laboratory of Nutrition of Aquatic Organisms-LANOA (EMA, FURG, Rio Grande, Brazil). The following formula used for converting nitrogen to protein was:

$$\text{Protein (\% of dry weight)} = [(0.1 \times \text{Vol} \times 0.014/\text{Sample}) \times 5.45] \times 100 \quad (3)$$

where Vol is the volume spent on titration and Sample is the dry weight of the sample [32].

2.7. Performance of the Shrimp

After shrimp storage, weekly biometrics were made to adjust the amount of feed offered. At the end of the experiment, all of the animals were weighed and counted. The performance variables collected at the end of the experiment were final average weight (g), final biomass yield (g) survival (%), feed conversion rate (FCR), specific growth rate (g week^{-1}), and productivity (kg m^{-3}).

The performance of the shrimp was analyzed from the weekly biometric measurements that included the following equations:

1. Final average weight (g): final biomass of live animals (g)/total number of animals;
2. Final biomass yield (g): final weight of all live animals (g);
3. Survival (%) = (final number of animals/initial number of animals) × 100;
4. Feed conversion rate (FCR) = feed offered (g)/(final biomass (g) − initial biomass (g));
5. Specific growth rate (g week^{-1}): weight gain (g)/number of weeks;
6. Productivity (kg m^{-3}): [(final biomass (kg) − initial biomass (kg)) × 100]/tank volume (L).

2.8. Statistical Analysis

The normality and homoscedasticity of the data were verified using the Shapiro–Wilk and Levene tests, respectively. Once the assumptions were met, ANOVA was performed followed by Tukey's post-hoc test or *t*-test. When ANOVA's assumptions were not satisfied, the Kruskall–Wallis nonparametric test was used. A minimum level of significance of 5% ($p \leq 0.05$) was applied in all analyzes.

3. Results

3.1. Lab-Scale Experiments

The final weight of the algae was higher ($p \leq 0.05$) then the initial weight of the algae at the beginning of the experiment. However, there was no significant difference ($p \leq 0.05$) in the final weight of the algae between the treatment groups, showing only an increase in the relative growth rate (RGR) in all treatments (Table 2).

Table 2. Initial and final biomass, relative growth rates, and protein contents (% of dry weight) (mean ± standard deviation) of *U. lactuca* growth in the treatments of BFT (in shrimp culture effluent with 400 mg L^{-1} of TSS), DEC (in effluent from shrimp culture after the settling period, with 30 mg L^{-1} of TSS) and VS. (cultivation of *U. lactuca* in standard von Stosch enrichment solution at a concentration of 10 mL L^{-1}) during the 35 days of the experimental period.

Parameters	Treatments		
	BFT	DEC	VS
Initial weight (g)	6.2 ± 0.1	6.2 ± 0.1	6.3 ± 0.1
Final weight (g)	9.1 ± 2.2	7.9 ± 1.1	8.3 ± 1.8
RGR (% day^{-1})	1.0 ± 0.7	0.7 ± 0.4	0.8 ± 0.6
Protein content (%)	18.4 ± 0.5 [a]	19.6 ± 1.4 [a]	10.8 ± 2.4 [b]

[a, b] = Different letters represent a significant difference ($p \leq 0.05$) between treatments after one-way ANOVA with Tukey's post-hoc test. RGR: relative growth rate.

The protein concentrations were evaluated at the end of cultivation and showed a significant difference ($p \leq 0.05$) between the treatment groups. The lower protein value was recorded in the experimental group where the algae were grown in von Stosch enrichment solution (Table 2).

The water quality parameters in the treatment groups are summarized in Table 3. The results show that the water temperature, dissolved oxygen, pH, salinity and alkalinity did not show any significant difference between the treatments ($p \leq 0.05$) (Table 3). The total suspended solids (TSS) concentrations were significantly lower ($p \leq 0.05$) in the DEC treatment where *U. lactuca* was cultivated in the shrimp effluent water with 30 mg L^{-1} of TSS. Turbidity, however, showed similar results between the DEC (with 30 mg L^{-1} of TSS) and BFT groups (with 400 mg L^{-1} of TSS).

Table 3. Water quality parameters (mean ± standard deviation) in the treatment groups where *U. lactuca* was cultivated in different shrimp effluent concentrations (BFT: with 400 mg L^{-1} of TSS; DEC: with 30 mg L^{-1} of TSS; and VS: cultivation of *U. lactuca* in von Stosch enrichment solution at a concentration of 10 mL L^{-1}) during the 35 days of the experimental period.

Parameters	Treatments		
	BFT	DEC	VS
Temperature (°C)	26.6 ± 0.8	26.6 ± 0.8	26.5 ± 0.7
D.O (mg L^{-1})	7.6 ± 0.2	7.8 ± 0.2	7.8 ± 0.2
pH	8.3 ± 0.1	8.4 ± 0.1	8.4 ± 0.1
Salinity (‰)	29.2 ± 2.9	29.1 ± 3.0	29.3 ± 2.3
Turbidity (NTU)	30.3 ± 37.4 [a]	3.3 ± 1.4 [a]	1.0 ± 0.4 [b]
Alkalinity (mg $CaCO_3$ L^{-1})	145.8 ± 6.0	140.3 ± 6.8	Nd
TSS (mg L^{-1})	228.0 ± 109.8 [b]	22.5 ± 6.8 [a]	Nd
TAN (mg L^{-1})	0.1 ± 0.0 [a]	0.1 ± 0.0 [a]	0.1 ± 0.0 [a]
Nitrite (mg L^{-1})	0.2 ± 0.1 [a]	0.2± 0.1 [a]	0.0± 0.0 [b]
Nitrate (mg L^{-1})	64.3 ± 4.6 [a]	62.2 ± 4.6 [a]	10.3 ± 2.9 [b]
Phosphate (mg L^{-1})	4.5 ± 0.8 [a]	4.0 ± 0.7 [a]	0.9 ± 0.2 [b]

[a, b] = represent a significant difference ($p \leq 0.05$) between treatments along the same lines, after one-way ANOVA with Tukey's post-hoc test. Nd: not determined.

The turbidity values did not differ statistically ($p \leq 0.05$) between the initial and final concentrations in the VS treatment where *U. lactuca* cultivation took place in von Stosch enrichment solution at a concentration of 10 mL L^{-1}. For the biofloc treatments (DEC and BFT), there was a decrease ($p \leq 0.05$) in turbidity concentration on the fourth day of sampling and throughout the experiment (Figure 2). There was no significant difference ($p \geq 0.05$) with the vs. treatment at the end of cultivation.

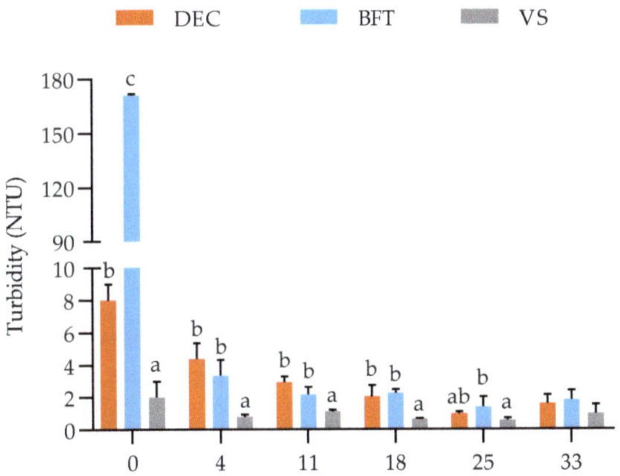

Figure 2. Weekly turbidity values in the DEC (with 30 mg L^{-1} of TSS), BFT (with 400 mg L^{-1} of TSS), and vs. (with von Stosch enrichment solution at a concentration of 10 mL L^{-1}) treatment groups during the 35 days of the experimental period. a, b, c = Different letters on the same day represent a significant difference ($p \leq 0.05$) between the treatments after one-way ANOVA with Tukey's post-hoc test.

3.2. Experiments in Greenhouse Conditions

The macroalgae biomass yields decreased in all the treatment groups compared to the initial biomass values and the difference was significant ($p \leq 0.05$). However, at the end of the experiment there was no significant difference ($p \leq 0.05$) in fresh biomass values between the treatment groups (Table 4).

Table 4. Relative growth rates (% day^{-1}) of *U. lactuca* (mean ± standard deviation) in the IMTA 1 (with 1 g Ulva L^{-1}), IMTA 2 (2 g Ulva L^{-1}), and IMTA 3 (3 g Ulva L^{-1}) treatment groups during the 35 days of cultivation.

Parameters	Treatments		
	IMTA 1	IMTA 2	IMTA 3
Initial fresh weight (g)	150.5 ± 0.3 [a]	300.5 ± 0.4 [b]	450.6 ± 0.4 [c]
Final fresh weight (g)	88.5 ± 31.6	120.9 ± 18.8	195.8 ± 69.8
RGR (% day^{-1})	−1.7 ± 1.2	−2.6 ± 0.4	−2.5 ± 0.9

[a, b, c] = indicate significant differences ($p \leq 0.05$) on the same line between the treatments after one-way ANOVA after Tukey's post-hoc test. RGR: relative growth rate.

There was a significant difference ($p \leq 0.05$) in the protein content of *U. lactuca* in the initial samples compared to the final samples taken from different treatment groups after 35 days of culture (Table 5). However, there was no significant differences between the treatments with biofloc in the protein content of *U. lactuca*.

Table 5. Initial and final protein concentration (% of dry weight) (mean ± standard deviation) of *U. lactuca* in the IMTA 1 (with 1 g Ulva L^{-1}), IMTA 2 (2 g Ulva L^{-1}), and IMTA 3 (3 g Ulva L^{-1}) treatment groups during the 35 days of cultivation.

	Initial Protein Content	Final Protein Content		
		IMTA 1	IMTA 2	IMTA 3
U. lactuca	11.8 ± 0.0 [b]	20.1 ± 0.7 [a]	20.1 ± 1.3 [a]	20.1 ± 1.0 [a]

[a, b] = Different lowercase letters represent significant differences ($p \leq 0.05$) between the initial and final values after one-way ANOVA after Tukey's post-hoc test.

During the 35 days of cultivation in non-climatized greenhouse conditions, the water quality parameters, such as temperature, dissolved oxygen, pH, salinity, phosphate, and nitrite showed no significant difference ($p \geq 0.05$) between the treatments (Table 6). However, the introduction of *U. lactuca* into the shrimp cultivation system resulted in significantly lower mean values of nitrate, turbidity, and settleable solids for treatments IMTA 1 (with 1 g Ulva L^{-1}), IMTA 2 (2 g Ulva L^{-1}), and IMTA 3 (3 g Ulva L^{-1}), compared to those found in MONO C (with no Ulva) (Table 6, Figures 3 and 4).

Table 6. Water quality parameters (mean ± standard deviation) of treatments: the MONO C (with no Ulva), IMTA 1 (with 1 g Ulva L^{-1}), IMTA 2 (2 g Ulva L^{-1}), and IMTA 3 (3 g Ulva L^{-1}) treatment groups during the 35 days of the experiment.

Parameters	Treatments			
	MONO C	IMTA 1	IMTA 2	IMTA 3
Temperature (°C)	26.3 ± 1.0	26.5 ± 0.8	26.5 ± 1.0	26.7 ± 0.9
D.O (mg L^{-1})	7.3 ± 0.2	7.3 ± 0.2	7.3 ± 0.2	7.3 ± 0.2
pH	8.1 ± 0.1	8.2 ± 0.1	8.2 ± 0.1	8.1 ± 0.1
Salinity (‰)	30.0 ± 0.3	29.8 ± 0.3	28.8 ± 0.6	29.6 ± 0.5
Turbidity (NTU)	243.6 ± 66.8 ^{b}	142.6 ± 43.0 ^{b}	117.5 ± 30.2 ^{a}	121.1 ± 45.4 ^{a}
Alkalinity (mg $CaCO_3$ L^{-1})	136.7 ± 12.4 ^{b}	152.0 ± 8.7 ^{a}	143.2 ± 9.8 ^{ab}	148.8 ± 8.7 ^{a}
TSS (mg L^{-1})	423.3 ± 109.7	295.3 ± 92.7	297.3 ± 92.1	302.0 ± 78.7
SS (ml L^{-1})	13.9 ± 7.80 ^{b}	5.5 ± 2.7 ^{a}	4.6 ± 2.3 ^{a}	6.4 ± 3.5 ^{a}
TAN (mg L^{-1})	0.3 ± 0.1	0.3 ± 0.1	0.2 ± 0.1	0.2 ± 0.1
Nitrite (mg L^{-1})	1.3 ± 0.7	1.7 ± 0.8	1.3 ± 0.6	1.5 ± 0.7
Nitrate (mg L^{-1})	62.0 ± 4.0 ^{b}	42.0 ± 2.0 ^{a}	51.3 ± 3.5 ^{ab}	50.7 ± 5.5 ^{ab}
Phosphate (mg L^{-1})	3.4 ± 0.8	4.6 ± 1.1	2.7 ± 0.4	3.2 ± 0.0

^{a,b} = Different lowercase letters represent a significant difference ($p \leq 0.05$) between the weeks of the same treatment, after performing one-way ANOVA followed by Tukey's post-hoc test. O.D (dissolved oxygen); TSS (total suspended solids); SS (settleable solids); TAN (total ammonium nitrogen).

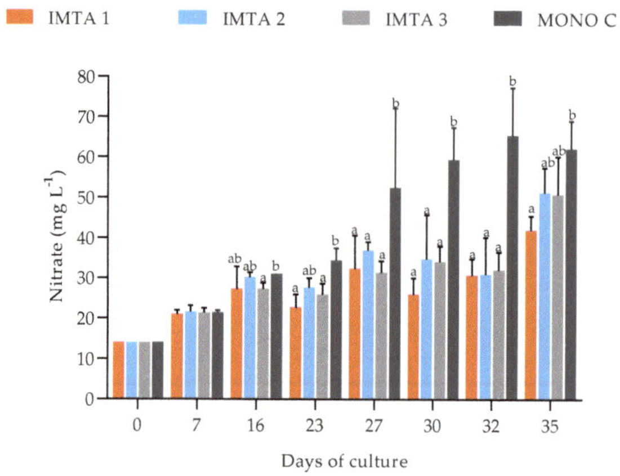

Figure 3. Variations in the nitrate concentrations in the IMTA 1 (with 1 g Ulva L^{-1}), IMTA 2 (2 g Ulva L^{-1}), IMTA 3 (3 g Ulva L^{-1}), and MONO C (with no Ulva) treatment groups during the 35 days of cultivation. a, b = Different lowercase letters represent a significant difference ($p \leq 0.05$) on the same days between the treatments after performing one-way ANOVA followed by Tukey's post-hoc test.

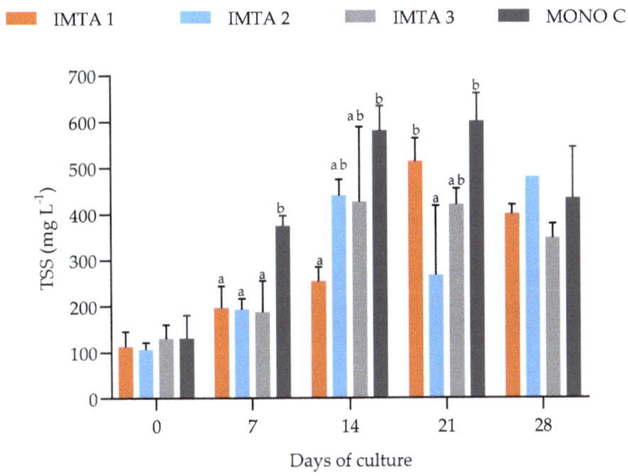

Figure 4. Variations in the total suspended solids (TSS) in the IMTA 1 (with 1 g Ulva L^{-1}), IMTA 2 (2 g Ulva L^{-1}), IMTA 3 (3 g Ulva L^{-1}), and MONO C (with no Ulva) treatment groups during the 35 days of cultivation. a, b = Different lowercase letters on the same day represent a significant difference ($p \leq 0.05$) between the treatments after ANOVA followed by Tukey's post-hoc test.

The alkalinity was higher in treatments with *U. lactuca* (Table 6), requiring a smaller amount of calcium hydroxide for its maintenance. The IMTA 1 (with 1 g Ulva L^{-1}) treatment required the least amount of calcium hydroxide (90 g in the entire crop), while the IMTA 2 (2 g Ulva L^{-1}), IMTA 3 (3 g Ulva L^{-1}), and MONO C (with no Ulva) treatments required 120, 105, and 157 g of calcium hydroxide to maintain the alkalinity, respectively.

In general, the total ammonium nitrogen (TAN) values showed no significant difference in the treatment groups including algal biomass *U. lactuca* ($p \leq 0.05$) (Table 6). However, TAN in the MONO C (with no Ulva) treatment group showed significantly ($p \leq 0.05$) higher values in the first week, with 1.1 ± 0.1 mg L^{-1} when compared to the treatments including the algae, with 0.7 ± 0.1; 0.7 ± 0.1, and 0.6 ± 0.0 mg L^{-1} in the treatments IMTA 1, IMTA 2 and IMTA 3, respectively, thus requiring the application of organic carbon, such as molasses, to control of nitrogen in the MONO C treatment.

In the control treatment (MONO C where there was no Ulva), in the first weeks of the experiment, the TSS (total suspended solids) concentrations raised above the established maximum limit of 500 mg L^{-1} (Figure 4). The clarifiers started to be used to keep optimal shrimp performance as recommended by Gaona et al. [34]. For this reason, during the experiment, the treatments of IMTA 1 (with 1 g Ulva L^{-1}), IMTA 2 (2 g Ulva L^{-1}), and IMTA 3 (3 g Ulva L^{-1}) needed 3, 3, and 5 h of clarification, respectively. These three treatment groups had smaller amounts of tailings compared to the treatment MONO C (with no Ulva) which needed 8 h of clarification.

There was no significant difference ($p \leq 0.05$) in the shrimp performance parameters between the treatment groups (Table 7).

Table 7. The performance of shrimp *L. vannamei* (mean ± standard deviation) in treatments of MONO C (with no Ulva) in the IMTA 1 (with 1 g Ulva L^{-1}), IMTA 2 (2 g Ulva L^{-1}), and IMTA 3 (3 g Ulva L^{-1}) treatment groups during the 35 days of culture.

Parameters	Treatments			
	MONO C	IMTA 1	IMTA 2	IMTA 3
Final average weight (g)	6.5 ± 0.5	6.8 ± 0.3	7.6 ± 0.3	7.4 ± 0.2
Survival (%)	99.3 ± 0.7	100.0 ± 0.0	99.3 ± 0.7	99.3 ± 0.7
FCR	2.0 ± 0.1	1.8 ± 0.1	1.7 ± 0.0	1.8 ± 0.1
WGW (g $week^{-1}$)	0.8 ± 0.3	0.9 ± 0.2	1.1 ± 0.2	1.0 ± 0.1
Final biomass (g)	187.3 ± 22.4	205.2 ± 16.0	239.2 ± 11.5	230.4 ± 9.0
Productivity (kg m^{-3})	1.9 ± 0.1	2.0 ± 0.1	2.3 ± 0.1	2.2 ± 0.1

FCR (food conversion rate); WGW (weekly weight gain).

4. Discussion

The introduction of species in integrated aquaculture is related to their ability to adapt according to system conditions and their interaction with other organisms [35]. In offshore systems, with low water turbidity and water renewal by currents, the integration of the macroalgae into the cultures shows better results. Verdian et al. [4] showed that in these systems the relative growth rate was 14.3 $\pm$ 4.3% day^{-1}, probably due to the easy adaptation of the macroalgae to the system because of the similarity with the natural environment. In land based biofloc systems, the environmental characteristics are different The system is characterized by a high organic load and nutrient accumulation throughout the production cycle. When cultured under unfavorable environmental conditions, such as high solids and nutrient loading in closed systems, such changes may influence the macroalgae's performance. Different management protocols should be applied for the best adaptation and performance of macroalgae in these systems.

Our laboratory experiment showed that different concentrations of solids in the biofloc system did not cause biomass loss of macroalgae (see Table 2). However, the system did not provide the best conditions to seaweed growth compared to offshore cultivation. The relative growth rate of the macroalgae was similar between the treatments, demonstrating the feasibility of using shrimp culture water in biofloc systems as culture medium. The laboratory environmental conditions were controlled, with a direct light source on the macroalgae and temperature regulation. Another way to favor macroalgae growth under these conditions was the use of transparent experimental units that allow greater light incidence, which may have influenced the processes of light absorption by the macroalgae even with the deposition of solids.

In the laboratory experimental conditions, there was also no increase in solids throughout the experiment. Since there were no shrimp present, the feces and feed waste were not being produced in the culture, which was characterized as a static system. In controlled and fixed conditions, the macroalgae were able to adapt to the different culture media and grow. In contrast, the conditions of the greenhouse experiment simulated real culture conditions, resulting in biomass loss (see Table 4). The integrated culture of shrimp and macroalgae in an intensive system and with a high feed intake generated an increase in solids. Even with intense aeration, the macroalgae culture structure allowed for greater deposition of organic matter on the macroalgae. This effect was smaller in the laboratory experiment, because the macroalgae were loose in the carboy. Such conditions can be stressful to the macroalgae, impacting their growth throughout the culture, which was verified in this experiment with the loss of biomass. A similar result was also observed by Legarda et al. [10], working in a closed system with integrated culture of the macroalgae *U. fasciata*, the shrimp *L. vannamei*, and the fish *Mugil liza* in a biofloc system, confirming the difficulty of adaptation of the macroalgae in this system.

The availability of nutrients in the biofloc system is advantageous for macroalgae cultures [9]. However, the presence of solids can be an inhibiting factor for macroalgae growth. In both experiments, solids were deposited on the macroalgae, and lower concentrations of suspended solids were found in the water. In the laboratory experiment, a decrease in turbidity was observed in the treatment with the highest concentration of solids (400 mg L^{-1}) on the fourth day of culture (Figure 2), showing that most of the solids were deposited on the macroalgae, even with constant aeration.

In the greenhouse experiment, a similar process was verified, with the aggravating factor that the production of solids was persistent due to the presence of the shrimp in the culture. The settleable solids (ml L^{-1}) showed a significant difference between the integrated culture and monoculture treatments. The lowest concentrations of settleable solids were observed in the *U. lactuca* treatment. As the individuals of this species are sessile organisms, they probably interfered with the dynamic movement of biofloc particles in the water column, causing the deposition of particles on them, unlike in the monoculture where there was no physical barrier for the particles to decant. Brito et al. [9] also found a decrease in solids levels due to the deposition of flocs on the photosynthetic macroalgae leaves, going from a suspended to a decanted solid. This result also induced the need to use the clarifier to control excess solids in the system. The treatments with the presence of *U. lactuca* required less clarification time compared to the shrimp monoculture treatments. However, removing the macroalgae from the system will cause solids to become suspended in the water again. Excess solids in the water can cause rapid oxygen depletion [36] and growth performance problems to the shrimp [5]. Some alternatives to control solids presented by Khanjani et al. [35] would be the use of organic consumers in integrated cultivation. Thus, the solids would be controlled without the use of clarifiers and there would not be a large amount of solids deposited on the macroalgae.

The deposition of solids on the macroalgae can be a stressing factor for the species, which may prevent the absorption of light and the performance of photosynthesis (see Figure 4). Such factors can trigger reproduction events such as the release of gametes or spores. These events can be initiated due to several environmental factors, such as high temperatures, the concentration of nutrients, and even the short life cycle of the species, thus resulting in the loss of biomass [19]. During the cultivation period, the presence of "ghost tissues" was also observed in *U. lactuca* as a sign of sporulation. The loss of macroalgae biomass was also verified by Legarda et al. [10] when cultivating *U. fasciata* in an integrated system, probably because of the different characteristics of the biofloc system compared to the natural environment where the macroalgae were collected.

The von Stosch nutrient solution used in the laboratory experiment is composed of balanced minerals and nutrients [13]. Despite the biomass gain, the relative growth rate of *U. lactuca* in this experiment in the laboratory condition was lower compared to the studies with a maximum of 16.9% day^{-1} for *U. prolifera*, when cultivated in the laboratory conditions with the nutrient medium F/2 [37]. The von Stosch medium used for the cultivation of *U. lactuca* and the concentration of the medium used may not be adequate to cause low growth of the algal biomass. In addition, the treatment water was not renewed nor were more nutrients added, which could have limited the growth of the algae. For the production of culture media, specific compounds of high economic value are needed, so alternative culture media and better management practices can facilitate the production of *U. lactuca*.

The increase in the protein content of *U. lactuca* cultivated in the biofloc system probably occurred because of the nutritional composition of the macroalgae changes according to the physical and chemical factors of the culture environment. For example, higher concentrations of nitrogen available in the culture system, such as in the biofloc system, provide an increase in tissue nitrogen. According to Duke et al. [38], greater availability of nitrogen in the medium results in its absorption and its transformation into protein, stored in the form of amino acids and pigments [32]. Treatments with biofloc effluent contained higher concentrations of nitrate and phosphate than treatment with von Stosch culture medium. This is due to the origin of the biofloc effluent, which came from a shrimp culture with 43 days of cultivation, with a gradual accumulation of nutrients. This high availability of nitrogen in the water favored the increase in the protein content of the macroalgae. The results obtained were superior to those observed by Fong et al. [39], who obtained protein concentrations ranging from 0.6 to 5.4% in algae grown in nutrient solutions in the laboratory. The high protein value of *U. lactuca* in the present study shows its importance for human food as a food supplement, for muscle tissue reconstruction, and in vegan foods, similar to the previous study conducted by Bleakley and Hayes [40].

Even with 20% biofloc inoculum in the greenhouse experiment, the ammonia concentrations in the first days of cultivation exceeded the concentration of 1 mg L^{-1} in the tanks without *U. lactuca*. Although ammonia concentrations were well controlled in the biofloc system, at times throughout the production cycle it may be necessary to add organic carbon to stimulate the development of heterotrophic bacteria in the system [26]. When using a low percentage of inoculum diluted in seawater, the bacteria can undergo adaptation in the system. Together with the feed supply and continuous excretion of the animals, these bacteria were not able to convert all the ammonia in the system and its concentration increased, with the use of molasses in shrimp monoculture being necessary to increase the number of heterotrophic bacteria, in this experiment. Such instability of the bacteria also occurred in the integrated treatment, but due to the absorption capacity by *U. lactuca*, there was no increase of ammonia concentration in the system. Castelar et al. [41] observed that the genus Ulva tends to have a preference for ammonia, making its assimilation faster and thus controlling nitrogen in the crop as a consequence.

Nitrate was another nitrogenous compound absorbed by *U. lactuca*. Its accumulation is constant throughout the production cycle in the biofloc system, with it reaching high concentrations. The *U. lactuca* absorbed the nitrate, as it is the nitrogenous compound with the highest availability in the system, with the best absorption result occurring in the treatment with the lowest density of *U. lactuca* (1 g L^{-1}). High densities, above 1 g L^{-1} are likely to increase intraspecific competition, due to *U. lactuca* overlap in the structure, and negatively affect nutrient absorption. Alencar et al. [15], testing different densities with *U. lactuca*, also observed that nutrient removal was impaired when the algae density and growth rate increased. Therefore, unlike the gradual accumulation of nitrate that occurred in the monoculture, the treatment with the density of 1 g L^{-1} of *U. lactuca* resulted in a lower concentration of nitrate at the end of the cultivation. Due to absorption by the macroalgae, this compound is used for the production of biomass and pigments.

In addition to nutrient absorption, the integration of *U. lactuca* into shrimp cultivation also interferes with other components in the system. For the biofloc system, calcium carbonate or calcium hydroxide is required to maintain alkalinity [30]. The integrated cultivation of *L. vannamei* and *U. lactuca* required a smaller number of corrections with $Ca(OH)_2$, maintaining a more stable level of alkalinity compared to the monoculture treatment where there was no *U. lactuca*. This possibly occurred due to the absorption of carbon dioxide from the medium by *U. lactuca* [36]. Chopin [11] comments on the decrease in water acidification due to the absorption of gases by macroalgae, acting on the greenhouse effect. The same pattern can occur in cropping systems integrated with *U. lactuca*.

The physical and chemical parameters of water quality were kept in the ideal ranges for *L. vannamei* cultivation, such as temperature [42], salinity [43], dissolved oxygen [44], pH [45], alkalinity [30], and TSS [5]. The shrimp cultivated in this study had a similar performance to that found in the literature for monoculture systems [26]. Therefore, the integration of *U. lactuca* in the system did not interfere in the performance of the shrimp, with it being environmentally advantageous. In the study of Brito et al. [9], a higher average final weight of the shrimp was observed when cultivated with the macroalgae, possibly due to nutritional advantages provided by the macroalgae to the shrimp. Although the macroalgae are grown in a separate floating structure from the shrimp, they are both grown in the same tank. Therefore, the reduction in the production of biomass served by Ulva can also be explained by its consumption by shrimp. In future studies, it is recommended to observe and monitor if the shrimp consume algae when they are integrated in the same tank as well as when they are in a separated cultivation structure.

In a conventional farming system, there is a significant loss of nitrogen that is not incorporated by the animals and becomes available in the water as a residue that can be toxic and the main source of environmental pollution [46]. The use of macroalgae for the absorption of nutrients from the system has been widely used due to greater sustainability and productivity gain [4]. The use of macroalgae *U. lactuca* at a density of 1 g L^{-1} in an integrated system with shrimp *L. vannamei* in biofloc was also viable due to the incorporation of nitrogen by the algae, resulting in a biomass with higher protein content, in addition to increasing the system productivity and sustainability.

5. Conclusions

The cultivation of macroalgae in biofloc promoted changes in water quality in both experiments. The concentration of total suspended solids decreased in both experiments with macroalgae integration.

When cultivated in an integrated system with shrimp, the addition of macroalgae at a density of 1 g L^{-1} in the system promoted the absorption of nitrogen, nitrate, and ammonia, generating lower final concentrations of these nutrients at the end of cultivation compared to shrimp monoculture conditions. In addition, the integration of the macroalgae into the system resulted in less use of inputs, such as molasses and calcium hydroxide.

Finally, the use of *U. lactuca* did not negatively affect the performance of the shrimp. Despite the loss of biomass under the conditions tested in the integrated system, the macroalgae *U. lactuca* showed potential for the consumption of nutrients available in the system.

Author Contributions: Conceptualization, A.C., L.H.P. and M.H.; methodology, A.C., G.T. and L.H.P.; software, A.C. and L.C.d.O.C.; validation, A.C., L.C.d.O.C., M.H., L.H.P. and G.T.; formal analysis, A.C. and L.C.d.O.C.; investigation, A.C., L.C.d.O.C., M.H., L.H.P. and G.T.; resources, G.T. and L.H.P.; data curation, A.C., L.C.d.O.C. and M.H.; writing—original draft preparation, A.C. and L.C.d.O.C.; writing—review and editing, L.C.d.O.C., M.H., L.H.P. and G.T.; visualization, A.C., L.C.d.O.C., M.H., L.H.P. and G.T.; supervision, G.T. and L.H.P.; project administration, L.C.d.O.C., M.H., L.H.P. and G.T.; funding acquisition, G.T. and L.H.P. All authors have read and agreed to the published version of the manuscript.

Funding: This research was funded by ASTRAL Project—H2020 grant Agreement 863034.

Institutional Review Board Statement: The experiment was approved by the Ethics and Animal Welfare Committee of FURG (Case number 23116.005895/2016-42).

Acknowledgments: The Authors are grateful to the ASTRAL project that has received funding from the European Union's Horizon 2020 research and innovation program under grant agreement No 863034. The Authors are also thankful to John Bolton of Biological Sciences at the University of Cape Town, South Africa for identification of the species and Amir Neori of National Center for Mariculture at Israel Oceanographic & Limnological Research Ltd, Eilat, Israel for valuable information and suggestions in the preparation and editing of this manuscript. Special thanks to GUABI Nutrition for donating the commercial diets.

Conflicts of Interest: The authors declare no conflict of interest.

References

1. Da Silva, K.R.; Wasielesky, W.; Abreu, P.C. Nitrogen and Phosphorus Dynamics in the Biofloc Production of the Pacific White Shrimp, *Litopenaeus vannamei*. *J. World Aquac. Soc.* **2013**, *44*, 30–41. [CrossRef]
2. Krummenauer, D.; Abreu, P.C.; Poersch, L.; Reis, P.A.C.P.; Suita, S.M.; dos Reis, W.G.; Wasielesky, W. The Relationship between Shrimp (*Litopenaeus vannamei*) Size and Biofloc Consumption Determined by the Stable Isotope Technique. *Aquaculture* **2020**, *529*, 735635. [CrossRef]
3. Ferreira, G.S.; Santos, D.; Schmachtl, F.; Machado, C.; Fernandes, V.; Bögner, M.; Schleder, D.D.; Seiffert, W.Q.; Vieira, F.N. Heterotrophic, Chemoautotrophic and Mature Approaches in Biofloc System for Pacific White Shrimp. *Aquaculture* **2021**, *533*, 736099. [CrossRef]
4. Verdian, A.H.; Effendi, I.; Budidardi, T.; Diatin, I. Production Performance Improvement of White Shrimp (*Litopenaeus vannamei*) Culture with Integrated Multi Trophic Aquaculture System in Seribu Islands, Jakarta, Indonesia. *Iran. J. Fish. Sci.* **2020**, *19*, 1415–1427. [CrossRef]
5. Gaona, C.A.P.; de Almeida, M.S.; Viau, V.; Poersch, L.H.; Wasielesky, W. Effect of Different Total Suspended Solids Levels on a *Litopenaeus vannamei* (Boone, 1931) BFT Culture System during Biofloc Formation. *Aquac. Res.* **2017**, *48*, 1070–1079. [CrossRef]
6. Ahmad, I.; Babitha Rani, A.M.; Verma, A.K.; Maqsood, M. Biofloc Technology: An Emerging Avenue in Aquatic Animal Healthcare and Nutrition. *Aquac. Int.* **2017**, *25*, 1215–1226. [CrossRef]
7. Azim, M.E.; Little, D.C. The Biofloc Technology (BFT) in Indoor Tanks: Water Quality, Biofloc Composition, and Growth and Welfare of Nile Tilapia (*Oreochromis niloticus*). *Aquaculture* **2008**, *283*, 29–35. [CrossRef]
8. Wasielesky, W.; Bezerra, A.; Poersch, L.; Hoffling, F.B.; Krummenauer, D. Effect of Feeding Frequency on the White Shrimp *Litopenaeus vannamei* during the Pilot-Scale Nursery Phase of a Superintensive Culture in a Biofloc System. *J. World Aquac. Soc.* **2020**, *51*, 1175–1191. [CrossRef]
9. Brito, L.O.; Arantes, R.; Magnotti, C.; Derner, R.; Pchara, F.; Olivera, A.; Vinatea, L. Water Quality and Growth of Pacific White Shrimp *Litopenaeus vannamei* (Boone) in Co-Culture with Green Seaweed *Ulva lactuca* (Linaeus) in Intensive System. *Aquac. Int.* **2014**, *22*, 497–508. [CrossRef]
10. Legarda, E.C.; da Silva, D.; Miranda, C.S.; Pereira, P.K.M.; Martins, M.A.; Machado, C.; de Lorenzo, M.A.; Hayashi, L.; do Nascimento Vieira, F. Sea Lettuce Integrated with Pacific White Shrimp and Mullet Cultivation in Biofloc Impact System Performance and the Sea Lettuce Nutritional Composition. *Aquaculture* **2021**, *534*, 736265. [CrossRef]

11. Chopin, T. Marine Aquaculture in Canada: Well-Established Monocultures of Finfish and Shellfish and an Emerging Integrated Multi-Trophic Aquaculture (IMTA) Approach Including Seaweeds, Other Invertebrates, and Microbial Communities. *Fisheries* **2015**, *40*, 28–31. [CrossRef]
12. Troell, M.; Joyce, A.; Chopin, T.; Neori, A.; Buschmann, A.H.; Fang, J.G. Ecological Engineering in Aquaculture—Potential for Integrated Multi-Trophic Aquaculture (IMTA) in Marine Offshore Systems. *Aquaculture* **2009**, *297*, 1–9. [CrossRef]
13. Mansilla, A.; Rodriguez, J.P.; Souza, J.M.C.; Rosenfeld, S.; Ojeda, J.; Yokoya, N.S. Growth Responses to Temperature, Salinity and Nutrient Variations, and Biomass Variation and Phenology of *Ahnfeltia plicata* (Rhodophyta, Ahnfeltiales): A Commercially Interesting Agarophyte from the Magellanic Region, Chile. *J. Appl. Phycol.* **2014**, *26*, 1133–1139. [CrossRef]
14. Kılınç, B.; Cirik, S.; Turan, G. Seaweeds for Food and Industrial Applications. In *Food Industry*; IntechOpen: London, UK, 2013; pp. 735–748.
15. de Alencar, J.R.; Junior, P.A.H.; Celino, J.J. Cultivo de Camarão Branco *Litopenaeus vannamei* (Boone, 1931) Com a Macroalga *Ulva lacuata* Linneaus (Chlorophyta) No Tratamento de Efluentes Em Sistema Fechado de Recirculação. *Rev. Biol. Ciênc. Terra* **2010**, *10*, 117–137.
16. Ramos, R.; Vinatea, L.; Santos, J.; Da Costa, R. Tratamiento de Efluentes Del Cultivo de *Litopenaeus vannamei* Mediante Procesos de Sedimentación, Filtración y Absorción. *Lat. Am. J. Aquat. Res.* **2010**, *38*, 188–200. [CrossRef]
17. El Baz, F.K.; El-Baroty, G.S.; Ibrahim, A.E.; Abd El Baky, H.H. Cytotoxicity, Antioxidants and Antimicrobial Activities of Lipids Extracted from Some Marine Algae. *J. Aquac. Res. Dev.* **2014**, *5*, 5–9. [CrossRef]
18. Turan, G.; Tekogul, H. The Turkish Mezzes Formulated with Protein-Rich Green Sea Vegetable (Chlorophyta), *Ulva rigida*, Cultured in Onshore Tank System. *J. Aquat. Food Prod. Technol.* **2014**, *23*, 447–452. [CrossRef]
19. Copertino, M.D.S.; Tormena, T.; Seeliger, U. Biofiltering Efficiency, Uptake and Assimilation Rates of *Ulva clathrata* (Roth) J. Agardh (Clorophyceae) Cultivated in Shrimp Aquaculture Waste Water. *J. Appl. Phycol.* **2009**, *21*, 31–45. [CrossRef]
20. Wei, Z.; You, J.; Wu, H.; Yang, F.; Long, L.; Liu, Q.; Huo, Y.; He, P. Bioremediation Using *Gracilaria lemaneiformis* to Manage the Nitrogen and Phosphorous Balance in an Integrated Multi-Trophic Aquaculture System in Yantian Bay, China. *Mar. Pollut. Bull.* **2017**, *121*, 313–319. [CrossRef]
21. del Río, M.J.; Ramazanov, Z.; García-Reina, G. *Ulva rigida* (Ulvales, Chlorophyta) Tank Culture as Biofilters for Dissolved Inorganic Nitrogen from Fishpond Effluents. *Hydrobiologia* **1996**, *326*, 61–66. [CrossRef]
22. Guiry, M.D.; Cunningham, E.M. Photoperiodic and Temperature Responses in the Reproduction of North-Eastern Atlantic *Gigartina acicularis* (Rhodophyta: Gigartinales). *Phycologia* **1984**, *23*, 357–367. [CrossRef]
23. Krummenauer, D.; Peixoto, S.; Cavalli, R.O.; Poersch, L.H.; Wasielesky, W. Superintensive Culture of White Shrimp, *Litopenaeus vannamei*, in a Biofloc Technology System in Southern Brazil at Different Stocking Densities. *J. World Aquac. Soc.* **2011**, *42*, 726–733. [CrossRef]
24. Unesco. *Chemical Methods for Use in Marine Environmental Monitoring*; Intergovernmental Oceanographic Commission: Paris, France, 1983.
25. Bendschneider, K.; Robinson, R.J. *A New Spectrophotometric Method for the Determination of Nitrite in Water*; Technical Report No. 8; Office of Naval Research: Seattle, WA, USA, 1952.
26. Samocha, T.M.; Patnaik, S.; Speed, M.; Ali, A.M.; Burger, J.M.; Almeida, R.V.; Ayub, Z.; Harisanto, M.; Horowitz, A.; Brock, D.L. Use of Molasses as Carbon Source in Limited Discharge Nursery and Grow out Systems for *Litopenaeus vannamei*. *Aquac. Eng.* **2007**, *36*, 184–191. [CrossRef]
27. Aminot, A.; Chaussepied, M. *Manuel des Analyses Chimiques en Milieu Marin*; Centre National Pour L'Exploitation des Océans: Paris, France, 1983.
28. Baumgarten, M.D.G.Z.; De Barros Rocha, J.M.; Niencheski, L.F.H. *Manual de Análises em Oceanografia Química*; Universidade Federal do Rio Grande: Rio Grande, Brazil, 1996.
29. American Public Health Association APHA; American Water Works Association; Water Pollution Control Association. *Standard Methods for the Examination of Water and Wastewater*, 21st ed.; American Public Health Association: Washington, DC, USA, 2005.
30. Furtado, P.S.; Poersch, L.H.; Wasielesky, W. Effect of Calcium Hydroxide, Carbonate and Sodium Bicarbonate on Water Quality and Zootechnical Performance of Shrimp *Litopenaeus vannamei* Reared in Biofloc Technology (BFT) Systems. *Aquaculture* **2011**, *321*, 130–135. [CrossRef]
31. Loureiro, R.R.; Reis, R.P.; Critchley, A.T. In Vitro Cultivation of Three *Kappaphycus alvarezii* (Rhodophyta, Areschougiaceae) Variants (Green, Red and Brown) Exposed to a Commercial Extract of the Brown Alga *Ascophyllum nodosum* (Fucaceae, Ochrophyta). *J. Appl. Phycol.* **2010**, *22*, 101–104. [CrossRef]
32. Baethgen, W.E.; Alley, M. A Manual Colorimetric Procedure for Measuring Ammonium Nitrogen in Soil and Plant Kjeldahl Digests. *Commun. Soil Sci. Plant Anal.* **1989**, *20*, 961–969. [CrossRef]
33. AOAC. *Official Methods of Analysis*; Association of Official Analytical Chemists: Arlington, VA, USA, 2005; p. 245.
34. Gaona, C.A.P.; Poersch, L.H.; Krummenauer, D.; Foes, G.K.; Wasielesky, W.J. Effect of Solids Removal on Production of Shrimp. *Int. J. Recirc. Aquac.* **2011**, *12*, 54–73.
35. Khanjani, M.H.; Zahedi, S.; Mohammadi, A. Integrated Multitrophic Aquaculture (IMTA) as an Environmentally Friendly System for Sustainable Aquaculture: Functionality, Species, and Application of Biofloc Technology (BFT). *Environ. Sci. Pollut. Res.* **2022**, *29*, 67513–67531. [CrossRef]

36. Granada, L.; Lopes, S.; Novais, S.C.; Lemos, M.F.L. Modelling Integrated Multi-Trophic Aquaculture: Optimizing a Three Trophic Level System. *Aquaculture* **2018**, *495*, 90–97. [CrossRef]
37. Luo, M.B.; Liu, F.; Xu, Z.L. Growth and Nutrient Uptake Capacity of Two Co-Occurring Species, *Ulva prolifera* and *Ulva linza*. *Aquat. Bot.* **2012**, *100*, 18–24. [CrossRef]
38. Duke, C.S.; Litaker, W.; Ramus, J. Effects of the Temperature, Nitrogen Supply and Tissue Nitrogen on Ammonium Uptake Rates of the Chlorophyte Seaweeds *Ulva Curvata* and *Codium decorticatum*. *J. Phycol.* **1989**, *25*, 113–120. [CrossRef]
39. Fong, P.; Boyer, K.E.; Desmond, J.S.; Zedler, J.B. Salinity Stress, Nitrogen Competition, and Facilitation: What Controls Seasonal Succession of Two Opportunistic Green Macroalgae? *J. Exp. Mar. Biol. Ecol.* **1996**, *206*, 203–221. [CrossRef]
40. Bleakley, S.; Hayes, M. Algal Proteins: Extraction, Application, and Challenges Concerning Production. *Foods* **2017**, *6*, 33. [CrossRef]
41. Castelar, B.; Reis, R.P.; dos Santos Calheiros, A.C. *Ulva lactuca* and *U. flexuosa* (Chlorophyta, Ulvophyceae) Cultivation in Brazilian Tropical Waters: Recruitment, Growth, and Ulvan Yield. *J. Appl. Phycol.* **2014**, *26*, 1989–1999. [CrossRef]
42. Wyban, J.; Walsh, W.A.; Godin, D.M. Temperature Effects on Growth, Feeding Rate and Feed Conversion of the Pacific White Shrimp (*Penaeus vannamei*). *Aquaculture* **1995**, *138*, 267–279. [CrossRef]
43. Khanjani, M.H.; Sharifinia, M.; Hajirezaee, S. Effects of Different Salinity Levels on Water Quality, Growth Performance and Body Composition of Pacific White Shrimp (*Litopenaeus vannamei* Boone, 1931) Cultured in a Zero Water Exchange Heterotrophic System. *Ann. Anim. Sci.* **2020**, *20*, 1471–1486. [CrossRef]
44. Schveitzer, R.; Arantes, R.; Costódio, P.F.S.; do Espírito Santo, C.M.; Arana, L.V.; Seiffert, W.Q.; Andreatta, E.R. Effect of Different Biofloc Levels on Microbial Activity, Water Quality and Performance of *Litopenaeus vannamei* in a Tank System Operated with No Water Exchange. *Aquac. Eng.* **2013**, *56*, 59–70. [CrossRef]
45. Wasielesky, W.; Krummenauer, D.; Lara, G.; Fóes, G. Cultivo de Camarões Em Sistema de Bioflocos: Realidade e perspectivas. *Rev. ABCC* **2013**, *15*, 16–26.
46. Poli, M.A.; Legarda, E.C.; de Lorenzo, M.A.; Pinheiro, I.; Martins, M.A.; Seiffert, W.Q.; do Nascimento Vieira, F. Integrated Multitrophic Aquaculture Applied to Shrimp Rearing in a Biofloc System. *Aquaculture* **2019**, *511*, 734274. [CrossRef]

Disclaimer/Publisher's Note: The statements, opinions and data contained in all publications are solely those of the individual author(s) and contributor(s) and not of MDPI and/or the editor(s). MDPI and/or the editor(s) disclaim responsibility for any injury to people or property resulting from any ideas, methods, instructions or products referred to in the content.

fishes

MDPI

Article

The Effects of Different Carbon Sources on Water Quality, Growth Performance, Hematology, Immune, and Antioxidant Status in Cultured Nile Tilapia with Biofloc Technology

Khalid Hussain Rind [1], Syed Sikandar Habib [2], Javed Ahmed Ujan [3], Francesco Fazio [4,*], Saira Naz [5], Aima Iram Batool [2], Mujeeb Ullah [6], Sobia Attaullah [6], Khayyam Khayyam [6] and Khalid Khan [6]

1 Department of Molecular Biology and Genetics, Shaheed Benazir Bhutto University Shaheed Benazirabad, Sindh 67450, Pakistan; khalid.rind@sbbusba.edu.pk
2 Department of Zoology, University of Sargodha, Sargodha 40100, Pakistan; sikandarzoo00@yahoo.com (S.S.H.); aima.iram@uos.edu.pk (A.I.B.)
3 Department of Zoology, Shah Abdul Latif University, Khairpur Mirs 66020, Pakistan; javed.ujan@salu.edu.pk
4 Department of Veterinary Sciences, University of Messina, Via Palatuccisnc, 98168 Messina, Italy
5 Institute of Molecular Biology and Biotechnology, University of Lahore Sargodha Campus, Sargodha 40100, Pakistan; saira.zoologist1991@gmail.com
6 Department of Zoology, Islamia College University, Peshawar 25120, Pakistan; mujibkhanicp@gmail.com (M.U.); sobia@icp.edu.pk (S.A.); khayyam@icp.edu.pk (K.K.); khalidkhan@icp.edu.pk (K.K.)
* Correspondence: ffazio@unime.it; Tel.: +39-0906766516

Abstract: The biofloc technology system (BFT) is considered to be one of the sustainable aquaculture systems, which is based on the principle of nutrient recycling with the addition of a carbon source to give dominance to heterotrophic microorganisms. The objective of this study was to evaluate the effect of sugar cane molasses and tapioca flour as carbon sources on the water quality, growth, hematology, immune status, and non-specific antioxidant status of Oreochromis juveniles. Methodologically, the experiment was carried out for 10 weeks on 225 juvenile Nile tilapia with initial body weights of 47.0 ± 1.3 g that were randomly distributed in 09 tanks (1000 L) with a stocking density of 25 tilapias per tank; the treatments were: BFT + SM (S molasses), BFT + TF tapioca flour (TF), and a control with no carbon source added. The control group was fed 100% feed, while the BFT experimental groups were fed microbial flocs along with 75% feed. The results revealed that the water quality parameters were affected by the carbon sources, but were adequate for normal fish welfare, and the biofloc volume was higher (28.94) with the TF carbon source. The growth performance, such as weight gain (98.61), survival (99.01), and improved feed conversion ratio (FCR) (1.69), was recorded in BFT + TF. Significant improvements in WBCs, HCT, HB, lymphocytes, plasma proteins, albumin, and non-specific immune factors (lysozyme activity, immunoglobulins levels, and ACH50) were observed in biofloc-reared fish with tapioca flour as the carbon source compared to the control and sugarcane molasses groups. Moreover, significant increases in catalase (CAT) and superoxide dismutase (SOD) were found in the biofloc-reared fish with different carbon sources. In conclusion, the use of BFT + TF was found to affect improving the water quality, growth, hematology, immunity, and antioxidant status of juvenile Tilapia.

Keywords: biofloc technology; carbon source; water quality; growth; immunity; antioxidant

Key Contribution: Tapioca meal effect as a carbon source improved the water quality parameters, growth, blood profile, immunity, and antioxidant status in juvenile Tilapia cultured in biofloc technology with tapioca meal.

Citation: Rind, K.H.; Habib, S.S.; Ujan, J.A.; Fazio, F.; Naz, S.; Batool, A.I.; Ullah, M.; Attaullah, S.; Khayyam, K.; Khan, K. The Effects of Different Carbon Sources on Water Quality, Growth Performance, Hematology, Immune, and Antioxidant Status in Cultured Nile Tilapia with Biofloc Technology. *Fishes* **2023**, *8*, 512. https://doi.org/10.3390/fishes8100512

Academic Editors: Wagner C. Valenti and Eduardo Luis Ballester

Received: 20 August 2023
Revised: 4 October 2023
Accepted: 12 October 2023
Published: 14 October 2023

Copyright: © 2023 by the authors. Licensee MDPI, Basel, Switzerland. This article is an open access article distributed under the terms and conditions of the Creative Commons Attribution (CC BY) license (https://creativecommons.org/licenses/by/4.0/).

1. Introduction

A major requirement for meeting the current fish protein demand for human consumption is the intensification of aquaculture [1]. Increasing the rearing density of fish increases the productivity per unit area [2]. A limited ability to control pathogens poses a major challenge to production intensification [3]. Sustainability in feed management is also considered to be a major component in intensifying any aquatic organism's production [4]. In addition, a sustainable cultural system that does not pollute the environment and utilizes limited natural resources is needed [5]. Such an ecological aquaculture system is biofloc technology, which ensures sustainable feed management. This system provides a high yield with limited water exchange [6]. The advantage includes maintaining a high C/N ratio; therefore, the microbial community can take up ammonium and enhance the management of health and biosecurity with a limited exchange of water [4].

The biofloc culture system assembles various suspended organic particles with useful microorganisms involved with polymeric extracellular substance, making it a heterogeneous system [7]. It is possible to increase the carbon ratio of the feed by adding various organic carbonaceous sources such as tapioca, glucose, corn, wheat, acetate, glycerol, and molasses, etc., to the aquaculture system or by altering the feed composition by adding additional organic carbon sources [8]. The organic carbon source significantly influences the composition of flocs, specifically, the kind of storage polymer used and its amount [9]. Tilapia and *Litopenaeus vannamei* were successfully farmed using biofloc technology, which substantiated to be better in terms of feed efficacy and water than traditional methods [5]. In the biofloc system, the type of carbon source used affects its management, nutritional value, microbial community, and the biofloc system organisms [8]. Wei et al. [10] described that adding diverse carbonaceous sources may influence the NH_4 elimination process, and simpler carbon sources (such as glucose and sucrose) may eradicate ammonia more rapidly than more complex carbon sources such as starch. Aquatic organisms have been treated with microorganisms and their cell components to enhance their immunity, growth, disease resistance, and antioxidant status [11]. Several bioactive compounds are present in bioflocs, such as polysaccharides, chlorophyll, fat-soluble vitamins, taurine, carotenoids, and phytosterols [12,13]. It is widely recognized that several microorganisms and their metabolites present in biofloc are immune-stimulants, which enhance immunity and are protective against several diseases [14]. Furthermore, bacterial species are typically linked with suspended particles in biofloc, facilitating with exogenous digestive enzymes and additional nutrients, thereby contributing to bacterial growth and survival [5]. Many studies have been conducted by utilizing several sources of carbons and their effects on several species cultured in biofloc technology, such as tapioca and plant starch for *Pelteobagrus vachelli* [8], Longan powder for Nile tilapia [9], and wheat bran and molasses for *Litopenaeus vannamei* [12]. Additional research is required to clarify the effects of various carbon sources on the water quality, growth, immunity, and stress response of species. The current study used tapioca flour (TF) and sugarcane molasses (SM), the best alternatives to the exhausted sources, which are readily available.

Globally, tilapia farming has increased in popularity and is the second-most farmed fish species [4]. It has a high growth rate and is a stress-tolerant species. Moreover, it is still under investigation to assess the effects of different carbon sources on the water quality, fish growth, haematology, immunity, and antioxidant status of biofloc-reared species. Therefore, the current research work was intended to assess the effect of various carbon sources (sugarcane molasses and tapioca flour) on the water characteristics, growth, hematology, and non-specific immune and antioxidant status of juvenile *Oreochromis niloticus* for the duration of 10 weeks.

2. Materials and Methods

2.1. Experimental Design

The research was performed in the Research Laboratory of Zoology, University of Lahore, Pakistan. Before the experiment began, 225 juvenile Nile tilapia with initial body weights of (27.0 ± 1.3 g) were collected from the local fish hatchery. The collected fish were acclimatized (indoor) for 14 days in a rectangular tank of a 2000 L volume. During this duration, commercial feed (Supreme Company, Lahore, Pakistan) at 3% of their body weight was provided three times daily (7:00, 12:00, and 17:00) under a light/dark period (12/12 h). According to a proximate analysis, the feed consisted of crude protein level (32%), ash (6.9%), and lipid (4.3%). A continuous aeration system (1.5 hp blower) was installed in the tank, and the temperature of the culture water was maintained using an electric heater at 27.0 ± 1 °C.

In the second stage of the study, a total of three treatments were designed, each with three replicates in 09 rectangular tanks (1000 L vol). The initial stocking density of the fish was 25 fish/tank, 1.3 kg/m^3, and body weight (47.0 ± 1.3 g). The experimental period of the study was 10 weeks. The control group was set in flow through a system with 30% water exchange on a daily basis and the fish of the control group were fed on commercial feed with 3% of their body weight [15]. However, the fish groups of the biofloc technology fed on bacterial flocs and commercial feed (75% daily feeding) with zero exchange of water [16]. The treatment protocols for the control and BFT groups were control: FT/30% water replacement, T1: BFT + SM, and T2: BFT + TF.

Fresh well water (salinity 15 ppt) was first added to all of the tanks prior to the start of the experiment. The concentrated biofloc (100 mL) was added to the BFT treatment groups from the old biofloc tanks. In addition, the C:N ratio of the BFT tanks was maintained at 15 [17] by a daily addition of the carbon sources two hours after feeding. In order to produce microbial flocs stock, 200 L of first-stage effluent was transferred into four conoid tanks and the total ammonium nitrogen (TAN) was estimated. Various carbon sources, including sugarcane molasses (SM) and tapioca flour (TF), were added to the tanks grounded on the calculation of Avnimelech [18], who presumed that 20 g of carbon source is necessary to transform 1 g of TAN. A biofloc was developed by adding carbonaceous materials to BFT tanks at a carbon-to-nitrogen ratio of 15 [17]. The carbon source amounts were calculated based on their characteristics [17]. Detail is given in Tables 1 and 2. The carbonated resources were weighed, tipped into 1-l plastic containers, and mixed thoroughly with the water of the culture tank. They were then dispersed consistently throughout the surface of the tank to promote the biofloc's growth. The continuous aerations were maintained and the experiment was conducted under 12 h darkness and 12 h light.

Table 1. Proximate investigation of the commercial fish feed and experimental flour utilized in experimental period.

Composition	Fish Feed	Sugarcane Molasses	Tapioca Flour
Crude protein (% dw)	32 ± 1.0	4.76 ± 0.12	10.78 ± 0.12
Dry matter (%)	84 ± 1.0	53.12 ± 0.38	90.13 ± 0.17
Crude lipid (% dw)	4.3 ± 1.0	0.2 ± 0.04	2.56 ± 0.02
Ash (% dw)	6.9 ± 1.0	7.87 ± 0.12	4.32 ± 0.05
Fiber (% dw)	1.8 ± 1.0	0.38 ± 0.06	6.12 ± 0.14
Carbohydrate (% dw)	26 ± 1.0	38.31 ± 0.21	65.21 ± 0.14

Table 2. Quantity of fish feed and carbon sources used in the research period.

Organic Materials	Daily Added (Grams)
Fish feed (32% protein)	16.1
Sugarcane molasses	9.8
Tapioca flour	8.6
C:N ratio	15:1

2.2. *Analysis of Water Quality*

The dissolve oxygen (DO) and water temperature were calculated daily in situ using an aquarium digital thermometer and DO meter (Jenway, London, UK). The pH values were measured after every three days using a pH meter (Beckman model-72). The salinity was measured with the help of an instrument (HQ30D Multi Meter HACH). Before feeding, ammonia (NH_4^+-N), nitrite nitrogen (NO^{2-}-N), and nitrate nitrogen (NO^{3-}-N) were sampled. An ammonia electrode (Model IS-570-NH_3, Germany) was used to measure the unionized ammonia (NH_4^+-N) on a weekly basis. The total suspended solids (TSS) were estimated with the help of a TSS-meter (TSS) (HM-COM-80). Nitrite nitrogen (NO^{2-}-N) and nitrate nitrogen (NO^{3-}-N) were determined weekly by using a freshwater 'master test kit'. An Imhoff cone was used to measure the volume of the biofloc, where the biofloc volume was recorded after 30 min of sedimentation of one liter of water in each BFT tank [18]; the number of observations was, therefore, 30 treatment^{-1}.

2.3. Growth Performance and Survival

The parameters, including the weight gain percentage (wt. gain %), survival rate, feed conversion ratio (FCR), and feed intake, were studied to determine the fish growth performance. From each group, $n = 10$ samples were studied biweekly.

$$\text{Wt. gain (\%)} = 100 \times (W_f - W_i/W_i)$$

where W_f is the final weight and W_i is the initial weight; feed intake (g diet kg^{-1} fish) = total weight of feed provided (g)/kg of fish; FCR = the total feed intake (g)/weight gain (g); and survival (%) = 100 × (final fish number/initial stocked number)

2.4. Blood Sampling

At the termination of the experiment, blood samples were collected from the healthy fish with no sign of infection. The fish (n = 5) from each group were anaesthetized by using clove oil (50 mg clove oil L^{-1}) [19] and the blood was drained from the caudal vein through a sterile syringe. Half (50%) of each blood sample was stored at 4 °C in an aqueous solution containing heparin (anticoagulant) and then used immediately to determine the hematological parameters. The other half of the blood was permitted to centrifuge at 1075× g for 10 min at 4 °C to acquire plasma. Further, for a biochemical and immunological assay analysis, the samples were stored at −80 °C.

2.5. Hematological Parameters

A Neubauer hemocytometer was used to count the White Blood Cells (WBC) and Red Blood Cells (RBCs) after dilution with phosphate-buffered saline. A hematocrit (HCT) determination was achieved by centrifuging the complete blood in heparinized microhematocrit capillary tubes for ten minutes at 3500× g. A cyanohemoglobin method was used to measure the haemoglobin (HB) concentration [20]. Giemsa-stained smears were used to determine the differential leucocyte count [21].

2.6. Humoral Non-Specific Immune Parameters

The determination of the total protein in the fish plasma samples was conducted through Biuret's method, as described by Gornall et al. [22]. The bromocresol green method [23] was used to estimate Albumin, and the globulin was determined as the change between the total albumin and protein.

Furthermore, the lysozyme activity was measured through a turbidimetric assay with some modification. Plasma aliquots of 25 µL were added to Micrococcus lysodeikticus (1 mL suspension) [24]. This suspension was prepared using 0.05 M of sodium phosphate buffer. The absorbance of the spectrophotometer (Spectrophotometer PD-303 UV, APEL, Japan) was set at 670 nm after 30 s and 180 s. The quantity of protein was estimated with the help of the micro protein determination technique, while the total immunoglobin was estimated by following Siwicki and Anderson [24]. With the help of a 12% polyethylene glycol solution, molecules of Ig were precipitated and the difference between before and after was considered as the Ig concentration.

To determine the alternative complement activity (ACH50), the RBC of sheep were used as the target, and, at 540 nm wavelength, the absorbance of the lysed cells was measured with the help of a spectrophotometer [25]. The ACH50 was determined for each treated group by measuring the volume of plasma producing 50% hemolysis as follows:

$$\text{ACH50 (unit mL}^{-1}) = 1/\text{Y} \times \text{(reciprocal of the plasma dilution)}$$

where Y is the amount of plasma (mL) giving 50% lysis.

2.7. Antioxidant Parameters

The catalase activity (CAT) was determined following the protocol of Luck [26]. Briefly a plasma sample of 10 µL was added to 1.24 mL of buffer (freshly prepared) containing 50 µL (H_2O_2) and 10 mL^{-1} of sodium–potassium phosphate buffer (0.15 M and pH 7). A change in absorbance was noted after 20 s (A1) and after 80 s (A2) of incubation at 240 nm against air. The catalase value was calculated as A1-A2/0.0008.

Superoxide dismutase (SOD) was calculated following Pedrajas et al.'s [27] method Briefly, a plasma sample of 20 µL was added to 945 µL of sodium carbonate buffer (0.05 M and pH 10) and 42 µL of epinephrine (30 mmol L^{-1} dissolved by adding 30 µL of HCL) The auto-oxidation of the epinephrine to adrenochrome inhibition was estimated at 480 nm after 30 and 80 s in an alkaline environment. In total, 40 µL of epinephrine and 960 µL of sodium carbonate buffer were used to prepare the control group.

$$\text{The inhibition (\%)} = 100 - [(\Delta\text{A control} - \Delta\text{A sample}/\Delta\text{A control}) \times 100]$$

$$\text{SOD activity in plasma (U/mL)} = \text{\% inhibition} \times 3.75$$

2.8. Statistical Analysis

A one-way ANOVA was used to statistically analyze the findings, and means were compared through a Duncan multiple range test at a significance level of 0.05. SPSS version 22.0 was used for the data analysis.

3. Results

3.1. Water Quality Characteristics

Detailed findings of the different water quality parameters are depicted in Table 3. After the data analysis, the water quality parameters (DO, pH, salinity, NH3, nitrite, nitrate, TSS, and biofloc volume) were significantly ($p < 0.05$) different. A lower DO level (5.58 mg/L) and pH (7.32) was observed in the BFT + SM group as compared to other groups. The changes in the water quality parameters and biofloc volume of tank water affected by different carbon sources for the 10-weeks experimental trial are depicted in Figure 1.

Table 3. Water quality parameters of juvenile *O. niloticus* (Nile tilapia) raised in biofloc tanks with different carbon sources for 10 weeks.

Parameters	Control	BFT + SM	BFT + TF
Water temperature (°C)	26.99 ± 0.23	27.01 ± 0.18	27.01 ± 0.11
Dissolved oxygen (mg L^{-1})	6.97 ± 0.073 ^{a}	5.58 ± 0.083 ^{c}	6.31 ± 0.073 ^{b}
pH	8.21 ± 0.057 ^{a}	7.32 ± 0.075 ^{c}	7.78 ± 0.054 ^{b}
Salinity (PPT)	1.47 ± 0.61 ^{b}	20.32 ± 11.21 ^{a}	18.11 ± 13.13 ^{a}
Unionized ammonia (mg L^{-1})	0.145 ± 0.016 ^{a}	0.052 ± 0.026 ^{b}	0.072 ± 0.025 ^{ab}
Nitrite (mg L^{-1})	0.256 ± 0.021 ^{a}	0.076 ± 0.017 ^{b}	0.136 ± 0.023 ^{b}
Nitrate (mg L^{-1})	3.81 ± 0.95 ^{b}	31.24 ± 4.62 ^{a}	23.33 ± 4.31 ^{a}
Total suspended solids (mg L^{-1})	31.3 ± 5.27 ^{b}	351.7 ± 70.44 ^{a}	314.6 ± 62.67 ^{a}
Biofloc volume (mL)	0.84 ± 0.13 ^{c}	18.47 ± 2.87 ^{b}	28.94 ± 4.12 ^{a}

Note: Significant difference is indicated with different superscripts on the mean in the same row ($p < 0.05$).

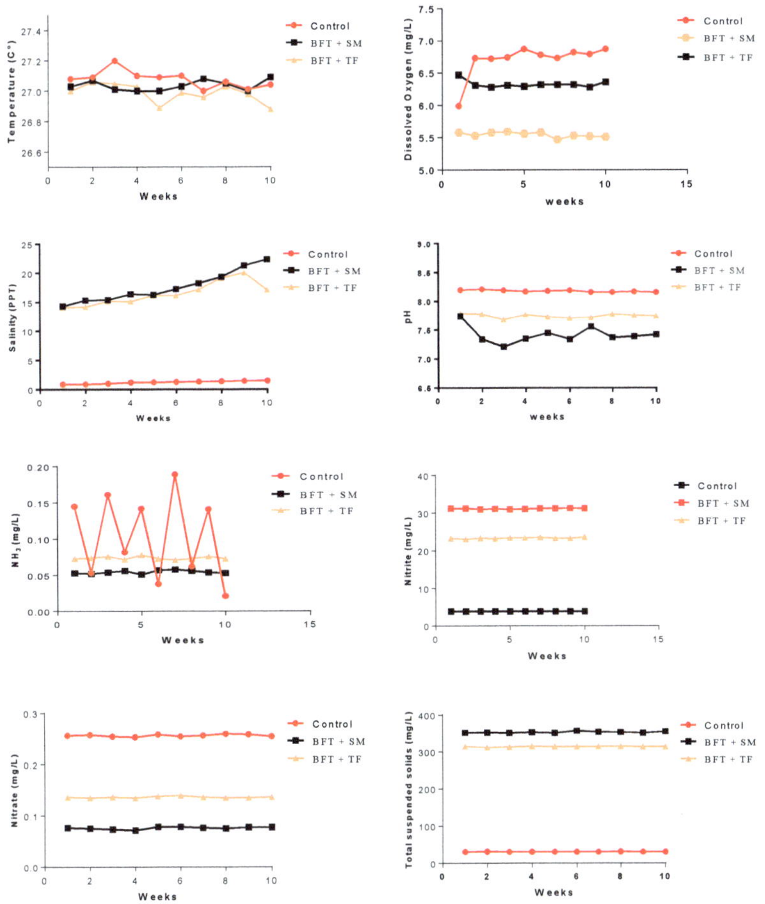

Figure 1. Variation in physicochemical parameters and biofloc volume of tanks water stocked by *O. niloticus* under biofloc system with different carbon sources for 10 weeks.

3.2. Fish Growth Performance

The results of the fish growth performance fed on different carbon sources (SM and TF) raised in biofloc tanks and the control group are given in Table 4. Significantly higher ($p < 0.05$) effects of biofloc technology with a carbon source (BFT + TF) were recorded in the fish weight gain as compared to the control and other group (BFT + SM). The feed intake was significantly greater ($p < 0.05$) in both the treated groups (BFT + TF and BFT + SM) than in the control group. The best FCR was found in the biofloc fish group with a carbon source (BFT + TF). The survival rate ranged from 95% to 99%, but, overall, the survival rate was also higher in the biofloc fish with the TF carbon source.

Table 4. Growth performance of juvenile *O. niloticus* (Nile tilapia) cultured in biofloc tanks with different carbon sources for 10 weeks.

Parameters	Control	BFT + SM	BFT + TF
Weight gain (%)	66.03 ± 2.82^{c}	87.15 ± 1.92^{b}	98.61 ± 3.71^{a}
Feed intake (g kg fish^{-1})	1001.21 ± 36.03^{a}	867.37 ± 23.17^{b}	869.31 ± 25.21^{b}

Table 4. *Cont.*

Parameters	Control	BFT + SM	BFT + TF
Feed conversion ratio (FCR)	2.35 ± 0.16 ^{a}	1.79 ± 0.06 ^{b}	1.69 ± 0.07 ^{c}
Survival (%)	95.83 ± 1.54	97.65 ± 1.02	99.01 ± 0.32

Note: Significant difference is indicated with different superscripts on the mean in the same row ($p < 0.05$).

3.3. Hematological Parameters

The fish hematological parameters, including (WBCs, HCT, and HB), were significantly higher ($p < 0.05$) in the fish reared in BFT + TF as compared to the other treated group (BFT + SM) and control (Table 5). This means the TF carbon source had a greater effect on the WBCs, HCT, and HB. The RBCs and monocyte values revealed no significant difference ($p > 0.05$) in both the carbon-treated and the control groups. However, the neutrophils value was significantly higher in the control group as compared to the experimental groups (Table 5).

Table 5. Hematological parameters of juvenile *O. niloticus* (Nile tilapia) cultured in biofloc tanks with different carbon sources for 10 weeks.

Hematological Parameters	Control	BFT + SM	BFT + TF
RBCs ($10^6/\mu L$)	1.32 ± 0.12	1.31 ± 0.11	1.33 ± 0.21
WBCs ($10^3/\mu L$)	27.41 ± 1.15 ^{c}	32.21 ± 1.23 ^{b}	37.43 ± 1.51 ^{a}
Hematocrit (%)	27.21 ± 0.39 ^{c}	33.11 ± 0.42 ^{b}	35.13 ± 0.46 ^{a}
Haemoglobin (%)	7.88 ± 0.32 ^{c}	9.03 ± 0.21 ^{b}	11.54 ± 0.25 ^{a}
Lymphocytes (%)	30.65 ± 0.84 ^{c}	34.52 ± 0.21 ^{b}	37.32 ± 0.24 ^{a}
Monocytes (%)	4.00 ± 0.21	4.88 ± 0.33	4.68 ± 0.31
Neutrophils (%)	62.49 ± 0.82 ^{a}	58.62 ± 0.61 ^{b}	56.16 ± 0.32 ^{c}

Note: Significant difference is indicated with different superscripts on the mean in the same row ($p < 0.05$).

3.4. Humoral Non-Specific Immune Parameters

The values of the total protein, humoral innate immunity, and albumin (lysozyme, immunoglobulin, and ACH50) increased significantly ($p < 0.05$) in the fish group reared in BFT with the TF carbon source compared to the control and other treated group (Table 6). The value of globulin revealed no significant difference ($p > 0.05$) between the treated groups (BFT + SM and BFT + TF), but significantly higher ($p < 0.05$) results compared to the control group.

Table 6. Non-specific immune parameters of juvenile *O. niloticus* (Nile tilapia) cultured in biofloc tanks with different carbon sources for 10 weeks.

Parameters	Control	BFT + SM	BFT + TF
Globulin ($g \times dL^{-1}$)	1.45 ± 0.07 ^{b}	1.86 ± 0.04 ^{a}	1.93 ± 0.04 ^{a}
Total protein ($g \times dL^{-1}$)	3.57 ± 0.05 ^{c}	3.84 ± 0.06 ^{b}	4.18 ± 0.08 ^{a}
Albumin ($g \times dL^{-1}$)	2.25 ± 0.02 ^{c}	2.31 ± 0.04 ^{b}	2.42 ± 0.04 ^{a}
Lysozyme ($U \times mg^{-1}$ protein)	24.00 ± 2.01 ^{c}	28.65 ± 1.42 ^{b}	31.43 ± 1.38 ^{a}
Immunoglobulin ($mg \times dL^{-1}$)	2.25 ± 0.16 ^{c}	3.02 ± 0.06 ^{b}	3.19 ± 0.08 ^{a}
ACH50 ($U \times mL^{-1}$)	18.85 ± 1.71 ^{c}	30.28 ± 1.03 ^{b}	37.12 ± 1.65 ^{a}

Note: Significant difference is indicated with different superscripts on the mean in the same row ($p < 0.05$).

3.5. Antioxidant Enzymatic Activities

From the current study, it was found that different carbon sources significantly affected ($p < 0.05$) the catalase (CAT) and superoxide dismutase (SAD) (Figure 2). However, higher enzymatic activities were obtained with BFT + TP than in the other groups.

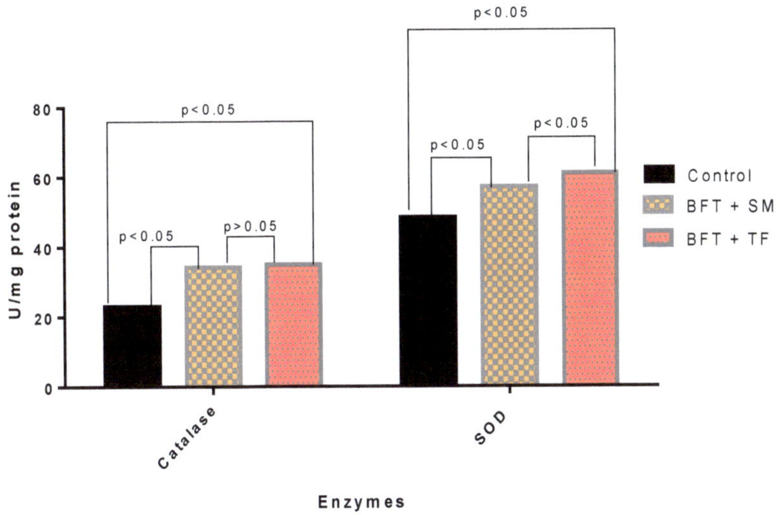

Figure 2. Antioxidant enzymes activities of juvenile *O. niloticus* (Nile tilapia) cultured in biofloc tanks with different carbon sources for 10 weeks.

4. Discussion

The composite carbon source in biofloc can enhance the nutritional value of the bacterial species in the system, which can further serve as an additional source of food for fish. Thus, it can play a significant role in the improvement of body weight. This study evaluated different carbon sources for *O. niloticus* reared in a biofloc system, and the results of the present research revealed that the rearing of juvenile Nile Tilapia in a biofloc system with variant carbon sources affects the water quality characteristics, growth performance, blood profile, non-specific immune response, and antioxidant status in no water exchange. Aquatic animals depend mainly on water quality to maintain their health and limit growth [28]. The water quality parameters (temperature, DO, pH, nitrite, nitrate, NH3, and TSS) observed in the present investigation were in the appropriate range, suitable for biofloc fish farming. The present study's findings are according to the water parameters reported by other studies [29,30]. Moreover, temperature is an essential factor affecting the formation and composition of biofloc [31] and is appropriate in the present research work. The decrease in microbial activity within the flocs led to deflocculation at lower temperatures (4 °C) than at higher temperatures (18–20 °C) [32]. According to Krishna and Van Loosdrecht [33], stable microbial flocs might be obtained at a temperature of (25–25 °C). In this study, the temperature range for both treatments was 27.01, slightly higher than the optimum level. The current research work also revealed a significant difference in the dissolved oxygen (DO) and pH levels in the control and treated groups. However, a lower DO level and pH were recorded in the BFT + SM group. This may have been due to higher respiration by the heterotrophic microbial community. As a result of the oxygen consumption by microbes and CO_2 emission, H_2CO_3 is produced in a limited water exchange system, reducing pH levels [2,10]. The DO level influences the structure of aerobic flocs due to its role in the metabolic activity of cells [34]. According to Martins et al. [35], filamentous bacteria were more numerous than zoogloeal bacteria at DO levels (less or equal to 1.02 mg/L). The floc volume index increases with DO levels above 3.5 mg/L [36,37]. The floc volume observed in the current study for the tapioca meal with biofloc was 28.94 at 6.31 mg/L DO level. The physiological function of tilapia is not adversely affected by pH 4.0–8.5 [38]. In this study, the pH value was 8.21 (control), (7.32) BFT + SM, and (7.78) BFT + TF. The treated groups had lower levels of nitrite and ammonia combined with higher levels of nitrate, which showed a greater bacteria abundance that oxidized nitrite and ammonia compared to the control group [39]. In addition, the control treatment involved constant water exchange, so lower levels of nitrogen compounds were expected in this treatment. According to Xu et al. [40], a change in TSS concentrations over time can serve as an indicator of the development of biofloc in aquaculture systems, which is consistent with the findings of the current study. A lower TSS in the current study was recorded in the control group compared to the treatment groups. It has been reported that during 14 weeks of raising *Labeo rohita* fingerlings, the highest TSS level (1.32840 mg/L) was recorded in a biofloc system [41]. Similarly, Azim and Little [42] observed the same observations for tilapia. However, in the current research, the TSS level did not exceed 351.7 mg/L. A TSS value higher than 1000 mg/L impacts the tilapia's health.

Additionally, biofloc microorganisms contribute to sustaining the water parameters, fecal waste, and uneaten feed metabolism, and, thus, decrease the nitrogenous compounds, particularly nitrite and NH_3 [43]. There is, however, an association between the reduction in ammonia concentrations and the development and formation of microbes in the biofloc [6]. According to Soliman and Abdel-Tawwab [29], the carbon source in biofloc technology improves microbial diversity, particularly ammonia-oxidizing bacteria, which decreases the NH_3 concentration. In current study, there was no evident variation in the nitrite and nitrate levels, which indicated that heterotrophic uptake was the key ammonia removal path in the system, and this is consistent with the studies that have been explained before. The biofloc volume recorded in the current study was appropriate for tilapia production [15].

Earlier studies have shown that, due to probiotic properties, the biofloc technology system increases fish growth performance and FCR [4,37,44,45]. Similarly, the current study results show that BFT with different carbon sources improves the fish growth performance more than the control. The BFT with TF significantly increased the fish growth and provided the best FCR. This shows that microbial floc with BFT + TF acts as a supplementary diet source that provides extra protein, vitamins, minerals, and polyunsaturated fatty acids [9,29,37]. The TF contain more than 90% carbohydrates, which might show better results for microbial growth as compared to SM. In biofloc technology, the FCR should be close to one, as reported in several studies such as Khanjani et al. [39], who observed an FCR of 0.99 while using starch as a carbon source in biofloc technology for Nile tilapia fingerlings. Similar findings were obtained by García-Ríos et al. [46] for cultured Nile tilapia fingerlings. However, in this study, the FCR value was found to be higher. This might have been due to the short experimental duration. The biofloc system represents a suitable environment for fish growth and feed utilization without affecting the water quality or survival of the fish, which revealed the significant effects of biofloc on fish growth, as reported by previous studies findings [12,43,47]. The carbon source is effective for the growth performance of reared species because it enhances the floc production, including its chemical composition and volume, and can also store different bioactive compounds (carotenoids, extracellular enzymes, polymers, and phytosterols [43,48]. Biofloc-reared fish such as rohu (*Labeo roita*) and tilapia (*Oreochromis niloticus*) have previously shown a better growth performance [49,50], which is according to the findings of the present study. The feed comprising different ingredients represents the major production cost in commercial aquaculture; as a result, improving the efficiency of fish nutrition is a key priority. Therefore, applying the BFT system to intensive tilapia culture can be advantageous.

Hematological parameters are essential for understanding abnormalities due to health status [51,52]. The findings of the current study showed that the WBCs, HCT, HB, and lymphocytes were largely affected by the carbon sources and their values were greater in BFT + TF than the other groups. The increased number of WBCs was due to lymphocytes. Fish health is closely associated with the number of leukocytes, which play a significant role in innate immunity during inflammation [53]. According to Mansour and Esteban [15], different carbon sources improve the number of WBCs, HB, and HCT in biofloc-cultured *O. niloticus*, which agrees with the current study's findings. According to the findings of many studies, stress, environmental conditions, carbon source type or amount, aquaculture system, microbial diversity, disease, and feeding treatment affect the blood profile of cultured fish [4,54–56].

It has been established that innate immunity is connected to increased levels of plasma proteins, albumin, and globulin, representing the significant proteins in plasma [57]. In the current study, all these proteins were significantly affected by the carbon sources and were higher in BFT + TF. Besides this, the lysozyme, immunoglobulin, and ACH50 levels were significantly higher in the fish cultured in biofloc technology with tapioca flour as a carbon source. The notable increase in these parameters is consistent with the findings of Mansour and Esteban [15]. They used wheat flour and rice bran as carbon sources for biofloc-cultured tilapia. Tapioca flour (TF) contains essential minerals such as iron and potassium. Besides this, it is also considered to be a good source of vitamin B, such as riboflavin (B2) and Niacin (B3) [58]. Vitamin B, potassium, and iron have long been known for their roles in immunity [59]. Therefore, an increase in the immunity of the fish reared in biofloc with tapioca flour can be linked to its nutritional value.

Moreover, Verma et al. [60] found that tapioca flour (TF) as a carbon source significantly increased the plasma proteins, globulin, and immunoglobulin levels in *Labeo rohita* reared in biofloc technology. Similar findings were revealed in the current study. The lysozyme produced by fish leukocytes causes bacterial cell wall lysis, which stimulates the complement system and facilitates the phagocytosis of many pathogens [61]. The complement activity plays a vital role in teleost's antibacterial defense mechanism [62]. Researchers might investigate in future studies whether the increases in the immune parameters are associated with improved defenses against disease or stressful situations.

The antioxidant status in the present study was also greater in the fish reared in the BFT with tapioca as a carbon source compared to the BFT with sugarcane molasses and control groups. In particular, the biofloc treatment significantly increased the activity of the CAT and SOD enzymes. The increased activity of antioxidants in the case of tapioca may be linked to the antioxidants present in this plant. Several studies have suggested that all possible tapioca extracts show significantly higher antioxidant activity [63,64]. The present research was according to the findings of other studies [15,18,45]. CAT and SOD are enzymes that prevent the oxidation of lipids. The catalytic reaction of SOD produces hydrogen peroxide from superoxide anion and is further decomposed by CAT to oxygen and water to prevent lipid peroxidation [65]. The increased SOD and CAT activity levels in the present study may be attributed to enhanced fish well-being and decreased oxidative stress.

5. Conclusions

In conclusion, the current study revealed that biofloc technology with different carbon sources (SM and TF) significantly increased the water quality, growth performance, blood profile, non-specific immunity, and antioxidant status of *O. niloticus* compared to the control group. The use of TF appeared to be more appropriate for the rearing of *O. niloticus* in biofloc than SM. Besides the other carbon sources, the current study's findings encourage biofloc fish farmers to consider TF as a carbon source for better results, because the BFT system was identified as an environmentally friendly alternative. This study provides new insight for future studies that can consider using TF as a carbon source on a larger scale with a long duration. This will enable us to understand better TF's effect on fish health and final yield.

Author Contributions: K.H.R. and S.S.H.: conceptualization, and investigation; J.A.U. and S.A.: data curation, and formal analysis; M.U. and A.I.B.: writing—original draft; S.A. and Khayyam Khayyam.: sampling, supervision and methodology; F.F. and Khalid Khan.: review and editing; S.N. and S.S.H.: funding, supervision and editing. Everyone involved in the manuscript has reviewed and approved it. All authors have read and agreed to the published version of the manuscript.

Funding: This study was not supported by external funding.

Institutional Review Board Statement: The current research work was sanctioned by the "Advanced Research Board of the Department of Zoology, The University of Lahore, sub campus Sargodha Punjab Pakistan" under the trial registration number (UOLS/FSH/0652) and the study was conducted according to the ethical deliberation of the European Legislation (Animal rights).

Data Availability Statement: Datasets from this study will be available upon reasonable request from the corresponding author.

Acknowledgments: We are pleased to express our gratitude to all the staff members of the Department of Zoology at the University of Sargodha (Pakistan) for their efforts and support during the experiment's planning and statistical analysis.

Conflicts of Interest: The authors have not disclosed any conflict of interest.

References

1. Tahergorabi, R.; Matak, K.E.; Jaczynski, J. Fish protein isolate: Development of functional foods with nutraceutical ingredients. *J. Funct. Foods* **2015**, *18*, 746–756. [CrossRef]
2. Boyd, C.E.; D'Abramo, L.R.; Glencross, B.D.; Huyben, D.C.; Juarez, L.M.; Lockwood, G.S.; McNevin, A.A.; Tacon, A.G.; Teletchea, F.; Tomasso, J.R., Jr.; et al. Achieving sustainable aquaculture: Historical and current perspectives and future needs and challenges. *J. World Aquac. Soc.* **2020**, *51*, 578–633. [CrossRef]
3. Ching, J.J.; Shuib, A.S.; Abdullah, N.; Majid, N.A.; Taufek, N.M.; Sutra, J.; Azmai, M.N.A. Hot water extract of *Pleurotus pulmonarius* stalk waste enhances innate immune response and immune-related gene expression in red hybrid tilapia *Oreochromis* sp. following challenge with pathogen-associated molecular patterns. *Fish Shellfish Immunol.* **2021**, *116*, 61–73. [CrossRef] [PubMed]
4. Habib, S.S.; Batool, A.I.; Rehman, M.F.U.; Naz, S. Comparative analysis of the haemato-biochemical parameters and growth characteristics of *Oreochromis niloticus* (Nile tilapia) cultured under different feed and habitats (biofloc technology and earthen pond system). *Aquac. Res.* **2022**, *53*, 6184–6192. [CrossRef]
5. Khanjani, M.H.; Sharifinia, M. Biofloc technology as a promising tool to improve aquaculture production. *Rev. Aquac.* **2020**, *12*, 1836–1850. [CrossRef]
6. Khanjani, M.H.; Sharifinia, M.; Hajirezaee, S. Recent progress towards the application of biofloc technology for tilapia farming. *Aquaculture* **2022**, *552*, 738021. [CrossRef]
7. Vethathirri, R.S.; Santillan, E.; Wuertz, S. Microbial community-based protein production from wastewater for animal feed applications. *Bioresou. Technol.* **2021**, *341*, 125723. [CrossRef]
8. Deng, M.; Chen, J.; Gou, J.; Hou, J.; Li, D.; He, X. The effect of different carbon sources on water quality, microbial community and structure of biofloc systems. *Aquaculture* **2018**, *482*, 103–110. [CrossRef]
9. Li, J.; Liu, G.; Li, C.; Deng, Y.; Tadda, M.A.; Lan, L.; Zhu, S.; Liu, D. Effects of different solid carbon sources on water quality, biofloc quality and gut microbiota of Nile tilapia (*Oreochromis niloticus*) larvae. *Aquaculture* **2018**, *495*, 919–931. [CrossRef]
10. Wei, Y.; Liao, S.A.; Wang, A.L. The effect of different carbon sources on the nutritional composition, microbial community and structure of bioflocs. *Aquaculture* **2016**, *465*, 88–93. [CrossRef]
11. Rohani, M.F.; Islam, S.M.; Hossain, M.K.; Ferdous, Z.; Siddik, M.A.; Nuruzzaman, M.; Padeniya, U.; Brown, C.; Shahjahan, M. Probiotics, prebiotics and synbiotics improved the functionality of aquafeed: Upgrading growth, reproduction, immunity and disease resistance in fish. *Fish Immunol.* **2022**, *120*, 569–589. [CrossRef] [PubMed]
12. Zhao, D.; Pan, L.; Huang, F.; Wang, C.; Xu, W. Effects of different carbon sources on bioactive compound production of biofloc, immune response, antioxidant level, and growth performance of *Litopenaeus vannamei* in zero-water exchange culture tanks. *J. World Aquac. Soc.* **2016**, *47*, 566–576. [CrossRef]
13. Elayaraja, S.; Mabrok, M.; Algammal, A.; Sabitha, E.; Rajeswari, M.V.; Zágoršek, K.; Ye, Z.; Zhu, S.; Rodkhum, C. Potential influence of jaggery-based biofloc technology at different C: N ratios on water quality, growth performance, innate immunity, immune-related genes expression profiles, and disease resistance against *Aeromonas hydrophila* in Nile tilapia (*Oreochromis niloticus*). *Fish Shellfish Immunol.* **2020**, *107*, 118–128. [PubMed]
14. Khanjani, M.H.; Sharifinia, M.; Emerenciano, M.G.C. A detailed look at the impacts of biofloc on immunological and hematological parameters and improving resistance to diseases. *Fish Shellfish Immunol.* **2023**, *137*, 108796. [CrossRef] [PubMed]
15. Mansour, A.T.; Esteban, M.Á. Effects of carbon sources and plant protein levels in a biofloc system on growth performance, and the immune and antioxidant status of Nile tilapia (*Oreochromis niloticus*). *Fish Shellfish Immunol.* **2017**, *64*, 202–209. [CrossRef]
16. Bakhshi, F.; Najdegerami, E.H.; Manaffar, R.; Tokmechi, A.; Rahmani Farah, K.; Shalizar Jalali, A. Growth performance, haematology, antioxidant status, immune response and histology of common carp (*Cyprinus carpio* L.) fed biofloc grown on different carbon sources. *Aquac. Res.* **2018**, *49*, 393–403. [CrossRef]
17. Avnimelech, Y. *Biofloc Technology: A Practical Guide Book*, 3rd ed.; World Aquaculture Society: San Diego, CA, USA, 2009.
18. Avnimelech, Y. Carbon/nitrogen ratio as a control element in aquaculture systems. *Aquaculture* **1999**, *176*, 227–235. [CrossRef]
19. Habib, S.S.; Naz, S.; Batool, A.I.; Rehman, F.U.; Ullah, M.; Kesbiç, O.S.; Maricchiolo, G.; Fazio, F. Effect of Different Anaesthetics on Hematology and Blood Biochemistry of *Labeo rohita*. *Aquac. Stud.* **2023**, *24*, AQUAST1195. [CrossRef]

20. Goldenfarb, P.B.; Bowyer, F.P.; Hall, E.; Brosious, E. Reproducibility in the hematology laboratory: The microhematocrit determination. *Am. J. Clin. Pathol.* **1971**, *56*, 35–39. [CrossRef]
21. Schütt, D.A.; Lehmann, J.; Goerlich, R.; Hamers, R. Haematology of swordtail, *Xiphophorus helleri*. I: Blood parameters and light microscopy of blood cells. *J. Appl. Ichthyol.* **1997**, *13*, 83–89. [CrossRef]
22. Gornall, A.G.; Bardawill, C.J.; David, M.M. Determination of serum proteins by means of the biuret reaction. *J. Biol. Chem.* **1949**, *177*, 751–766. [CrossRef]
23. Doumas, B.T.; Watson, W.A.; Biggs, H.G. Albumin standards and the measurement of serum albumin with bromcresol green. *Clin. Chim. Acta* **1971**, *31*, 87–96. [CrossRef]
24. Siwicki, A.K.; Anderson, D.P. Nonspecific Defence Mechanisms Assay in Fish: II. Potential Killing Activity of Neutrophils and Macrophages, Lysozyme Activity in Serum and Organs. In *Disease Diagnosis and Prevention Methods*; Siwicki, A.K., Anderson, D.P., Waluga, J., Eds.; Wydawnictwo Instytutu Rybactwa Strodladowego: Olsztyn, Poland, 1993; pp. 105–111.
25. Sun, S.; Ye, J.; Chen, J.; Wang, Y.; Chen, L. Effect of dietary fish oil replacement by rapeseed oil on the growth, fatty acid composition and serum non-specific immunity response of fingerling black carp, *Mylopharyngodon piceus*. *Aquac. Nutr.* **2011**, *17*, 441–450.
26. Luck, H. Catalase. In *Method of Enzymatic Analysis*; Bergmayer, M.V., Ed.; Verlag Chemic/Academic Press: New York, NY, USA, 1974; p. 885.
27. Pedrajas, J.R.; Peinado, J.; Lopez-Barea, J. Oxidative stress in fish exposed to model xenobiotics. Oxidatively modified forms of Cu, Zn-superoxide dismutase as potential biomarkers. *Chem. Biol. Interact.* **1995**, *98*, 267–282. [PubMed]
28. El-Sayed, A.F.M. Use of biofloc technology in shrimp aquaculture: A comprehensive review, with emphasis on the last decade. *Rev. Aquac.* **2021**, *13*, 676–705.
29. Soliman, A.M.; Abdel-Tawwab, M. Effects of different carbon sources on water quality, biofloc quality, and the productivity of Nile tilapia reared in biofloc-based ponds. *Ann. Anim. Sci.* **2022**, *22*, 1281–1289.
30. Khanjani, M.H.; Alizadeh, M.; Sharifinia, M. Effects of different carbon sources on water quality, biofloc quality, and growth performance of Nile tilapia (*Oreochromis niloticus*) fingerlings in a heterotrophic culture system. *Aquac. Int.* **2021**, *29*, 307–321.
31. Hostins, B.; Braga, A.; Lopes, D.L.; Wasielesky, W.; Poersch, L.H. Effect of temperature on nursery and compensatory growth of pink shrimp *Farfantepenaeus brasiliensis* reared in a super-intensive biofloc system. *Aquac. Eng.* **2015**, *66*, 62–67.
32. Wilen, B.M.; Nielsen, J.L.; Keiding, K.; Nielsen, P.H. Influence of microbial activity on the stability of activated sludge flocs. *Colloids Surf. B Biointerfaces* **2000**, *18*, 145–156.
33. Krishna, C.; Van Loosdrecht, M.C.M. Effect of temperature on storage polymers and settleability of activated sludge. *Water Res.* **1999**, *33*, 2374–2382.
34. De Schryver, P.; Crab, R.; Defoirdt, T.; Boon, N.; Verstraete, W. The basics of bio-flocs technology: The added value for aquaculture. *Aquaculture* **2008**, *277*, 125–137.
35. Martins, A.M.P.; Heijnen, J.J.; van Loosdrecht, M.C.M. Effect of dissolved oxygen concentration on sludge settleability. *Appl. Microbiol. Biotechnol.* **2003**, *62*, 586–593.
36. Shamsuddin, M.; Hossain, M.B.; Rahman, M.; Kawla, M.S.; Shufol, M.B.A.; Rashid, M.M.; Asadujjaman, M.; Rakib, M.R.J. Application of Biofloc Technology for the culture of *Heteropneustes fossilis* (Bloch) in Bangladesh: Stocking density, floc volume, growth performance, and profitability. *Aquac. Int.* **2022**, *30*, 1047–1070.
37. Luo, G.; Gao, Q.; Wang, C.; Liu, W.; Sun, D.; Li, L.; Tan, H. Growth, digestive activity, welfare, and partial cost-effectiveness of genetically improved farmed tilapia (*Oreochromis niloticus*) cultured in a recirculating aquaculture system and an indoor biofloc system. *Aquaculture* **2014**, *422*, 1–7.
38. Mustapha, M.K.; Atolagbe, S.D. Tolerance level of different life stages of Nile tilapia *Oreochromis niloticus* (Linnaeus, 1758) to low pH and acidified waters. *J. Basic Appl. Zool.* **2018**, *79*, 1–6.
39. Khanjani, M.H.; Alizadeh, M.; Mohammadi, M.; Sarsangi Aliabad, H. Biofloc system applied to Nile tilapia (*Oreochromis niloticus*) farming using different carbon sources: Growth performance, carcass analysis, digestive and hepatic enzyme activity. *Iran. J. Fish. Sci.* **2021**, *20*, 490–513.
40. Xu, W.J.; Morris, T.C.; Samocha, T.M. Effects of C/N ratio on biofloc development, water quality, and performance of *Litopenaeus vannamei* juveniles in a biofloc-based, high-density, zero-exchange, outdoor tank system. *Aquaculture* **2016**, *453*, 169–175.
41. Kamilya, D.; Debbarma, M.; Pal, P.; Kheti, B.; Sarkar, S.; Singh, S.T. Biofloc technology application in indoor culture of *Labeo rohita* (Hamilton, 1822) fingerlings: The effects on inorganic nitrogen control, growth and immunity. *Chemosphere* **2017**, *182*, 8–14.
42. Azim, M.E.; Little, D.C. The biofloc technology (BFT) in indoor tanks: Water quality, biofloc composition, and growth and welfare of Nile tilapia (*Oreochromis niloticus*). *Aquaculture* **2008**, *283*, 29–35.
43. Wang, G.; Yu, E.; Xie, J.; Yu, D.; Li, Z.; Luo, W.; Qiu, L.; Zheng, Z. Effect of C/N ratio on water quality in zero-water exchange tanks and the biofloc supplementation in feed on the growth performance of crucian carp, *Carassius auratus*. *Aquaculture* **2015**, *443*, 98–104.
44. Adineh, H.; Naderi, M.; Hamidi, M.K.; Harsij, M. Biofloc technology improves growth, innate immune responses, oxidative status, and resistance to acute stress in common carp (*Cyprinus carpio*) under high stocking density. *Fish Shellfish Immunol.* **2019**, *95*, 440–448. [CrossRef]
45. Long, L.; Yang, J.; Li, Y.; Guan, C.; Wu, F. Effect of biofloc technology on growth, digestive enzyme activity, hematology, and immune response of genetically improved farmed tilapia (*Oreochromis niloticus*). *Aquaculture* **2015**, *448*, 135–141. [CrossRef]

46. García-Ríos, L.; Miranda-Baeza, A.; Coelho-Emerenciano, M.G.; Huerta-Rábago, J.A.; Osuna-Amarillas, P. Biofloc technology (BFT) applied to tilapia fingerlings production using different carbon sources: Emphasis on commercial applicatons. *Aquaculture* **2019**, *502*, 26–31. [CrossRef]
47. Arnold, S.J.; Coman, F.E.; Jackson, C.J.; Groves, S.A. High-intensity, zero water-exchange production of juvenile tiger shrimp *Penaeus monodon*: An evaluation of artificial substrates and stocking density. *Aquaculture* **2009**, *293*, 42–48. [CrossRef]
48. Kishawy, A.T.; Sewid, A.H.; Nada, H.S.; Kamel, M.A.; El-Mandrawy, S.A.; Abdelhakim, T.M.; El-Murr, A.E.I.; Nahhas, N.E.; Hozzein, W.N.; Ibrahim, D. Mannanoligosaccharides as a carbon source in Biofloc boost dietary plant protein and water quality, growth, immunity and *Aeromonas hydrophila* resistance in Nile tilapia (*Oreochromis niloticus*). *Animals* **2020**, *10*, 1724. [CrossRef]
49. Mahanand, S.S.; Moulick, S.; Rao, P.S. Water quality and growth of Rohu, *Labeo rohita*, in a biofloc system. *J. Appl. Aquac.* **2013**, *25*, 121–131. [CrossRef]
50. Crab, R.; Kochva, M.; Verstraete, W.; Avnimelech, Y. Bio-flocs technology application in over-wintering of tilapia. *Aquac. Eng.* **2009**, *40*, 105–112. [CrossRef]
51. Habib, S.S.; Naz, S.; Khalid, S.; Kanwal, R.; Ameer, I.; Khan, S.N.A.; Rehman, A.U.; Kousar, M.; Khan, S.U.; Nazir, N. Effect of white button mushroom (*Agaricus bisporus*) on immunity and haematological parameters of *Oreochromis niloticus*. *Pak. J. Zool.* **2022**, *54*, 785–790. [CrossRef]
52. Amenyogbe, E.; Yang, E.J.; Xie, R.T.; Huang, J.S.; Chen, G. Influences of indigenous isolates Pantoea agglomerans RCS2 on growth, proximate analysis, haematological parameters, digestive enzyme activities, serum biochemical parameters, antioxidants activities, intestinal morphology, disease resistance, and molecular immune response in juvenile's cobia fish (*Rachycentron canadum*). *Aquaculture* **2022**, *551*, 737942.
53. Rebl, A.; Goldammer, T. Under control: The innate immunity of fish from the inhibitors' perspective. *Fish Shellfish Immunol.* **2018**, *77*, 328–349. [CrossRef]
54. Habib, S.S.; Naz, S.; Nawaz, S.; Ameer, I.; Khatoon, A.; Rehman, H.U.; Jawad, S.M.; Ali, H. Comparative analysis of hematological parameters of some farmed and wild fish species. *Pak. J. Zool.* **2021**, *54*, 591–598. [CrossRef]
55. Habib, S.S.; Fazio, F.; Naz, S.; Arfuso, F.; Piccione, G.; Rehman, H.U.; Achakzai, W.M.; Uddin, M.N.; Rind, K.H.; Rind, N.A. Seasonal variations in haematological parameters and body composition of *Labeo rohita* (Rohu) and *Cirrhinus mrigala* (Mrigal carp) in River Indus, District Dera Ismail Khan, Pakistan. *Turk. J. Fish. Aquat. Sci.* **2021**, *21*, 435–441. [CrossRef] [PubMed]
56. Burgos-Aceves, M.A.; Lionetti, L.; Faggio, C. Multidisciplinary haematology as prognostic device in environmental and xenobiotic stress-induced response in fish. *Sci. Total Environ.* **2019**, *670*, 1170–1183. [CrossRef] [PubMed]
57. Akrami, R.; Gharaei, A.; Mansour, M.R.; Galeshi, A. Effects of dietary onion (*Allium cepa*) powder on growth, innate immune response and hemato–biochemical parameters of beluga (*Huso huso* Linnaeus, 1754) juvenile. *Fish Shellfish Immunol.* **2015**, *45*, 828–834. [CrossRef]
58. Chinnadurai, V.; Viswanathan, P.; Kalimuthu, K.; Vanitha, A.; Ranjitha, V.; Pugazhendhi, A. Comparative studies of phytochemical analysis and pharmacological activities of wild and micropropagated plant ethanol extracts of *Manihot esculenta*. *Biocatal. Agric. Biotechnol.* **2019**, *19*, 101166. [CrossRef]
59. Chirra, M.; Newton, H.S.; Gawali, V.S.; Wise-Draper, T.M.; Chimote, A.A.; Conforti, L. How the potassium channel response of T lymphocytes to the tumor microenvironment shapes antitumor immunity. *Cancers* **2022**, *14*, 3564. [CrossRef]
60. Verma, A.K.; Rani, A.B.; Rathore, G.; Saharan, N.; Gora, A.H. Growth, non-specific immunity and disease resistance of *Labeo rohita* against *Aeromonas hydrophila* in biofloc systems using different carbon sources. *Aquaculture* **2016**, *457*, 61–67.
61. de Mello, M.M.M.; de Faria, C.D.F.P.; Zanuzzo, F.S.; Urbinati, E.C. β-glucan modulates cortisol levels in stressed pacu (*Piaractus mesopotamicus*) inoculated with heat-killed *Aeromonas hydrophila*. *Fish Shellfish Immunol.* **2019**, *93*, 1076–1083. [CrossRef]
62. Uribe, C.; Folch, H.; Enríquez, R.; Moran, G.J.V.M. Innate and adaptive immunity in teleost fish: A review. *Vet. Med.* **2011**, *56*, 486. [CrossRef]
63. Omar, N.F.; Hassan, S.A.; Yusoff, U.K.; Abdullah, N.A.P.; Wahab, P.E.M.; Sinniah, U.R. Phenolics, flavonoids, antioxidant activity and cyanogenic glycosides of organic and mineral-base fertilized cassava tubers. *Molecules* **2012**, *17*, 2378–2387. [CrossRef]
64. Homayouni, H.; Kavoosi, G.; Nassiri, S.M. Physicochemical, antioxidant and antibacterial properties of dispersion made from tapioca and gelatinized tapioca starch incorporated with carvacrol. *LWT* **2017**, *77*, 503–509. [CrossRef]
65. El-Beltagi, H.S.; Mohamed, H.I. Reactive oxygen species, lipid peroxidation and antioxidative defense mechanism. *Not. Bot. Horti Agrobot. Cluj-Napoca* **2013**, *41*, 44–57. [CrossRef]

Disclaimer/Publisher's Note: The statements, opinions and data contained in all publications are solely those of the individual author(s) and contributor(s) and not of MDPI and/or the editor(s). MDPI and/or the editor(s) disclaim responsibility for any injury to people or property resulting from any ideas, methods, instructions or products referred to in the content.

fishes

MDPI

Article

Biological Responses of Oyster *Crassostrea gasar* Exposed to Different Concentrations of Biofloc

Léa Carolina de Oliveira Costa [1], Andrezza Carvalho [2], Mariana Holanda [3], Jorge Santos [2], Lucélia Borges [2], Bruna Guterres [2], Je Nam Junior [2], Virginia Fonseca [2], Larissa Muller [2], Luis Romano [2], Silvia Botelho [2], Marcelo Pias [2], Juliane Ventura [2] and Luís H. Poersch [2,*]

1 Instituto Federal de Educação Ciência e Tecnologia do Pará, Rodovia BR 316, KM 61, Castanhal 68740970, PA, Brazil; lea.costa@ifpa.edu.br

2 Marine Aquaculture Station, Institute of Oceanography, Federal University of Rio Grande—FURG, Rua do Hotel n○2, Cassino, Rio Grande 96210030, RS, Brazil; andrezzachagas@hotmail.com (A.C.); jorgerenato_82@hotmail.com (J.S.); lu.fito@gmail.com (L.B.); guterres.bruna@gmail.com (B.G.); jenamjunior@hotmail.com (J.N.J.); vikavetp@gmail.com (V.F.); larissamullern@gmail.com (L.M.); luisalbertoromano0@gmail.com (L.R.); silviacb.botelho@gmail.com (S.B.); pias.marcelo@gmail.com (M.P.); juliane.lima@furg.br (J.V.)

3 Escuela de Ciencias del Mar, Pontificia Universidad Católica de Valparaíso, Valparaíso 2373223, Chile; mariana.holandalima@gmail.com

* Correspondence: lpoersch@gmail.com

Abstract: Oysters have the potential to be a part of more sustainable farming systems, such as multitrophic systems integrated into biofloc systems, due to their filtration activity, which enables them to act as organic consumers. However, the stress experienced by animals in a system with a high organic load can compromise their productive performance. The objective of this study was to evaluate the biological responses of *Crassostrea gasar* oysters when exposed to different concentrations of total suspended solids in biofloc systems. The oysters were exposed to four different concentrations of solids for 28 days. Hall effect sensors were installed on the outside of the shells to detect the movement of the oyster valves. Also, biochemical and histological analyses were conducted to assess the biological responses of the oysters to exposure to varying levels of solids. A difference in valve opening detected by the Hall sensors was observed from the second week of culture, indicating a relationship between shell closure and higher concentrations of suspended solids present in the system. In terms of biochemical analysis, a significant increase in lipid damage was observed in treatments with medium and high levels of total suspended solids compared with the control group. Conversely, no changes were observed in the gill structure of the oysters caused by the concentrations of suspended solids in the system when compared with the control. According to the analyses of gill activity and biochemistry, it is suggested that *C. gasar* should be cultured with total suspended solids at less than 200 mg/L. Oysters cultivated in a biofloc system keep their shells closed when subjected to high concentrations of total suspended solids; concentrations of total suspended solids below 200 mg/L do not induce oxidative stress, changes in behavior or histological alterations in *C. gasar* oysters cultivated in a biofloc system.

Keywords: nutrients; *Crassostrea gasar*; biofloc; density; biosensor; bioremediation

Key Contribution: Oyster cultivated in a biofloc system keep their shells closed when subjected to high concentrations of total suspended solids; Concentrations of total suspended solids below 200 mg/L do not induce oxidative stress, changes in behavior or histological alterations in oyster *C. gasar* cultivated in a biofloc system.

Citation: Costa, L.C.d.O.; Carvalho, A.; Holanda, M.; Santos, J.; Borges, L.; Guterres, B.; Nam Junior, J.; Fonseca, V.; Muller, L.; Romano, L.; et al. Biological Responses of Oyster *Crassostrea gasar* Exposed to Different Concentrations of Biofloc. *Fishes* **2023**, *8*, 586. https://doi.org/10.3390/fishes8120586

Academic Editors: Wagner C. Valenti and Eduardo Luis Ballester

Received: 17 October 2023
Revised: 11 November 2023
Accepted: 14 November 2023
Published: 29 November 2023

Copyright: © 2023 by the authors. Licensee MDPI, Basel, Switzerland. This article is an open access article distributed under the terms and conditions of the Creative Commons Attribution (CC BY) license (https://creativecommons.org/licenses/by/4.0/).

1. Introduction

The integration of filter-feeding species, such as the oyster *Crassostrea gasar*, into multi-trophic biofloc cultures provides mutual benefits: increased productivity, a reduced organic load in the system, and a constant natural food source for the oysters. The inclusion of mollusks in integrated farming has gradually increased, with a view of using more sustainable and productive crops. According to the FAO [1], mollusks are the second most produced group, with the *Crassostrea* genus in first place, representing a group of economic interest with rising production. In addition to profitability, the use of more sustainable systems has been gaining ground in aquaculture, with Lima et al. [2] showing that the integration of the *Crassostrea* sp. oyster in the cultivation of white shrimp provided lower concentrations of settleable solids and better performance of the species compared with monoculture. The key to a successful integrated system lies in achieving the appropriate species composition as well as ensuring that waste from the feeding species can be utilized by species that consume organic material and that the selected species are tolerant to the conditions of the culture system.

When it comes to integrating species such as oysters that consume fine particulate organic material into biofloc system cultures, uncertainties arise regarding the physiological responses of these bivalves to high solid concentrations in this system. It is not definitively known whether the increased availability of solids might diminish the oysters' filtration potential or even induce stress that could lower their performance. This is because studies involving the optimal concentrations of total suspended solids in bivalve cultures are still limited [2]. The biological responses caused by environmental stress can affect various levels of organization, ranging from the cellular level to the entire population [3]. This occurs as an attempt to overcome challenges and aid the organism in returning to its normal physiological state [4]. However, these responses can lead to negative consequences, including reduced resilience, impacts on growth and reproduction, and heightened susceptibility to diseases.

However, defining the state of well-being of sessile bivalve mollusks is not a simple task. Indicators can be as conspicuous as the mortality of individuals or as subtle as alterations in the activities of biomolecules, such as enzymes. Enhancing the understanding of the physiological and behavioral reactions of mollusks is crucial for mitigating losses in aquaculture systems [5].

In this context, to comprehend the condition of animals in aquaculture environments, a range of analyses can be employed as indicators of environmental comfort [6]. Monitoring the oyster valve opening behavior through biosensors, oxidative stress biomarkers, and the histological examination of oyster gills are some of the methods utilized to observe animal behavior and health. The utilization of biosensors allows for the analysis of organism behavior in response to environmental changes, providing a swift, sensitive, and cost-effective approach [7–9]. The opening and closing movements of bivalve shells are closely linked to vital activities and environmental conditions [8], making the monitoring of this movement instrumental in understanding behavioral patterns influenced by environmental characteristics [10,11].

The development of biosensors based on the behavioral analysis of bivalve mollusks has been explored; these biosensors are sensitive tools for detecting pollutants and physicochemical conditions of interest, as well as their impacts on ecosystem services [12]. In terms of behavioral analysis, Le Moullac et al. [13] emphasized the measurement of valve activity (shell opening and closing) as a potentially valuable tool for biologically monitoring water quality and comprehending metabolic optimization strategies.

Likewise, Andrewartha and Elliott [14] introduced biosensors centered around sentinel animals, combined with the monitoring of environmental variables, as prospective key technologies for managing aquaculture farms. Generally, changes in behavior associated with bivalves' exposure to harmful algae [15–17] and other contaminants include a decrease in mean opening amplitude and an increase in transition frequency. To monitor the amplitude of shell opening in bivalve mollusks, valvometry techniques are employed. Among these techniques, Hall effect sensors in conjunction with magnets offer several advantages, such as durability, lightness, easy attachment, and reduced stress on the animals, which facilitates the measurement of their movements [16].

In aquaculture, oxidative stress in cultured aquatic organisms can be induced by various conditional factors [18]. Molecular and biochemical biomarkers have been widely employed to demonstrate the biological responses of organisms to specific environmental conditions [19]. They are used to highlight alterations in biomolecules and other aspects of oxidative stress resulting from reactive oxygen species or shifts in redox balance in experimental animals [20]. Sies and Jones [21] conceptualized oxidative stress as an imbalance between oxidants and antioxidants in favor of oxidants. This imbalance can lead to changes in redox signaling and control as well as molecular damage. During stressful events, reactive oxygen species are generated [22,23]; however, aerobic organisms possess an antioxidant defense system that enables them to manage these species by either neutralizing/intercepting them or repairing the damage they cause [24,25].

Another way of assessing the condition of a cultured animal is through histological analysis, which allows the structure of tissues to be observed through microscopic examination of their components [26]. One of the advantages of this analysis is its ability to provide information on the state of health of the organism at the cellular level [27]. Many tissues can be used in these analyses, but the gills stand out as one of the most important organs for aquatic species.

This study aimed to evaluate the biological responses of *C. gasar* oysters when exposed to different concentrations of total suspended solids in biofloc culture systems on the basis of behavioral, biochemical, and histological analyses of the bivalves. The results obtained will demonstrate the best concentration of solids for the cultivation of *C. gasar* oysters in biofloc systems.

2. Materials and Methods

2.1. Oyster Origin

The oysters were obtained from a commercial farm in the state of Santa Catarina, Brazil. Upon arrival at the Marine Aquaculture Station (EMA-FURG), they underwent a four-week acclimatization period in the laboratory. During this phase, they were housed in 500 L water tanks filled with filtered seawater (salinity 27) and maintained with continuous aeration. Biofloc water from a marine shrimp farm with a total suspended solids (TSSs) concentration of 100 mg/L was introduced to feed the oysters. The water was completely renewed twice a week. Initially, the oysters' mean weight was 54.18 $\pm$ 13.45 g, and they measured 68.2 $\pm$ 4.44 mm in mean height, 50.4 $\pm$ 3.50 mm in mean length, and 24.1 $\pm$ 5.02 mm in mean width.

2.2. Experimental Design

Water recirculation systems were established comprising one 500 L water reservoir (considered the macrocosm) and three 100 L boxes where the oysters were distributed. The 500 L reservoir was equipped with a pump (Sarlo Better—1000 L/h, São Caetano do Sul, São Paulo, Brazil) that circulated water to the three 100 L boxes, and through gravity, the water returned to the 500 L reservoir. Throughout the experiment, constant water circulation was maintained in the boxes, and aeration was provided via blowers and micro-perforated hoses. Each set consisting of three 100 L boxes and one 500 L reservoir represented one treatment. A suspended screen was placed in each 100 L box to keep the oysters elevated from the bottom of the box (Figure 1). Four times a day, the water in the 100 L boxes was agitated to resuspend the solids that had settled on the box structures.

The oysters were subjected to varying concentrations of total suspended solids (TSSs) over a 28-day period in order to expose them to the concentrations of solids that can be found in the natural environment [28] and the higher concentrations found in biofloc shrimp farming [29]. The experiment comprised four treatments involving the exposure of oysters to three distinct TSS concentrations of water sourced from the cultivation of marine shrimps in a mature biofloc system (with the presence of nitrate). The treatments were as follows: control treatment—oysters were fed with microalgae; low treatment—maintained at a low TSS concentration, approximately 100 mg/L TSSs; medium treatment—maintained at a nominal medium TSS concentration ranging from above 100 to 200 mg/L TSSs; and high treatment—maintained at a nominal high TSS concentration exceeding 200 mg/L TSSs. Each 100 L box represented one repetition of a treatment, resulting in a total of four treatments, each with three repetitions. Each experimental unit initially contained 30 oysters.

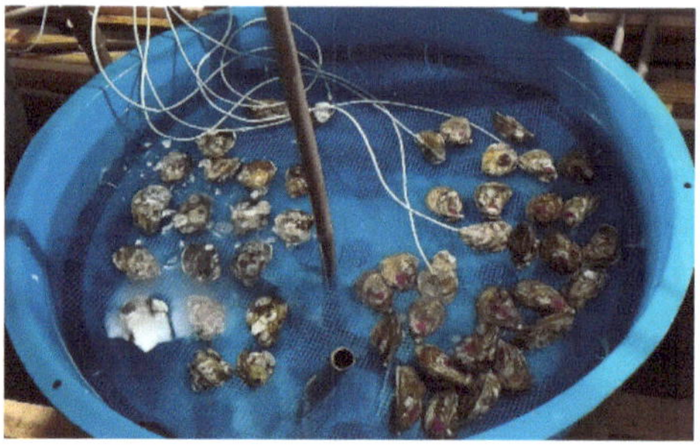

Figure 1. Experimental units (100 L) containing the oysters stored on the screens.

To feed the oysters in the control treatment, the diatom *Chaetoceros muelleri* was cultivated in f/2 media [30] with a salinity of 28, a temperature of 20 °C, and a light intensity of 40 µmol photons/m^2/s^1 in batch-type cultivation. Cell growth was monitored daily using a Neubauer chamber and an optical microscope. Daily counts of the remaining microalgae's density in the control experimental units were also conducted, and if necessary, aliquots of the microalgae culture were added to maintain a density of 16×10^4 cells/mL in the units. Every two weeks, three oysters were collected from each experimental unit for biochemical analysis. At the end of the experiment, two oysters were collected from each experimental unit for histological analysis. For biochemical and histological analysis, the oysters' adductor muscles were cut, the shells were opened, the visceral mass was removed from the shell, and only the gills were collected from the visceral mass for analysis. For biochemical analysis, the samples were stored at −80 °C, and for the histological analysis, the samples were stored in a 20% saline formalin solution.

2.3. Water Quality

Total suspended solids (TSSs) were measured twice a week using gravimetry, following a methodology adapted from AOAC [31] with adjustments made in accordance with treatment definitions. As needed, concentrated flake water was introduced to maintain the concentrations in the biofloc treatments. This matured floc, containing nitrate, originated from an ongoing marine shrimp culture and was stored in a 1000 L water tank with continuous aeration. The dissolved oxygen level was 8.6 ± 0.68 mg/L, the total alkalinity was 240 ± 20.8 mg $CaCO_3$/L, and the TSS concentration was 1460 ± 15.2 mg/L.

In the 100 L boxes where the oysters were stored, the following analyses were carried out twice a week. Settleable solids (SSs) were measured by an Imhoff Cone [32], and turbidity was measured by a portable turbidimeter (2100P, Hach®, Loveland, CO, USA). Temperature and oxygen were measured daily in the early morning and late afternoon using a multiparameter probe e (model Pro-20, YSI Inc., Yellow Springs, OH, USA). The pH reading was also taken daily in the morning by a bench pH meter (Seven2Go S7 Básico, Mettler Toledo, São Paulo, Brazil). Oyster mortality was observed daily.

Salinity, total alkalinity, ammonia, nitrite, nitrate, and phosphate were checked twice a week. Salinity was measured with an optical refractometer. Alkalinity was measured according to APHA [33], ammonia was measured according to UNESCO [34], and nitrite was measured according to Strickland and Parsons [35]. When total alkalinity values were below 150 mg $CaCO_3$/L, they were adjusted with calcium hydroxide according to Furtado et al. [36]. Nitrate and phosphate were measured 1x/week according to the methodology of Aminot and Chaussepied [37].

2.4. Hall Sensor

Oyster valve activity was monitored using Hall effect sensors, following the methodology outlined by Guterres et al. [8]. The Hall effect sensor was affixed to the upper valve using epoxy resin (Araldite 5 min) [11], while neodymium magnets were attached to the lower valve using cyanoacrylate glue [10,15,16]. The choice of glue was informed by existing literature, along with the study conducted by Hartmann et al. [11], which evaluated the performance (mechanical strength and fixation time) of different substances in constructing biosensors based on bivalve behavioral analysis. The sensors were made waterproof using epoxy resin and linked to the electronic system via 3-way cables.

Positioned on opposing sides of the shell (see Figure 2), these sensors registered the oyster's opening by detecting their movement away from each other. The opening amplitudes for each individual were expressed as percentages, considering the maximum and minimum values recorded throughout the experiment [38]. The sensors were linked to an Arduino MEGA board and a computer to enable real-time recording of oyster openings throughout the experiment. The simultaneous connection of all sensors to the acquisition system was facilitated by employing two multiplexer boards, each with 16 channels. Within each experimental unit, 4 oysters were equipped with sensors, resulting in 12 sensor-equipped oysters per treatment and a total of 48 oysters in the experiment. Data were recorded using an SD memory card module.

Figure 2. Positioning of the Hall sensor and magnet on the right and left valves, respectively, of *C. gasar* oysters.

The acquisition system captured two readings per second (2 Hz) in bytes. These bytes were then converted into percentages of open state (%) on the basis of the individual oyster's lowest (0%) and highest (100%) observed values. To analyze the extensive volume of readings gathered during the experiment effectively, the average openings were computed every 1 min. Consequently, the behavioral data collected over the course of 24 h were represented by 1440 observations.

To facilitate data organization, this work designated oyster opening levels as follows: closed for 0% opening, slightly open for openings ranging from 0.1% to 25%, moderately open for 25.1% to 50%, open for 50.1% to 75%, and fully open for 75.1% to 100% openings.

2.5. Biochemical Analyses

After freezing in an ultrafreezer, the gills of the *C. gasar* oysters were homogenized according to Bebianno et al. [39] in a chilled buffer solution (TRIS-HCl 50 mM, EDTA 1 mM, sucrose 0.5 M, DTT 1 mM, KCl 0.15 M, PMSF 0.1 M), adjusted to pH 7.6, at a ratio of 1:5 (w/v). Subsequently, the tissue was centrifuged for 30 min at 4 °C and 10,000× g, and the resulting supernatant was collected and stored at −80 °C for the analysis of enzyme activity, including glutathione S-transferase (GST), catalase (CAT), and superoxide dismutase (SOD) activity, along with total antioxidant capacity against peroxyl radicals (ACAP) and peroxidized lipids (TBARS). For the assessment of reduced glutathione (GSH) and protein-associated sulfhydryl groups (P-SH), the same solution was employed, excluding DTT, and the same protocol as the other analyses was followed.

To determine the total protein in the tissues (gills), the Bradford method [40] was adopted using a Sensiprot commercial kit and a microplate reader (BioTek LX 800, Winooski, Vermont, USA). For the analysis of GST enzyme activity, the methodology of Habig et al. [41] was adopted, in which the increase in the CDNB-GSH product formed from the consumption of GSH in the presence of the extracts of the tissue samples and 0.1 M potassium phosphate buffer solution at pH 7.0 was monitored. The analysis was performed in a microplate reader at a temperature of 25 °C with a wavelength of 340 nm. The units used were nmol of CDNB per mg of protein per minute.

For catalase (CAT), the methodology described by Beutler [42] was adopted, in which the consumption of hydrogen peroxide (H2O2) in the presence of the samples was recorded in a spectrophotometer at 240 nm, pH 8.0, and a temperature of 25 °C. The units adopted were μmol of H_2O_2 per min and per mg of protein, which are expressed as CAT units.

The analysis of reduced glutathione (GSH) levels was performed using the protocol described by Sedlak and Lindsay [43]. Samples were adjusted to a concentration of 2 mg of protein per mL, followed by the addition of 50% (*w*/*v*) trichloroacetic acid (TCA) and centrifugation at 20,000× *g* for 10 min at 4 °C to precipitate the proteins. The resulting supernatant was collected for GSH analysis and mixed with 0.4 M Tris-base buffer at pH 8.9 along with DTNB. Following a 15 min incubation, the mixture was read at 405 nm for subsequent analysis. The units used for measurement were μmol of GSH per mg of protein. The pellet was collected, resuspended in homogenization buffer, and then treated with 0.2 M Tris-base at pH 8.2 and DTNB. After another 15 min incubation, the mixture was read at 405 nm to assess P-SH levels, with the unit of measurement of μmol of P-SH groups per mg of protein.

Lipid peroxidation levels (TBARS) were determined using the methodology outlined by Oakes and Van Der Kraak [44]. This method involves the reaction of lipid peroxidation byproducts with thiobarbituric acid (TBA) under conditions of high temperature and acidity, leading to the generation of a chromogen that can be detected through fluorimetry (excitation at 520 nm and emission at 580 nm). The units of measurement used were nmol of malondialdehyde (MDA) per mg of protein.

The assessment of total antioxidant capacity against peroxyl radicals (ACAP) was conducted in accordance with the methodology established by Amado et al. [45]. This involved quantifying the reactive oxygen species (ROS) in the samples, both treated and untreated, with a peroxyl radical generator. The samples were adjusted to a concentration of 2 mg of protein per mL and exposed to peroxyl radicals, which reacted with a fluorescent substrate (H2DCF-DA). Using fluorimetry, with excitation at 485 nm and emission at 520 nm, readings were taken on a microplate reader at 5 min intervals over a 35 min period. The results were expressed as an area relative to the difference between the area of ROS with ABAP (2,2-azobis (2-methylpropionamidine) dichloride) and the area of ROS without ABAP. In terms of results interpretation, a larger relative area indicated a lower antioxidant capacity.

The SOD (superoxide dismutase) activity in the oyster gills was evaluated following the methods described by McCord and Fridovich [46] and Bainy et al. [47]. The reduction of cytochrome c absorbance at a wavelength of 550 nm was measured in a potassium phosphate buffer at pH 7.8. The specific activity of the SOD enzyme is expressed as SOD activity per mg of total protein.

2.6. Histological Analysis

At the end of the experiment, 6 oyster samples were collected from each treatment, resulting in a total of 24 oysters. After opening the oysters, the visceral mass was extracted and preserved in 20% formalin saline. Subsequently, only the gills were removed from the visceral mass for analysis. The samples underwent a dehydration process using increasing ethanol concentrations in sequence, followed by clarification in xylene, and were subsequently embedded in Paraplast® using an automatic tissue processor (PT 05, LUPETEC, São Carlos, Brazil).

Sections with a thickness of 5 µm were prepared using a microtome (MRP03, LUPETEC, Brazil), stained with hematoxylin and eosin (H&E), and examined using an optical microscope (ZEISS Primo Star) at magnifications of 4× and 40×. The analysis of the gills was qualitative, involving the observation of structures and the identification of any damage. For histochemical assessment, specialized staining with periodic acid-Schiff (PAS) dye was employed, whereby PAS-positive structures were stained in magenta/red [26].

2.7. Statistical Analysis

The statistical analysis of water quality parameters, including pH, turbidity, alkalinity, salinity, ammonia, nitrite, nitrate, and phosphate, was conducted using one-way ANOVA. Assumptions of normality and homoscedasticity were verified using the Shapiro–Wilk and Levene tests, respectively [48]. If these assumptions were not met, the data were subjected to logarithmic transformation. For oxygen, temperature, suspended solids (SS), and total suspended solids (TSS) parameters, the non-parametric Kruskal–Wallis test was employed.

Regarding the statistical analysis of oyster valve activity monitored by the Hall sensor, two separate analyses were carried out. Initially, descriptive statistics were performed, and subsequently, a non-parametric Kruskal–Wallis analysis was conducted to assess the mean opening values within each treatment. The second analysis involved creating a frequency table for the data, categorizing them according to their opening level within each treatment and per week of the experiment. The non-parametric Kruskal–Wallis analysis was then performed to compare the values of each opening level across the weeks.

For the statistical analysis of biochemical parameters, including GST, CAT, SOD, GSH, P-SH, TBARS, and ACAP, a two-way ANOVA utilizing parametric statistics was conducted. This analysis considered both the cultivation time and the treatment as factors. The assumptions underlying the analysis were assessed using the Shapiro–Wilk and Levene tests. Whenever necessary, mathematical transformations were applied to the data. Significant differences were determined at a significance level of $p < 0.05$, followed by Newman–Keuls post hoc testing for multiple comparisons.

Across the experiment results, data are presented as mean values accompanied by standard deviation. The statistical software employed for analysis was Statistica 12. Microsoft Excel was utilized for the generation of tables, while PrismGraph software was utilized for the creation of graphs. A consistent significance level of 5% was adopted for all conducted tests.

3. Results

3.1. Water quality

The parameters of dissolved oxygen, salinity, alkalinity, and ammonia did not show significant differences between the different treatments applied in this experiment (Table 1).

Temperature, pH, nitrite, nitrate, and phosphate exhibited significant differences between the treatments. For pH and nitrite, the control treatment displayed significant differences ($p < 0.05$) compared with the low, medium, and high treatments, but no differences were found between the latter three treatments. Temperature values were significantly lower in the medium and high treatments. Nitrate levels showed the lowest means in the control, low, and medium treatments, while the high treatment had the highest mean, which was not significantly different from the medium treatment.

Phosphate levels were higher in the medium and high treatments compared with the control and low treatments. The low treatment showed no significant differences compared with the control for these parameters, but its mean values were lower than those of the medium and high treatments. The high treatment had higher mean values than the medium treatment, in line with the experimental design.

Table 1. Water quality parameters with mean values and standard deviation over the 28 days of cultivation in the treatments: control—oysters fed with microalgae; low treatment—100 mg/L total suspended solids (TSSs); medium treatment—from 100 to 200 mg/L TSSs; high treatment—above 200 mg/L TSSs.

	Treatments			
Parameters	**Control**	**Low**	**Medium**	**High**
Oxygen (mg/L)	6.83 ± 0.67	6.65 ± 0.62	6.84 ± 0.64	6.79 ± 0.63
Temperature (°C)	20.76 ± 1.83 ^{a}	20.64 ± 1.51 ^{a}	19.40 ± 1.79 ^{b}	20.07 ± 1.43 ^{b}
pH	8.21 ± 0.05 ^{b}	8.31 ± 0.06 ^{a}	8.31 ± 0.05 ^{a}	8.32 ± 0.05 ^{a}
SS (mL/L)	0.00 ± 0.00 ^{a}	0.06 ± 0.12 ^{a}	1.78 ± 2.00 ^{b}	5.65 ± 4.62 ^{c}
TSS (mg/L)	83.04 ± 29.18 ^{a}	71.27 ± 44.74 ^{a}	189.88 ± 120.15 ^{b}	291.38 ± 98.26 ^{c}
Turbidity (NUT)	22.74 ± 18.17 ^{a}	33.30 ± 35.02 ^{a}	108.29 ± 72.43 ^{b}	259.22 ± 142.88 ^{c}
Alkalinity (mg $CaCO_3$/L)	146.11 ± 10.54	167.22 ± 25.51	173.33 ± 21.94	173.33 ± 41.53
Salinity	33.63 ± 2.39	32.38 ± 1.41	31.63 ± 1.69	32.00 ± 1.07
Ammonia (mg/L)	0.17 ± 0.13	0.09 ± 0.04	0.08 ± 0.04	0.11 ± 0.04
Nitrite (mg/L)	0.65 ± 0.60 ^{b}	0.07 ± 0.06 ^{a}	0.12 ± 0.18 ^{a}	0.13 ± 0.11 ^{a}
Nitrate (mg/L)	9.56 ± 4.42 ^{a}	10.87 ± 4.85 ^{a}	17.20 ± 5.55 ^{ab}	20.76 ± 8.56 ^{b}
Phosphate (mg/L)	2.64 ± 1.14 ^{a}	3.80 ± 0.87 ^{a}	6.76 ± 2.05 ^{b}	6.50 ± 0.84 ^{b}

Different letters represent a significant difference ($p \leq 0.05$) between treatments.

3.2. Hall Sensor

For mean valve opening values across each week within each treatment, no significant differences ($p > 0.05$) were observed between treatments during week 1 and week 4. However, significant differences ($p < 0.05$) emerged during weeks 2 and 3 of the experiment, as indicated in Table 2.

During week 2, a significant difference ($p < 0.05$) was observed. The control and medium treatments had a greater number of fully open oysters compared with the high treatment. In week 3, differences ($p < 0.05$) between treatments were observed across three ranges of oyster opening: closed, slightly open, and fully open. Notably, the high treatment exhibited a higher occurrence of closed oysters than the control treatment. Furthermore, the medium treatment had a higher frequency of slightly open oysters compared with the high treatment, while the high treatment had a lower frequency of fully open oysters compared with the control treatment.

Overall, it is evident that closed oysters predominated in all treatments during the four weeks of cultivation. The second most frequent opening range was fully open, followed by open. The medium open level represented the opening range with the lowest number of records in terms of percentage.

3.3. Biochemical Analysis

The activities of the enzymes GST and CAT did not exhibit significant differences ($p \geq 0.05$) between the various treatments throughout the experiment, as depicted in Figures 3 and 4, respectively.

Table 2. The frequency distribution (in %) of valve opening values for *C. gasar* oysters over the four weeks of culture in the control, low, medium, and high treatments. Control treatment: oysters fed with microalgae. Low treatment: 100 mg/L total suspended solids (TSSs). Medium treatment: ranging from 100 to 200 mg/L TSSs. High treatment: exceeding 200 mg/L TSSs.

		Opening Range of Oyster Valves				
Treatment	**Week**	**Closed**	**Slightly Open**	**Moderately Open**	**Open**	**Fully Open**
Control	1	64.54 ± 7.15	0.20 ± 0.45	0.90 ± 0.67	7.06 ± 4.89	27.30 ± 7.25
Low	1	73.23 ± 5.88	0.02 ± 0.02	1.90 ± 1.23	6.78 ± 2.93	18.07 ± 7.79
Medium	1	64.32 ± 7.81	0.24 ± 0.48	0.95 ± 0.71	7.51 ± 5.20	26.97 ± 7.89
High	1	60.86 ± 14.36	0.00 ± 0.00	0.89 ± 0.53	6.40 ± 2.80	31.84 ± 14.02
Control	2	39.35 ± 22.57	0.00 ± 0.00	0.18 ± 0.29	3.87 ± 3.48	56.60 ± 20.75 [ab]
Low	2	64.80 ± 26.72	0.07 ± 0.07	0.69 ± 0.60	4.45 ± 2.95	30.00 ± 25.67 [bc]
Medium	2	31.01 ± 10.75	0.00 ± 0.00	0.13 ± 0.28	4.37 ± 3.64	64.40 ± 9.03 [a]
High	2	65.04 ± 20.13	0.11 ± 0.13	5.39 ± 10.55	18.00 ± 25.16	11.47 ± 9.71 [c]
Control	3	66.97 ± 15.14 [a]	0.03 ± 0.02 [ab]	1.17 ± 1.02	3.77 ± 3.25	28.05 ± 12.88 [a]
Low	3	79.97 ± 4.93 [ab]	0.04 ± 0.01 [ab]	0.56 ± 0.71	2.99 ± 2.77	16.44 ± 7.64 [ab]
Medium	3	79.67 ± 4.57 [ab]	0.12 ± 0.05 [a]	1.23 ± 0.77	4.83 ± 2.95	14.14 ± 2.09 [ab]
High	3	85.10 ± 6.29 [b]	0.01 ± 0.01 [b]	0.19 ± 0.12	5.66 ± 8.08	9.04 ± 3.91 [b]
Control	4	56.99 ± 27.22	0.02 ± 0.04	1.45 ± 0.13	10.83 ± 5.04	30.71 ± 28.23
Low	4	65.06 ± 14.52	0.04 ± 0.04	0.94 ± 0.73	6.30 ± 2.66	27.66 ± 15.98
Medium	4	54.10 ± 15.54	0.04 ± 0.07	1.38 ± 0.68	10.60 ± 7.24	33.88 ± 15.81
High	4	52.43 ± 17.72	0.00 ± 0.00	0.57 ± 0.33	8.78 ± 7.76	38.22 ± 10.58

Different lowercase letters in the same row represent significant differences ($p < 0.05$) between treatments.

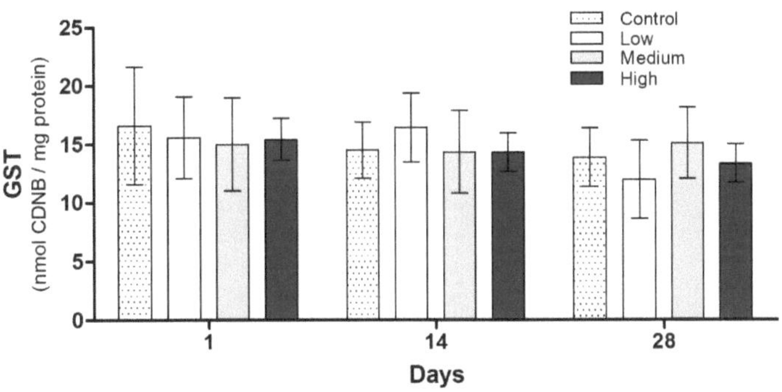

Figure 3. Specific activity of the enzyme GST (Glutathione S-transferase) in the gills of *C. gasar* oysters in the following treatments: control—oysters fed with microalgae; low treatment—100 mg/L total suspended solids (TSSs); medium treatment—100 to 200 mg/L TSSs; high treatment—TSSs exceeding 200 mg/L over a 28-day experimental period. The bars represent the mean enzyme activity values, and the vertical lines above the bars represent the standard deviation of the values.

Regarding the SOD enzyme, a significant difference ($p < 0.05$) was observed between the treatments on day 28. No differences between treatments were evident on days 1 and 14. On day 28, SOD activity displayed higher values for the high treatment than for the control. Notably, the low and medium treatments did not demonstrate significant differences ($p > 0.05$) from each other or when compared with the high and control treatments (Figure 5).

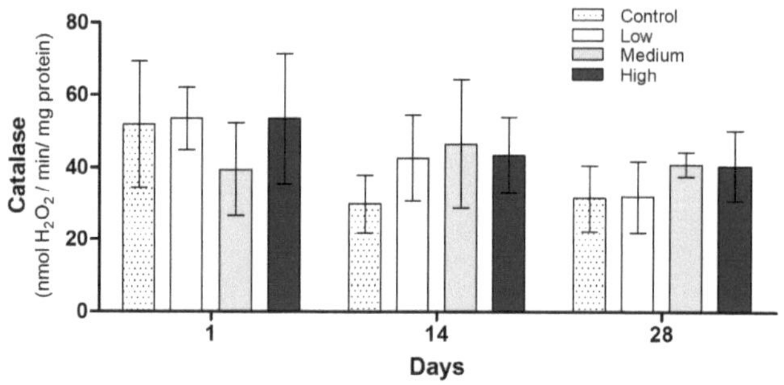

Figure 4. Specific activity of catalase (µm/mg protein) in the gills of *C. gasar* oysters in the following treatments: control—oysters fed with microalgae; low treatment—100 mg/L total suspended solids (TSSs); medium treatment—100 to 200 mg/L TSSs; high treatment—TSSs exceeding 200 mg/L over a 28-day experimental period. The bars represent the mean enzyme activity values, and the vertical lines above the bars represent the standard deviation of the values.

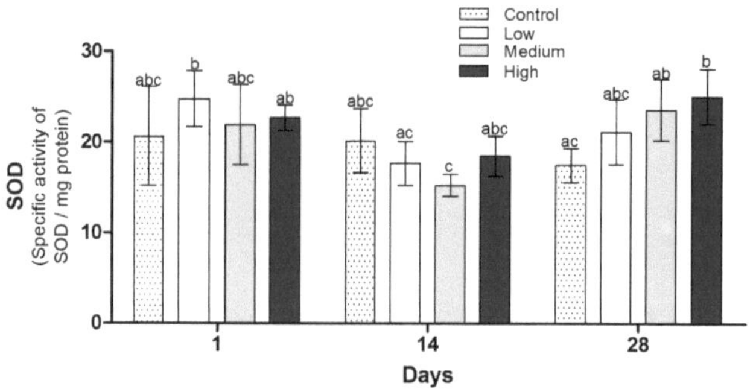

Figure 5. Specific activity of the SOD enzyme (specific SOD activity/mg protein) in the gills of *C. gasar* oysters in the following treatments: control—oysters fed with microalgae; low treatment—100 mg/L total suspended solids (TSSs); medium treatment—100 to 200 mg/L TSSs; high treatment—TSSs exceeding 200 mg/L over a 28-day experimental period. The bars represent the mean values, and the vertical lines above the bars represent the standard deviation of the values. Lowercase letters for different days and treatments indicate significant differences ($p < 0.05$).

The concentration of GSH did not show significant differences ($p > 0.05$) between the different treatments during the experiment (Figure 6). No significant differences ($p > 0.05$) emerged between treatments or culture days with respect to the concentration of sulfhydryl groups in the gills of *C. gasar* oysters (Figure 7). Regarding lipid peroxidation, significant differences ($p < 0.05$) between treatments were observed. No significant differences ($p > 0.05$) in lipid damage were observed between treatments on days 1 and 28. However, on day 14, a substantial increase in lipid damage was noted in the medium and high treatments in comparison to the control group. By contrast, the low treatment did not show significant differences ($p > 0.05$) on day 14 (Figure 8).

In relation to the total antioxidant capacity against peroxyl radicals, a significant difference ($p < 0.05$) emerged solely on day 28 between the treatments. Specifically, the high treatment exhibited a greater mean area compared with the control treatment. By contrast, the low and medium treatments did not display differences from each other or from the control or high treatments (Figure 9). A higher mean area in the high treatment signified a reduced total antioxidant capacity.

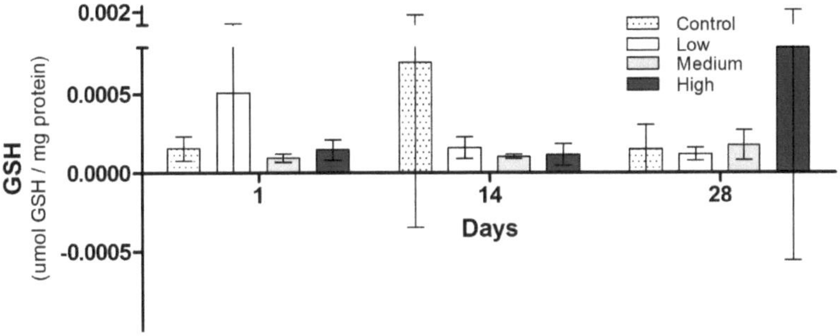

Figure 6. Concentration of reduced glutathione (GSH) (in nmol/mg protein) in the gills of *C. gasar* oysters in the following treatments: control—oysters fed with microalgae; low treatment—100 mg/L total suspended solids (TSSs); medium treatment—100 to 200 mg/L TSSs; high treatment—TSSs exceeding 200 mg/L over a 28-day experimental period. The bars represent the mean values, and the vertical lines above the bars represent the standard deviation of the values.

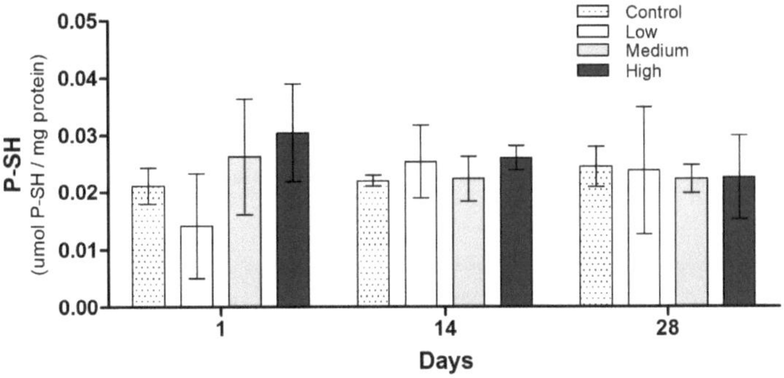

Figure 7. *P*-SH concentration (in nmol P-SH/mg protein) in the gills of *C. gasar* oysters in the following treatments: control—oysters fed with microalgae; low treatment—100 mg/L total suspended solids (TSSs); medium treatment—100 to 200 mg/L TSSs; high treatment—TSSs exceeding 200 mg/L over a 28-day experimental period. The bars represent the mean values, and the vertical lines above the bars represent the standard deviation of the values.

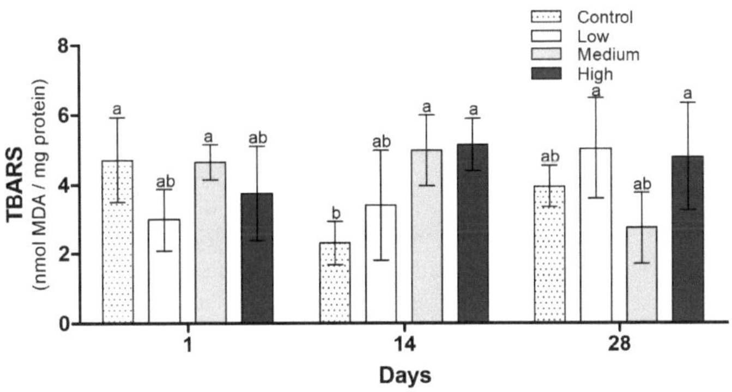

Figure 8. Lipid peroxidation (TBARS) (in mmol/mg tissue) in the gills of *C. gasar* oysters in the following treatments: control—oysters fed with microalgae; low treatment—100 mg/L total suspended solids (TSSs); medium treatment—100 to 200 mg/L TSSs; high treatment—TSSs exceeding 200 mg/L over a 28-day experimental period. The bars represent the mean values, and the vertical lines above the bars represent the standard deviation of the values. Lowercase letters for different days and treatments indicate significant differences ($p < 0.05$).

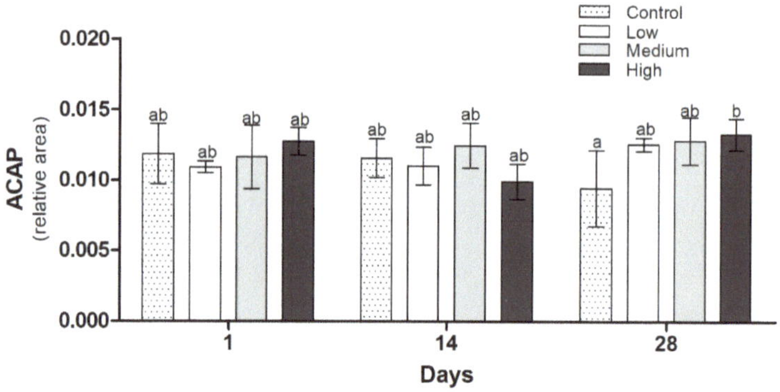

Figure 9. Antioxidant capacity (relative area) in the gills of *C. gasar* oysters in the following treatments control—oysters fed with microalgae; low treatment—100 mg/L total suspended solids (TSSs) medium treatment—100 to 200 mg/L TSSs; high treatment—TSSs exceeding 200 mg/L over four weeks of the experiment. The bars represent the mean values, and the vertical lines above the bars represent the standard deviation of the values. Lowercase letters for different days and treatments indicate significant differences ($p < 0.05$).

3.4. Histological Analysis

Following 28 days of culture, oysters from the control treatment exhibited well-defined gill lamellae characterized by intact edges, organized cells, and the presence of infiltrated hemocytes (Figure 10).

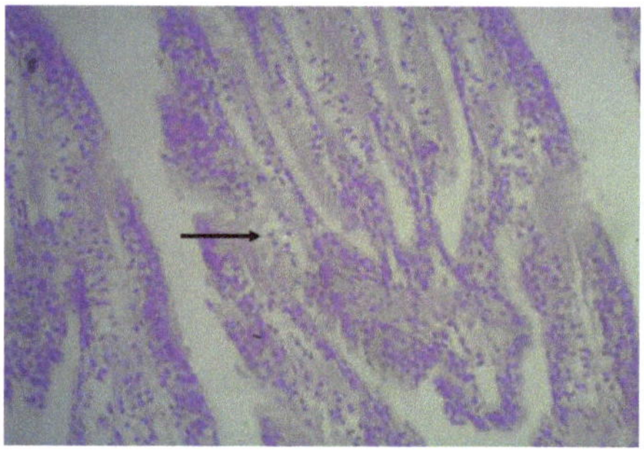

Figure 10. Gill filaments of *C. gasar* oysters after 28 days of culture in the control treatment stained with hematoxylin and eosin, magnification 40×.

After the same cultivation duration, oysters from the low treatment showed well-defined gill lamellae with intact edges and organized cells. The hyperplasia of the epithelium within the gill lamellae was more pronounced at the filament ends, accompanied by the presence of infiltrated hemocytes (Figure 11A). In the medium treatment, the gill lamellae structure exhibited poor organization yet intact filaments. Notably, hyperplasia of the gill lamellae's epithelium was evident, particularly at the filament ends, alongside the presence of infiltrated hemocytes (Figure 11B). In the high treatment, the gill lamellae structure was less organized, with filaments lacking consistent height and width patterns. The hyperplasia of the gill lamellae's epithelium was pronounced, and infiltrated hemocytes were observed within the gill filaments (Figure 11C).

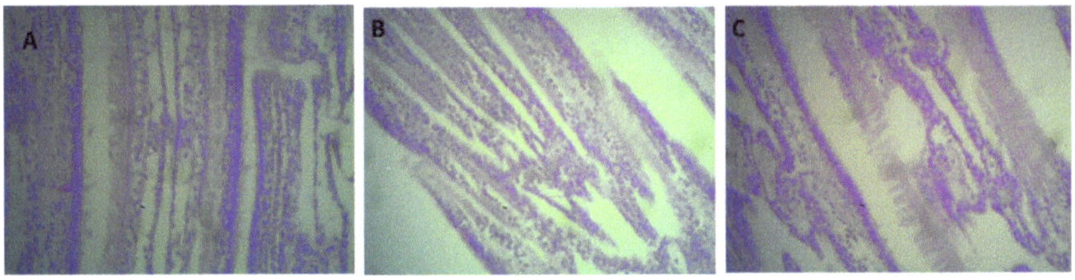

Figure 11. Gill filaments of *C. gasar* oysters after 28 days of culture in biofloc treatments, stained with hematoxylin and eosin, magnification 40×. (**A**): Low treatment; (**B**): medium treatment; (**C**): high treatment.

Concluding the experiment, the gills of oysters from both the medium and high treatments displayed notable hyperplasia of the gill epithelium and the presence of infiltrated hemocytes. In the low and control treatments, infiltrated hemocytes and filament hyperplasia were also observable, albeit with lower intensity.

For the PAS-stained samples, in the control treatment, no PAS-positive cells were observed, and the gill architecture was preserved. The gills of the treatments with biofloc presented gill structures with hyperplasia of the lamellar epithelium and isolated PAS-positive cells of a mucosecretory aspect (Figure 12).

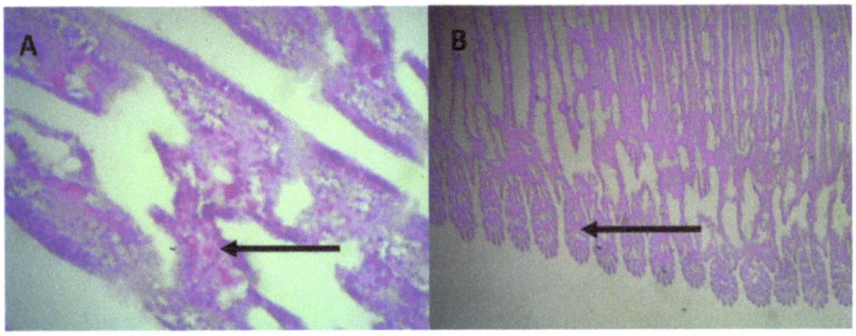

Figure 12. Gill filaments of *C. gasar* oysters after 28 days of culture in biofloc treatments, stained with PAS. (**A**): Magnification 40×; (**B**): magnification 4×.

4. Discussion

To maintain a biofloc system with marine shrimps, the recommended TSS concentration falls within 100 to 300 mg/L [29]. Schveitzer et al. [49] indicated that a TSS reduction to 200 mg/L in biofloc systems led to decreased nitrification rates. In addition, although 44.8% of a biofloc volume is composed of particles smaller than 48 µm, which is favorable for filtering oysters [50], the guidelines for oyster cultivation suggest lower TSS concentrations. Barillé et al. [28] investigated the impact of high seston (suspended material) concentrations on food selection by *Crassostrea* oysters. They found that seston concentrations below 90 mg/L were regulated by pseudofecal production, while levels above 90 mg/L exhibited reduced filtration and rejection rates, indicative of physical constraints hindering food acquisition. Consequently, reconciling optimal solids concentrations for shrimp and microorganisms with tolerable levels for oysters in an integrated biofloc system appears to pose challenges given the conflicting concentration requirements.

Usually, elevated levels of suspended solids can potentially lead to health issues in cultured species, although tolerance thresholds vary among species [51,52]. Cultivated organisms may experience hindered gill function due to excessive particulate matter in the water, rendering them more vulnerable to hypoxia. This scenario applies similarly to oysters.

The monitoring of valve movements in bivalves serves as a common approach to discerning behavioral patterns that might signify the presence of contaminants or alterations in environmental conditions [10,11]. For instance, Lombardi et al. [53] noted that oysters tended to keep their valves closed when exposed to unfavorable salinity levels. Various factors contribute to the modulation of valve opening and closing behaviors, encompassing hypoxia [54], food availability [55], pollutants [56,57], acoustic disturbances [58], and other environmental influences. In this context, total suspended solids may impact the valve behaviors of *C. gasar*.

Within the scope of the present experiment, a notable observation was that oysters subjected to the control treatment exhibited a higher average valve opening than those in the biofloc treatments. This disparity indicates a certain constraint on valve opening in biofloc cultures, irrespective of the TSS concentration. This implies that TSSs could be a limiting parameter in *C. gasar* cultivation. These results corroborate a study by Lima et al. [2], who tested different concentrations of settleable solids in oyster seeds and juveniles and found low survival at high concentrations (10 and 20 ml/L of settleable solids). Filter-feeding bivalves employ valve opening to filter water, respire, and feed. Accordingly, well-opened valves denote feeding periods, while partially opened or closed valves suggest decreased or halted respiratory filtration [59–61].

Schreck et al. [62] proposed that a common initial response of many marine organisms to environmental stressors is to modify their behavior, either to evade lethal consequences or minimize the metabolic costs related to maintaining physiological equilibrium under stressful conditions. If organisms are unable to escape unfavorable conditions, these circumstances can have cascading effects on various other behaviors [63]. In the context of the current experiment, during the initial days of cultivation, the TSS concentrations did not appear to significantly impact the opening behavior of the oysters. However, in the second and third weeks, the effects became prominently evident, and by the fourth week, the biochemical effects started manifesting in the high treatment.

Stressful conditions within culture environments can trigger physiological transformations [64], including changes in metabolism, the initiation of a pro-oxidative state involving modifications in the antioxidant defense system, and the potential generation of reactive oxygen species (ROS), which may or may not be accompanied by oxidative damage. Particularly in confined and intensive aquatic environments like those established in biofloc systems, the oxidative equilibrium of aquatic organisms can be adversely affected. Numerous stressors have been investigated in biofloc systems, including salinity [65], temperature [66], and pH [67], among others [18,68,69].

Regarding the SOD enzyme activity, it became evident that the high treatment induced the production of superoxide anion radicals (O_{2-}). Although no signs of lipid damage were observed during this period, a subsequent reduction in total antioxidant capacity was noticeable. Thus, it can be postulated that within the initial two weeks of culture, the elevated TSSs in the high treatment did not prompt oxidative stress in *C. gasar* oysters. However, after this period, spanning 28 days of culture, discernible alterations in SOD enzyme activity appeared, signifying the emergence of a pro-oxidative scenario in *C. gasar* oysters. In a study by Zanette et al. [70], investigations involving distinct salinities revealed no modifications in CAT and GST enzyme activities over a span of 17 days in *C. gigas* oysters' gills. However, when exposed to diesel oil under varying salinity conditions, just one week of diesel exposure was adequate to elicit biochemical changes in the SOD and GST enzymes. This underscores the influence of the type of contaminant on the pace and nature of biochemical responses.

Organisms' antioxidant defense mechanisms can be categorized into prevention, interception, and repair [71]. When confronted with stressors, an organism's primary line of defense entails employing preventive strategies to counter the exposure. For instance, plankton employ a preventive measure during photooxidative stress by positioning themselves beneath the sea surface to avoid direct solar radiation [72]. In the case of oysters, it is plausible that, prior to engaging enzymatic defense mechanisms, they employ physical defense strategies. One such strategy might involve minimizing valve opening as a means of curtailing exposure. This adjustment could serve to limit the contact between the gills and the surrounding water, thereby mitigating the potential adverse effects associated with elevated TSS concentrations. However, as the experiment progressed, the oysters in the high treatment ceased to display distinguishable differences in valve opening compared with the other treatments. Consequently, water ingress into the pallial cavity resumed. This marked a turning point at which enzymatic activity alterations and a decline in total antioxidant capacity manifested, indicative of a pro-oxidative scenario.

A potential influencing factor on the enzymatic defense responses of the organisms in the current experiment was the presence of a biofloc system and the subsequent utilization of biofloc by the oysters. Some studies have indicated increased antioxidant effects in cultured organisms exposed to biofloc compared with those raised in clear water conditions [18].

In the context of this study, the absence of observable shifts in antioxidant defenses and macromolecular damage in the low and medium treatments throughout the experiment could be attributed to the antioxidative properties of biofloc. In cases of lower concentrations, such as those found in the low and medium treatments, the TSS levels did not prove detrimental to the gill health of the oysters concerning the analyzed enzymatic and non-enzymatic defense mechanisms. This implies that the presence of biofloc may have conferred a protective effect against oxidative stress induced by elevated TSS concentrations.

On the 28th day of the culture period, a decline in the overall antioxidant capacity of the oyster gills became evident in the high treatment. This shift in the normal pattern compared with the control treatment during the same period signifies a departure from the expected antioxidant response. Simultaneously, an elevation in SOD activity was observed, indicating the organism's reaction to dismutase superoxide anion radicals, which are generated due to the presence of oxygen. Essentially, gas exchange persisted, but it appeared to have had detrimental effects on the oysters in the high treatment.

As a result, it is highly plausible that the reduction in the total antioxidant capacity could be attributed to other antioxidant defense mechanisms that were not directly assessed in this study. This is because antioxidant defense systems tend to collaborate [73], and their activities can be either boosted or hindered by stressors [74]. In light of this, an approach that evaluates antioxidant capacity in a broader sense, encompassing various antioxidant defenses and offering comprehensive insights into an organism's resilience against ROS-induced toxicity [45], seems better suited to elucidating the redox state of oyster gills.

Interestingly, in the context of bivalves, the presence of food sources like microalgae has been shown to mitigate oxidative stress induced by external factors [75], possibly owing to the antioxidant properties of microalgae. Notably, in the current study, there was a significant decrease ($p < 0.05$) in lipid damage in the control treatment on day 14 compared with day 1, and this reduction was also observed on day 14 in the control treatment compared with the medium and high treatments, aligning with the findings of Li et al. [75].

Pathologies affecting bivalve gills span from inflammatory responses to tissue necrosis. Hyperplasia, a defensive inflammatory pathology, represents an immune reaction in bivalves [76]. Interestingly, such hyperplasia was observed in treatments involving biofloc, irrespective of the concentration of solids. However, the morphological changes in the gills of *C. gasar* oysters did not display notably more pronounced alterations compared with those in the control treatment. As Au [77] suggested, gill histopathological changes are generally indicative of responsiveness rather than specificity to pollutant exposure, implying that they can signify a broad spectrum of contaminants, signifying environmental toxicity.

In terms of the gill lamellae structure of *C. gasar* oysters exposed to biofloc, the mild effects observed in the medium and high treatments suggest the influence of higher solids concentrations on the bivalves. Nevertheless, these effects did not compromise the overall integrity of the gill structure. Notably, the presence of biofloc in the low, medium, and high treatments did not result in significant damage to the gill morphology of *C. gasar* oysters during the 28-day culture period under the conditions of this experiment.

However, a noticeable variation was observed in mucus production in the gills of oysters subjected to biofloc treatments. Through the use of the PAS method, the presence of PAS-positive cells was discerned in the brachial filaments of oysters from the biofloc treatments, indicative of the activity of mucosecretory cells in these treatment groups. This observation underscores the potential influence of biofloc on the mucous production of the oysters' gills.

A study conducted by Salas-Yanquin et al. [78] provided relevant insights into the relationship between mollusks and the production of mucus in response to environmental conditions. Their analysis of different ash concentrations in water and the subsequent increase in mucus production in mussels highlighted the ability of mollusks to separate ingested material from material to be expelled as pseudofeces, often involving substantial amounts of mucus.

In the context of the present experiment, the observed increase in mucus production associated with the biofloc treatments aligns with the presence of cells displaying mucus production. This phenomenon may be attributed to the requirement for greater pseudofeces production in bivalves within the biofloc treatments. The bioflocculation process could potentially necessitate enhanced pseudofeces production by bivalves, contributing to the selection of food items.

The findings of Garrido et al. [79] further support this notion. Their observations of the food particle selection process in bivalves through endoscopy revealed that particles reaching the mantle cavity are intercepted by gill filaments and transported toward the food groove in small mucous aggregates facilitated by the frontal cilia. Particles slated for rejection, to be eliminated as pseudofeces, are expelled from the mantle cavity through mucous channels that actively engage in mucus secretion. This process's demand for high mucus production could explain the presence of a greater quantity of mucosecretory cells within the gills of oysters exposed to biofloc treatments in the present experiment.

Another explanation for the presence of mucocytes could be the interaction with biofloc, since the diversity of microorganisms in this system is high, especially when compared with the control treatment. The production of mucus in aquatic organisms is for protection against pathogenic organisms.

David et al. [80] analyzed changes in the gills of Mytella falcata collected in polluted regions of the Santos estuary. Among other histopathological changes, they observed morphological alterations in the epithelium and an increase in the number of mucous cells, possibly as an attempt to prevent the entry of pollutants through the gill filaments to the entire organism. Zannella et al. [81] reinforced that the production of antimicrobial substances, mainly peptides or polypeptides, is an ancient mechanism of innate immunity. These substances are produced by different types of cells and secretions and are synthesized at the time of infection.

5. Conclusions

The initial defense response of *C. gasar* oysters to elevated concentrations of biofloc involved the regulation of valve closure behavior, followed by subsequent biochemical reactions. Hemocytes were consistently present in the gills, serving as an active defense mechanism for the oysters across all treatments in this study. Furthermore, the gill morphology of the oysters remained relatively stable in the biofloc treatments, with no significant structural changes. Exposure to total suspended solids (TSSs) concentrations comparable to those in the high treatment prompted alterations in the valve opening behavior of *C. gasar* oysters from the second week of exposure onward. Prolonged exposure exceeding 14 days to such TSS concentrations could impact the antioxidant capacity of *C. gasar* oysters while also influencing the modulation of superoxide dismutase (SOD) enzyme activity under similar conditions.

Total suspended solids (TSSs) concentrations below 200 mg/L, under conditions comparable to those in this study, did not induce oxidative stress, alterations in behavior, or histological changes in *C. gasar* oysters cultured over a 28-day biofloc period. However, higher TSS concentrations exceeding 200 mg/L prompted shifts in valve opening behavior and led to oxidative stress in *C. gasar* oysters, making them unsuitable for cultivation.

Author Contributions: Conceptualization, L.C.d.O.C.; L.H.P.; B.G.; J.N.J.; M.P.; S.B.; L.R.; J.V. and M.H.; methodology, A.C.; M.H.; J.S.; L.B.; L.R.; V.F.; B.G.; J.N.J.; L.M.; J.V. and L.H.P.; formal analysis, L.C.d.O.C.; L.M.; L.R.; V.F.; L.B. and J.V.; investigation, L.C.d.O.C.; A.C.; M.H.; L.B.; J.S.; B.G.; S.B.; J.N.J. and L.M.; resources, L.H.P.; M.P.; S.B.; J.V. and L.R.; data curation, L.C.d.O.C.; L.M.; V.F.; B.G; J.N.J. and L.R.; writing—original draft preparation, L.C.d.O.C.; M.H. and A.C.; writing—review and editing, A.C.; M.H.; B.G; J.N.J.; L.M.; J.V.; V.F.; L.B.; M.P.; S.B.; J.S. and L.H.P.; supervision, L.H.P. and M.P.; project administration, L.H.P.; funding acquisition, L.H.P. All authors have read and agreed to the published version of the manuscript.

Funding: This research was funded by ASTRAL Project—H2020 grant Agreement 863034.

Data Availability Statement: Data are contained within the article.

Acknowledgments: The Authors are grateful to the ASTRAL project that has received funding from the European Union's Horizon 2020 research and innovation program under grant agreement No 863034. Special thanks to Brazilian Council of Research (CNPq) and Coordination for the Improvement of Higher Level or Education Personnel (CAPES). L.H. Poersch received productivity research fellowship from CNPq.

Conflicts of Interest: The authors declare that they have no conflict of interest.

References

1. FAO. *The State of the World Fisheries and Aquaculture—2022 (SOFIA)*; FAO: Rome, Italy, 2022; ISBN 9789251072257.
2. Lima, P.C.M.; Silva, A.E.M.; Silva, D.A.; Silva, S.M.B.C.; Brito, L.O.; Gálvez, A.O. Effect of stocking density of Crassostrea sp. in a multitrophic biofloc system with Litopenaeus vannamei in nursery. *Aquaculture* **2021**, *530*, 735913. [CrossRef]
3. Iwama, G.K.; Afonso, L.O.B.; Todgham, A.; Ackerman, P.; Nakano, K. Are hsps suitable for indicating stressed states in fish? *J. Exp. Biol.* **2004**, *207*, 15–19. [CrossRef]
4. Afonso, L.O.B. Identifying and managing maladaptive physiological responses to aquaculture stressors. *Fish Physiol.* **2020**, *38*, 163–191. [CrossRef]
5. Coates, C.J.; Söderhäll, K. The stress–immunity axis in shellfish. *J. Invertebr. Pathol.* **2021**, *186*, 107492. [CrossRef]
6. Guo, X.; He, Y.; Zhang, L.; Lelong, C.; Jouaux, A. Immune and stress responses in oysters with insights on adaptation. *Fish Shellfish Immunol.* **2015**, *46*, 107–119. [CrossRef]
7. Gerhardt, A.; Janssens de Bisthoven, L.; Soares, A.M.V. Evidence for the Stepwise Stress Model: Gambusia holbrooki and Daphnia magna under Acid Mine Drainage and Acidified Reference Water Stress. *Environ. Sci. Technol.* **2005**, *39*, 4150–4158. [CrossRef] [PubMed]
8. de Vargas Guterres, B.; da Silveira Guerreiro, A.; Sandrini, J.Z.; da Costa Botelho, S.S. Feasibility of visual signals on the construction of biosensors based on behavioral analysis of Perna perna mussels. *Ecol. Inform.* **2020**, *59*, 101118. [CrossRef]
9. Newton, T.; Gregory Cope, W. Biomarker Responses of Unionid Mussels to Environmental Contaminants. In *Freshwater Bivalve Ecotoxicology*; CRC Press: Boca Raton, FL, USA, 2006; pp. 257–284. [CrossRef]
10. Comeau, L.A.; Babarro, J.M.F.; Longa, A.; Padin, X.A. Valve-gaping behavior of raft-cultivated mussels in the Ría de Arousa, Spain. *Aquac. Rep.* **2018**, *9*, 68–73. [CrossRef]
11. Hartmann, J.T.; Beggel, S.; Auerswald, K.; Geist, J. Determination of the most suitable adhesive for tagging freshwater mussels and its use in an experimental study of filtration behaviour and biological rhythm. *J. Molluscan Stud.* **2016**, *82*, 415–421. [CrossRef]
12. Lummer, E.-M.; Auerswald, K.; Geist, J. Fine sediment as environmental stressor affecting freshwater mussel behavior and ecosystem services. *Sci. Total Environ.* **2016**, *571*, 1340–1348. [CrossRef]
13. Le Moullac, G.; Soyez, C.; Lyonnard, P.; Chabrier, S.; Milhade, L.; Gueguen, Y.; Beliaeff, B. Non-invasive functional exploration techniques for bivalves with applications to pearl oyster Pinctada margaritifera. *Rev. Aquac.* **2020**, *12*, 1783–1791. [CrossRef]
14. Andrewartha, S.J.; Elliott, N.G. Aquaculture Sentinels: Smart-farming with Biosensor Equipped Stock. *J. Aquac. Res. Dev.* **2015**, *7*, 393. [CrossRef]
15. Basti, L.; Nagai, K.; Shimasaki, Y.; Oshima, Y.; Honjo, T.; Segawa, S. Effects of the toxic dinoflagellate Heterocapsa circularisquama on the valve movement behaviour of the Manila clam Ruditapes philippinarum. *Aquaculture* **2009**, *291*, 41–47. [CrossRef]
16. Nagai, K.; Honjo, T.; Go, J.; Yamashita, H.; Oh, S.J. Detecting the shellfish killer Heterocapsa circularisquama (Dinophyceae) by measuring bivalve valve activity with a Hall element sensor. *Aquaculture* **2006**, *255*, 395–401. [CrossRef]

17. Tran, D.; Haberkorn, H.; Soudant, P.; Ciret, P.; Massabuau, J.-C. Behavioral responses of Crassostrea gigas exposed to the harmful algae Alexandrium minutum. *Aquaculture* **2010**, *298*, 338–345. [CrossRef]
18. da Silva Martins, Á.C.; Artigas Flores, J.; Porto, C.; Wasielesky Junior, W.; Monserrat, J.M. Antioxidant and oxidative damage responses in different organs of Pacific white shrimp Litopenaeus vannamei (Boone, 1931) reared in a biofloc technology system *Mar. Freshw. Behav. Physiol.* **2015**, *48*, 279–288. [CrossRef]
19. Cogo, A.J.; Siqueira, A.F.; Ramos, A.C.; Cruz, Z.M.; Silva, A.G. Utilização de enzimas do estresse oxidativo como biomarcadoras de impactos ambientais. *Nat. Line* **2009**, *7*, 37–42.
20. Valavanidis, A.; Vlahogianni, T.; Dassenakis, M.; Scoullos, M. Molecular biomarkers of oxidative stress in aquatic organisms in relation to toxic environmental pollutants. *Ecotoxicol. Environ. Saf.* **2006**, *64*, 178–189. [CrossRef] [PubMed]
21. Sies, H.; Jones, D.P. Oxidative stress. In *Encyclopedia of Stress*; Fink, G., Ed.; Elsevier: Amsterdam, The Netherlands, 2007; pp. 45–48.
22. Jones, D.P. Extracellular Redox State: Refining the Definition of Oxidative Stress in Aging. *Rejuvenation Res.* **2006**, *9*, 169–181. [CrossRef]
23. Kunwar, A.; Priyadarsini, K.I. Free radicals, oxidative stress and importance of antioxidants in human health. *J. Med. Allied Sci.* **2011**, *1*, 53–60.
24. Comporti, M.; Signorini, C.; Leoncini, S.; Gardi, C.; Ciccoli, L.; Giardini, A.; Vecchio, D.; Arezzini, B. Ethanol-induced oxidative stress: Basic knowledge. *Genes Nutr.* **2010**, *5*, 101–109. [CrossRef]
25. Urso, M.L.; Clarkson, P.M. Oxidative stress, exercise, and antioxidant supplementation. *Toxicology* **2003**, *189*, 41–54. [CrossRef] [PubMed]
26. Genten, F.; Terwinghe, E.; Danguy, A. *Atlas of Fish Histology*; Science Publishers: Enfield, NH, USA, 2008.
27. Costa, P.M.; Carreira, S.; Costa, M.H.; Caeiro, S. Development of histopathological indices in a commercial marine bivalve (Ruditapes decussatus) to determine environmental quality. *Aquat. Toxicol.* **2013**, *126*, 442–454. [CrossRef] [PubMed]
28. Barillé, L.; Prou, J.; Héral, M.; Razet, D. Effects of high natural seston concentrations on the feeding, selection, and absorption of the oyster Crassostrea gigas (Thunberg). *J. Exp. Mar. Bio. Ecol.* **1997**, *212*, 149–172. [CrossRef]
29. Gaona, C.A.P.; de Almeida, M.S.; Viau, V.; Poersch, L.H.; Wasielesky, W. Effect of different total suspended solids levels on a Litopenaeus vannamei (Boone, 1931) BFT culture system during biofloc formation. *Aquac. Res.* **2017**, *48*, 1070–1079. [CrossRef]
30. Guillard, R.R.L. Culture of phytoplankton for feeding marine invertebrates. In *Culture of Marine Invertebrate Animals*; Smith, W.L., Chanley, M.H., Eds.; Plenum: New York, NY, USA, 1975; pp. 29–60.
31. AOAC. *Official Methods of Analysis of the AOAC*, 14th ed.; AOAC International: Washington, DC, USA, 1999.
32. Avnimelech, Y. Feeding with microbial flocs by tilapia in minimal discharge bio-flocs technology ponds. *Aquaculture* **2007**, *264*, 140–147. [CrossRef]
33. APHA. *Standard Methods for the Examination of Water and Waste Water*, 16th ed.; American Public Health Association, AWWA, WPFC: New York, NY, USA, 1989.
34. UNESCO. *Chemical Methods for Use in Marine Environmental Monitoring. Manual and Guides 12*; Intergovernamental Oceanographic Commissiony: Paris, France, 1983.
35. Strickland, J.D.H.; Parsons, T.R. *A Practical Handbook of Seawater Analysis*, 2nd ed.; Fisheries research board of Canada: Ottawa, ON, Canada, 1972.
36. Furtado, P.S.; Poersch, L.H.; Wasielesky, W. The effect of different alkalinity levels on Litopenaeus vannamei reared with biofloc technology (BFT). *Aquac. Int.* **2015**, *23*, 345–358. [CrossRef]
37. Aminot, A.; Chaussepied, M. *Manuel des Analyses Chimiques en Milieu Marin*; Centre National Pour L'exploitation des Océans: Brest, France, 1983.
38. Clements, J.C.; Poirier, L.A.; Pérez, F.F.; Comeau, L.A.; Babarro, J.M.F. Behavioural responses to predators in Mediterranean mussels (Mytilus galloprovincialis) are unaffected by elevated pCO_2. *Mar. Environ. Res.* **2020**, *161*, 105148. [CrossRef]
39. Bebianno, M.J.; Mello, A.C.P.; Serrano, M.A.S.; Flores-Nunes, F.; Mattos, J.J.; Zacchi, F.L.; Piazza, C.E.; Siebert, M.N.; Piazza, R.S.; Gomes, C.H.A.M.; et al. Transcriptional and cellular effects of paracetamol in the oyster Crassostrea gigas. *Ecotoxicol. Environ. Saf.* **2017**, *144*, 258–267. [CrossRef]
40. Bradford, M.M. A rapid and sensitive method for the quantitation of microgram quantities of protein utilizing the principle of protein-dye binding. *Anal. Biochem.* **1976**, *72*, 248–254. [CrossRef]
41. Habig, W.H.; Pabst, M.J.; Jakoby, W.B. Glutathione S-Transferases. *J. Biol. Chem.* **1974**, *249*, 7130–7139. [CrossRef] [PubMed]
42. Beutler, E. The preparation of red cells for assay. In *Red Cell Metabolism: A Manual of Biochemical Methods*; Beutler, E., Ed.; Grune & Straton: New York, NY, USA, 1975; pp. 8–18.
43. Sedlak, J.; Lindsay, R.H. Estimation of total, protein-bound, and nonprotein sulfhydryl groups in tissue with Ellman's reagent. *Anal. Biochem.* **1968**, *25*, 192–205. [CrossRef] [PubMed]
44. Oakes, K.D.; Van Der Kraak, G.J. Utility of the TBARS assay in detecting oxidative stress in white sucker (Catostomus commersoni) populations exposed to pulp mill effluent. *Aquat. Toxicol.* **2003**, *63*, 447–463. [CrossRef]
45. Amado, L.L.; Garcia, M.L.; Ramos, P.B.; Freitas, R.F.; Zafalon, B.; Ferreira, J.L.R.; Yunes, J.S.; Monserrat, J.M. A method to measure total antioxidant capacity against peroxyl radicals in aquatic organisms: Application to evaluate microcystins toxicity. *Sci. Total Environ.* **2009**, *407*, 2115–2123. [CrossRef] [PubMed]
46. McCord, J.M.; Fridovich, I. Superoxide Dismutase. *J. Biol. Chem.* **1969**, *244*, 6049–6055. [CrossRef]

47. Bainy, A.C.D.; Saito, E.; Carvalho, P.S.M.; Junqueira, V.B.C. Oxidative stress in gill, erythrocytes, liver and kidney of Nile tilapia (Oreochromis niloticus) from a polluted site. *Aquat. Toxicol.* **1996**, *34*, 151–162. [CrossRef]
48. Zar, J.H. *Biostatistical Analysis*, 5th ed.; Prentice Hall: Upper Saddle River, NJ, USA, 2010.
49. Schveitzer, R.; Arantes, R.; Costódio, P.F.S.; do Espírito Santo, C.M.; Arana, L.V.; Seiffert, W.Q.; Andreatta, E.R. Effect of different biofloc levels on microbial activity, water quality and performance of Litopenaeus vannamei in a tank system operated with no water exchange. *Aquac. Eng.* **2013**, *56*, 59–70. [CrossRef]
50. Ekasari, J.; Angela, D.; Waluyo, S.H.; Bachtiar, T.; Surawidjaja, E.H.; Bossier, P.; De Schryver, P. The size of biofloc determines the nutritional composition and the nitrogen recovery by aquaculture animals. *Aquaculture* **2014**, *426*, 105–111. [CrossRef]
51. Avnimelech, Y. Bio-filters: The need for an new comprehensive approach. *Aquac. Eng.* **2006**, *34*, 172–178. [CrossRef]
52. Hargreaves, J.A. Photosynthetic suspended-growth systems in aquaculture. *Aquac. Eng.* **2006**, *34*, 344–363. [CrossRef]
53. Lombardi, S.A.; Harlan, N.P.; Paynter, K.T. Survival, Acid—Base Balance, and Gaping Responses of the Asian Oyster Crassostrea ariakensis and the Eastern Oyster Crassostrea virginica During Clamped Emersion and Hypoxic Immersion. *J. Shellfish Res.* **2013**, *32*, 409–415. [CrossRef]
54. Porter, E.T.; Breitburg, D.L. Eastern oyster, Crassostrea virginica, valve gape behavior under diel-cycling hypoxia. *Mar. Biol.* **2016**, *163*, 218. [CrossRef]
55. Ballesta-Artero, I.; Witbaard, R.; Carroll, M.L.; van der Meer, J. Environmental factors regulating gaping activity of the bivalve Arctica islandica in Northern Norway. *Mar. Biol.* **2017**, *164*, 116. [CrossRef] [PubMed]
56. Comeau, L.A.; Sonier, R.; Guyondet, T.; Landry, T.; Ramsay, A.; Davidson, J. Behavioural response of bivalve molluscs to calcium hydroxide. *Aquaculture* **2017**, *466*, 78–85. [CrossRef]
57. Liao, C.-M.; Jau, S.-F.; Lin, C.-M.; Jou, L.-J.; Liu, C.-W.; Liao, V.H.-C.; Chang, F.-J. Valve movement response of the freshwater clam Corbicula fluminea following exposure to waterborne arsenic. *Ecotoxicology* **2009**, *18*, 567–576. [CrossRef] [PubMed]
58. Charifi, M.; Sow, M.; Ciret, P.; Benomar, S.; Massabuau, J.-C. The sense of hearing in the Pacific oyster, Magallana gigas. *PLoS ONE* **2017**, *12*, e0185353. [CrossRef] [PubMed]
59. Møhlenberg, F.; Riisgård, H.U. Filtration rate, using a new indirect technique, in thirteen species of suspension-feeding bivalves. *Mar. Biol.* **1979**, *54*, 143–147. [CrossRef]
60. Newell, C.R.; Wildish, D.; MacDonald, B. The effects of velocity and seston concentration on the exhalant siphon area, valve gape and filtration rate of the mussel Mytilus edulis. *J. Exp. Mar. Biol. Ecol.* **2001**, *262*, 91–111. [CrossRef]
61. Riisgård, H.U.; Larsen, P.S. Physiologically regulated valve-closure makes mussels long-term starvation survivors: Test of hypothesis. *J. Molluscan Stud.* **2015**, *81*, 303–307. [CrossRef]
62. Schreck, C.B.; Olla, B.L.; Davis, M.W. Behavioral responses to stress. In *Fish Stress and Health in Aquaculture*; Cambridge University Press: Cambridge, UK, 1997; pp. 145–170.
63. Clements, J.C.; Comeau, L.A. Use of High-Frequency Noninvasive Electromagnetic Biosensors to Detect Ocean Acidification Effects on Shellfish Behavior. *J. Shellfish Res.* **2019**, *38*, 811. [CrossRef]
64. Romano, N.; Zeng, C. Toxic Effects of Ammonia, Nitrite, and Nitrate to Decapod Crustaceans: A Review on Factors Influencing their Toxicity, Physiological Consequences, and Coping Mechanisms. *Rev. Fish. Sci.* **2013**, *21*, 1–21. [CrossRef]
65. Brol, J.; Müller, L.; Prates, E.C.A.; de Farias, B.S.; Pedrosa, V.F.; de Almeida Pinto, L.A.; Sant'anna Cadaval, T.R.; Tesser, M.B.; Wasielesky, W.; Ventura-Lima, J. Dietary chitosan supplementation in Litopenaeus vannamei reared in a biofloc system: Effect on antioxidant status facing saline stress. *Aquaculture* **2021**, *544*, 737034. [CrossRef]
66. de Souza, D.M.; Borges, V.D.; Furtado, P.; Romano, L.A.; Wasielesky, W.; Monserrat, J.M.; de Oliveira Garcia, L. Antioxidant enzyme activities and immunological system analysis of Litopenaeus vannamei reared in biofloc technology (BFT) at different water temperatures. *Aquaculture* **2016**, *451*, 436–443. [CrossRef]
67. Martins, G.B.; da Rosa, C.E.; Tarouco, F.d.M.; Robaldo, R.B. Growth, water quality and oxidative stress of Nile tilapia *Oreochromis niloticus* (L.) in biofloc technology system at different pH. *Aquac. Res.* **2019**, *50*, 1030–1039. [CrossRef]
68. da Silva Martins, Á.C.; Artigas Flores, J.; Porto, C.; Romano, L.A.; Wasielesky Junior, W.; Caldas, S.S.; Primel, E.G.; Külkamp-Guerreiro, I.; Monserrat, J.M. Antioxidant effects of nanoencapsulated lipoic acid in tissues and on the immune condition in haemolymph of Pacific white shrimp Litopenaeus vannamei (Boone, 1931). *Aquac. Nutr.* **2018**, *24*, 1255–1262. [CrossRef]
69. Pellegrin, L.; Nitz, L.F.; Pinto, D.d.S.B.; Copatti, C.E.; Wasielesky, W.; Garcia, L. Effects of suspended solids in the survival and haematological parameters of pacu juveniles (Piaractus mesopotamicus) in a biofloc technology culture system. *Aquac. Res.* **2022**, *53*, 276–284. [CrossRef]
70. Zanette, J.; de Almeida, E.A.; da Silva, A.Z.; Guzenski, J.; Ferreira, J.F.; Di Mascio, P.; Marques, M.R.F.; Bainy, A.C.D. Salinity influences glutathione S-transferase activity and lipid peroxidation responses in the Crassostrea gigas oyster exposed to diesel oil. *Sci. Total Environ.* **2011**, *409*, 1976–1983. [CrossRef]
71. Sies, H. Strategies of antioxidant defense. *Eur. J. Biochem.* **1993**, *215*, 213–219. [CrossRef]
72. Sies, H.; Berndt, C.; Jones, D.P. Oxidative stress. *Annu. Rev. Biochem.* **2017**, *86*, 715–748. [CrossRef] [PubMed]
73. Winston, G.W.; Regoli, F.; Dugas, A.J.; Fong, J.H.; Blanchard, K.A. A Rapid Gas Chromatographic Assay for Determining Oxyradical Scavenging Capacity of Antioxidants and Biological Fluids. *Free Radic. Biol. Med.* **1998**, *24*, 480–493. [CrossRef]
74. Sabatini, S.E.; Rocchetta, I.; Nahabedian, D.E.; Luquet, C.M.; Eppis, M.R.; Bianchi, L.; Ríos de Molina, M.d.C. Oxidative stress and histological alterations produced by dietary copper in the fresh water bivalve Diplodon chilensis. *Comp. Biochem. Physiol. Part C Toxicol. Pharmacol.* **2011**, *154*, 391–398. [CrossRef]

75. Li, Z.; Chang, X.; Hu, M.; Fang, J.K.-H.; Sokolova, I.M.; Huang, W.; Xu, E.G.; Wang, Y. Is microplastic an oxidative stressor *Evidence from a meta-analysis on bivalves. J. Hazard. Mater.* **2022**, *423*, 127211. [CrossRef]
76. Basti, L.; Endo, M.; Segawa, S.; Shumway, S.E.; Tanaka, Y.; Nagai, S. Prevalence and intensity of pathologies induced by the toxi dinoflagellate, Heterocapsa circularisquama, in the *Mediterranean mussel*, *Mytilus galloprovincialis. Aquat. Toxicol.* **2015**, *163*, 37–5([CrossRef] [PubMed]
77. Au, D.W.T. The application of histo-cytopathological biomarkers in marine pollution monitoring: A review. *Mar. Pollut. Bul* **2004**, *48*, 817–834. [CrossRef] [PubMed]
78. Salas-Yanquin, L.P.; Navarro, J.M.; Pechenik, J.A.; Montory, J.A.; Chaparro, O.R. Volcanic ash in the water column: Physiologica impact on the suspension-feeding bivalve *Mytilus chilensis. Mar. Pollut. Bull.* **2018**, *127*, 342–351. [CrossRef]
79. Garrido, M.V.; Chaparro, O.R.; Thompson, R.J.; Garrido, O.; Navarro, J.M. Particle sorting and formation and eliminatior of pseudofaeces in the bivalves *Mulinia edulis* (siphonate) and *Mytilus chilensis* (asiphonate). *Mar. Biol.* **2012**, *159*, 987–100([CrossRef]
80. de Oliveira David, J.A.; Salaroli, R.B.; Fontanetti, C.S. Fine structure of *Mytella falcata* (Bivalvia) gill filaments. *Micron* **2008**, *39* 329–336. [CrossRef] [PubMed]
81. Zannella, C.; Mosca, F.; Mariani, F.; Franci, G.; Folliero, V.; Galdiero, M.; Tiscar, P.G.; Galdiero, M. Microbial Diseases of Bivalve Mollusks: Infections, Immunology and Antimicrobial Defense. *Mar. Drugs* **2017**, *15*, 182. [CrossRef]

Disclaimer/Publisher's Note: The statements, opinions and data contained in all publications are solely those of the individual author(s) and contributor(s) and not of MDPI and/or the editor(s). MDPI and/or the editor(s) disclaim responsibility for any injury to people or property resulting from any ideas, methods, instructions or products referred to in the content.

fishes

MDPI

Article

Circularity Assessment in Aquaculture: The Case of Integrated Multi-Trophic Aquaculture (IMTA) Systems

Daniel Checa [1,*], Brett M. Macey [2], John J. Bolton [3], Marissa Brink-Hull [3], Pauline O'Donohoe [4], Alessandro Cardozo [5], Luis Henrique Poersch [5] and Inmaculada Sánchez [1,*]

1 LEITAT Technological Center, 08225 Terrassa, Spain
2 Department of Forestry, Fisheries and the Environment, Government of South Africa, Cape Town 8001, South Africa; bmacey@dffe.gov.za
3 Department of Biological Sciences, University of Cape Town, Rondebosch, Cape Town 7701, South Africa; john.bolton@uct.ac.za (J.J.B.); marissa.brink-hull@uct.ac.za (M.B.-H.)
4 Aquaculture Fisheries Ecosystems Advisory Services (FEAS), Marine Institute, Rinville, Co. Galway, H91 R673 Oranmore, Ireland; pauline.odonohoe@marine.ie
5 Marine Aquaculture Station, Institute of Oceanography, Federal University of Rio Grande—FURG, Rio Grande 96210030, Brazil; ocalessandro79@gmail.com (A.C.); lpoersch@gmail.com (L.H.P.)
* Correspondence: dcheca@leitat.org (D.C.); isanchez@leitat.org (I.S.); Tel.: +34-93-788-2300 (I.S.)

Abstract: Aquaculture is a strategic sector that aims to meet the increased demands for healthy food for current and future populations. However, this progression needs to be sustainable, which can potentially be achieved by the implementation of circular practices. Integrated multi-trophic aquaculture (IMTA) systems promote the incorporation of circular principles. Nevertheless, the lack of harmonized definitions and standards impedes the quantification of these circular attributes. This study aims to explore the potential principles embedded in IMTA and the existing alternatives to quantify circularity. Two basic pillars (nutrient management and resource use efficiency) were identified as the most relevant circularity attributes for IMTA systems and were quantified through aquaculture-specific indicators. Bioremediation indicators, together with the efficiency indicators in terms of feed, water, energy, and infrastructure materials used, were selected to evaluate the circularity performance of four IMTA trials in three aquaculture facilities in Ireland, Brazil, and South Africa. Salmon, white shrimp, tilapia, abalone, and sea urchins were studied and cultivated together in various combinations with several low-trophic species in these IMTA trials to evaluate the improvement in circularity compared with corresponding monoculture conditions. The results showed an increase in circularity of up to 90% in terms of water recirculation, as well as bioremediation, which was improved by 80%–90%, providing evidence for the potential role of IMTA in the circularity transition.

Keywords: aquaculture; IMTA; circularity assessment; nutrients; bioremediation; resource use

Key Contribution: Specific methodology is developed to evaluate multi-trophic aquaculture (IMTA) systems from a circularity perspective. The benefits of IMTA compared to monoculture are quantified in terms of the bioremediation of nutrients and the efficient use of resources.

Citation: Checa, D.; Macey, B.M.; Bolton, J.J.; Brink-Hull, M.; O'Donohoe, P.; Cardozo, A.; Poersch, L.H.; Sánchez, I. Circularity Assessment in Aquaculture: The Case of Integrated Multi-Trophic Aquaculture (IMTA) Systems. *Fishes* **2024**, *9*, 165. https://doi.org/10.3390/fishes9050165

Academic Editors: Wagner C. Valenti, Eduardo Luis Ballester and Domitília Matias

Received: 14 March 2024
Revised: 22 April 2024
Accepted: 1 May 2024
Published: 4 May 2024

Copyright: © 2024 by the authors. Licensee MDPI, Basel, Switzerland. This article is an open access article distributed under the terms and conditions of the Creative Commons Attribution (CC BY) license (https://creativecommons.org/licenses/by/4.0/).

1. Introduction

Population growth leads to an increased demand for food, while the pressure on the environment due to the intensification of food sector activities highlights the need for a more efficient use of natural resources. Aquaculture has a key role to play in feeding the growing population, but the sustainability of production systems must be ensured to meet the increasing demand for healthy aquatic food [1]. However, addressing sustainability in aquaculture is complex not only due to the existence of multiple frameworks for its evaluation but also the great variety of production systems to which sustainability approaches and tools can be applied [2].

To mitigate the negative effects on the environment, extensively studied ecological approaches promote aquaculture production designs that reduce nutrient discharges [3]. In addition to reducing environmental impacts, increasing production efficiency is the other main aspect by which the circular economy drives sustainable development, as recognized in policies put in place at the European level [4–6]. To improve environmental performance, as part of the strategy for the sustainable development of aquaculture, the European Commission recommends adopting the circular approach to waste management including treating waste as a resource [7].

Although the principles of the "circular economy" have been expressed in many ways depending on the point of view [8], a definition based on examples would be a good approach to harmonize the concept. In this sense, the circular economy is a model of production and consumption where cleaner and more competitive practices focus on saving and recovering resources [4]. As part of the recommendations provided by Balsells et al. [9], a harmonized definition of the circular economy should address the importance of biological flows and the role of aquaculture in producing renewable biological resources.

In addition to the definition, the range of examples is also very extensive for aquaculture, and no previous studies are available that provide the state of the art of the real and current implementation of the circular economy in the sector [10]. The concept of circular aquaculture can be addressed in many ways [11], but generally, circularity in aquaculture includes the adoption of practices regarding waste management [12–14], the recycling of nutrients [15–18], or the incorporation of novel ingredients in feeds derived from the bio-economy [19,20].

Evidence for the alignment of aquaculture systems with the circular economy comes from the study of resource use efficiency and nutrient management, which are measured as the capacity to collect and use excess nutrients and uneaten feed fractions. In this regard, integrated multi-trophic aquaculture (IMTA) is a circular production system where different species such as fish, shellfish, and seaweed are strategically integrated/linked to create a symbiotic relationship that contributes to reducing the impacts on the environment [21,22]. Within IMTA systems, uneaten feed and nutrient losses can be recaptured by other organisms and converted into valuable nutrients for harvestable seafood and crops [23]. Fish excrete nitrogenous and phosphorus waste, the dissolved component of which can be utilized by cultivated seaweed species as a nutrient source for growth, thus reducing nutrient levels in the water and preventing eutrophication [13]. Additionally, filter-feeding organisms such as mussels and oysters can help in the removal of excess nutrients as particulate matter by efficiently filtering water [24]. Successful integrated systems require not only that the appropriate species are selected but also that waste from the fed species can be efficiently utilized by tolerant species [25]. For effective IMTA, bioremediation is interpreted as a circular attribute of the system, as it promotes the recycling of nutrients. Bioremediation capacity plays a crucial role in IMTA systems by utilizing the natural metabolic activities of various organisms to remediate and improve water quality [26–29]. This approach promotes a circular economy within aquaculture, where waste from one component becomes a valuable resource for another [30].

Following the premise that "what gets measured gets managed" [31], the evaluation of circularity performance at the farm level could provide useful information to manage and maximize the bioremediation capacity of IMTA systems. At the same time, the potential nutrient mitigation service achieved by IMTA may be of interest for possible economic benefits, but this is limited, among other reasons, by the lack of common standards to verify this mitigation. Standardized measurements of circularity could provide evidence for how well circular economy principles are applied to products [32].

A good approach for the evaluation of aquaculture products is provided by Valentí et al. [33], who defined quantitative indicators that address the three dimensions of sustainability (economic, environmental, and social) through 56 indicators. This framework aims to evaluate relevant aspects of sustainability by quantifying the efficiency of natural resource use, which would perfectly link to the evaluation of some circularity attributes. Complementary aspects such as natural resource depletion and the useful service lifetime should be considered essential indicators [34] at the product or material level. However, there is no agreement regarding the most appropriate framework for measuring circularity indicators [35,36].

Linder et al. compared different product-level circularity metrics, concluding that none of the existing initiatives scored highly across the criteria of validity, reliability, transparency, and generality [31]. As part of the metric studied by Linder et al. [31], the Material Circularity Indicator (MCI) was identified as an appropriate approach for the evaluation of product circularity. Although the revised version (2019) included biological cycles, the MCI was originally developed to measure the circularity of technical products [37]. Even so, it is relatively difficult to associate concepts such as utility (referring to durability or usage intensity) and lifespan to biological products. Moreover, the MCI refers to the use phase, which has a different interpretation depending on the focus and objective of the circularity analysis. Regarding aquaculture, feed is used (ingested) and assimilated by the fish, and then feed flow is transformed into nutrients that are released into the environment in the form of excretion and feces, in addition to the uneaten fraction. The concept of waste from the use phase within the MCI approach would refer not to feed but to the release of nutrients, although, if the focus is at the nutrient level, circularity can be addressed though the evaluation of nutrients entering and leaving the system during the aquaculture production period.

In addition to the fact that most circular economy (CE) metrics focus on the technical cycle and materials from non-renewable resources [35], none of the approaches reviewed at the farm level fulfill the particularities of aquaculture (even less of integrated multi-trophic production). The diversity of aquaculture systems makes it challenging to determine the circular profile of the sector using a single approach.

Life cycle assessment (LCA) is a very appropriate approach for measuring the environmental footprint for products, processes, or services. The International Organization for Standardization (ISO) provides the guidelines to evaluate environmental impacts through indicators such as carbon footprint, acidification, and land use, among others (ISO 14040 [38], 14044 [39]). When LCA is applied to aquaculture processes, the environmental impact assessment potentially reflects circularity in terms of nutrient management or resource use (through the study of the material resource depletion impact category). Furthermore, LCA studies with a broader approach ("from cradle to gate") would potentially inform circularity beyond nutrients at the production level, encompassing the whole value chain. From a life cycle assessment perspective, the EU Environmental Footprint initiative [40] defines a circular footprint formula (CFF) to estimate emissions from processes involving recycling and energy recovery. However, as with the MCI, the CFF could be difficult to apply to measure the circularity of aquaculture production.

Given this context, the purpose of this work is to explore the potential principles embedded in IMTA through the definition of specific indicators that are adapted for and focused on IMTA production. This paper presents the specific case of three different IMTA production systems (known as IMTA laboratories) that were evaluated by applying sector-specific circularity indicators.

2. Materials and Methods

2.1. IMTA Laboratories

This study included four trials that were addressed through 3 IMTA laboratories. The term laboratory (lab) refers to large-scale demonstration sites in real environments, where new species and different combinations were investigated in the context of the EU ASTRAL project (GA 863034) (https://cordis.europa.eu/project/id/863034, accessed on 1 March 2024), which aims to develop new, sustainable, profitable, and resilient value chains for IMTA production within the framework of existing, emerging, and potential Atlantic markets.

Table 1 shows the IMTA systems that were evaluated following the methodology detailed in Section 2.2.

Table 1. Study scenarios.

Case Study	Monoculture	IMTA	
Irish IMTA lab	Salmon marine cages	Salmon marine cages with seaweed longlines and urchin and oysters in baskets	
Brazilian IMTA lab	White shrimp in ponds	White shrimp with tilapia and seaweed in closed BFT	
South African IMTA lab	Semi-closed abalone system	Semi-closed abalone and seaweed system	
	Semi-closed urchin system	Semi-closed urchin and seaweed system	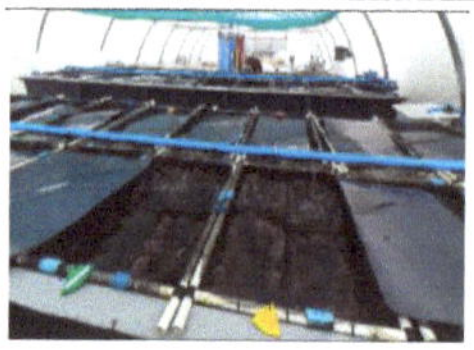

2.1.1. Irish IMTA Lab

The research at this IMTA lab, managed by the Marine Institute, was focused on multi-trophic production involving salmon (*Salmo salar*), seaweed (two kelp species, *Alaria esculenta* and *Saccharina. latissima*), urchin (*Paracentrotus lividus*), and oyster (*Ostrea edulis*). The primary objective of this lab was to investigate the capacity of these combined species to create a self-sustaining ecosystem that efficiently utilizes and recycles nutrients, thereby minimizing eutrophication and other negative effects on benthic zones. Bioremediation processes within this IMTA model were examined to understand the natural capabilities of the system to mitigate nutrient release. Salmon, known for producing nutrient-rich waste, served as the primary fed species in this IMTA setup. Salmon waste, which is particularly rich in nitrogen and phosphorus, becomes a valuable resource for secondary extractive species. Seaweeds were cultivated to absorb and utilize these excess soluble nutrients, contributing to the purification of the water and their own biomass production. Simultaneously, urchins and oysters were integrated into the system. Urchins were grown using the seaweed produced on-site, while oysters acted as filter feeders, removing particulate matter and enhancing water quality. The presence of the extractive species complements the nutrient cycling process and further contributes to maintaining a balanced ecosystem. Figure 1 shows the laboratory configuration located in Bertraghboy Bay (Connemara, Galway, Ireland).

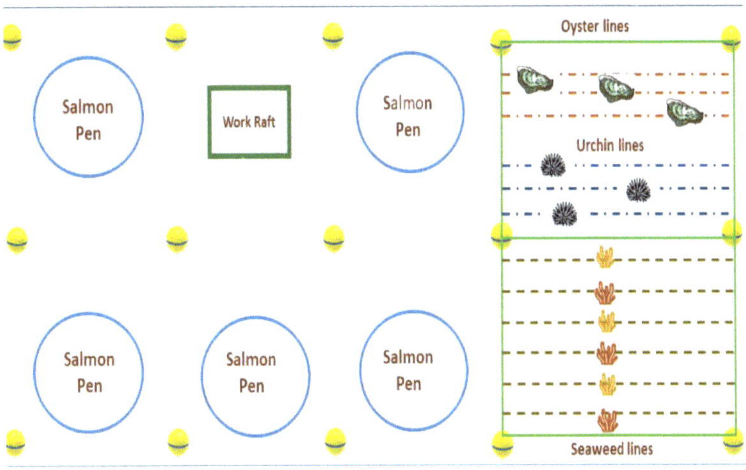

Figure 1. Irish IMTA lab design: the low-trophic grid with oysters, seaweeds, and urchins lies adjacent to the salmon pens within the licensed aquaculture site.

The circularity assessment of the Irish IMTA lab focused on the evaluation of the experimental trial, with a production of 50 t of salmon (FW/y), 2.1 t of seaweed (FW/y), 75.6 kg of oysters (FW/y), and 64 kg of urchins (FW/y).

2.1.2. Brazilian IMTA Lab

The research at this IMTA lab, carried out by the Federal University of Rio Grande (FURG) located in the city of Rio Grande in Brazil, with Biofloc technology (BFT), focused on the optimization of white shrimp (*Litopenaeus vannamei*) production in a Recirculating Aquaculture System (RAS) (Figure 2).

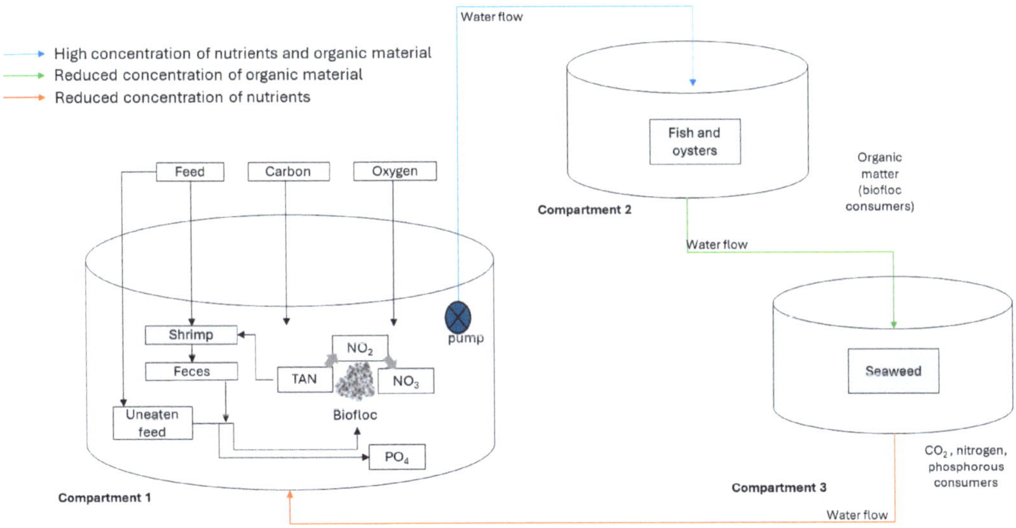

Figure 2. Brazilian IMTA lab design.

BFT entails the cultivation of dense populations of microorganisms in the water, creating a dynamic environment where the nitrogen excreted by fed species is converted into microorganism biomass (biofloc) under aeration and carbon fertilization. The biofloc growth is stimulated by the addition of active carbon (molasses) and constant aeration. The experiments carried out by this lab aimed at determining the most effective biomass ratios among marine white shrimp, tilapia, and algae, particularly evaluating the impacts of tilapia (*Oreochromis niloticus*) and sea lettuce (*Ulva lactuca*) in regulating organic material and nutrients. Moreover, the lab tested the performance of the system when fish meal was substituted by a fish meal analogue (FMA) in the shrimp diet formulation. The FMA used in this study was developed by Guabi Nutrition and Animal Health S.A. and consisted of a balanced blend of terrestrial animal by-products supplemented with amino acids, minerals, and commercial vitamins (Table 2). The animals were fed 2 times per day with the isoprotein and isoenergetic diets (feed pellets with 38% crude protein and 7% lipids) and FMA at a 50/50 ratio.

Table 2. Fish meal analog properties (used to prepare the shrimp diet).

FMA Composition	%
Crude protein	47.4
Lipids	16.3
Linoleic acid	2.1
Phospholipids	3.5
Cholesterol	0.3
HUFA	1.7
Fiber	0.9
Ash	16.2
Calcium	5.7
Total phosphorus	3.0
Total lysine	3.9
Total cysteine	0.9
Total methionine	1.4

This IMTA was designed to maximize the efficiency of nutrient utilization and circularity within the system [41,42]. White shrimp served as the primary species and produced organic waste and nutrients as by-products of their metabolic processes. These nutrients included nitrogen and phosphorus compounds. Nutrient-rich water was recirculated among the shrimp, tilapia, and seaweed tanks. Tilapia is a marketable species capable of consuming excess bioflocs, and *Ulva* absorbs excess dissolved nutrients, including nitrogen and phosphorus, from water [43], contributing to the overall nutrient balance [44]. This not only aids in water quality management but also provides an additional valuable product that can be harvested for various applications, such as biofuel, or as a nutrient-rich supplement for aquaculture feed [43].

The Brazilian IMTA lab circularity assessment was focused on the evaluation of the experimental trial, with a production of 4.3 kg/m^3 of FW of shrimp, 13.4 kg/m^3 of FW of tilapia, and 2 kg/m^3 of FW of seaweed.

2.1.3. South African IMTA Lab

IMTA research at the lab in South Africa consisted of two experimental systems (Figure 3). The first system is a fully commercial integrated abalone–*Ulva* system at Buffeljags Abalone farm. The farm is managed by Viking Aquaculture and is located approximately 200 km east of Cape Town on a pristine stretch of coastal land near the remote settlement of Buffeljags on the Cape southwest coast. The second system is an urchin-*Ulva* pilot commercial-scale experimental system on the same farm. Data from the experimental trials conducted in the abalone and urchin systems were used to perform the circularity assessments.

The abalone (*Haliotis midae*)-*Ulva lacinulata* IMTA system consists of seven modular platforms, each comprising four clusters. Each cluster is made up of a 150 m^3 D-ended *Ulva* paddle raceway and multiple abalone raceway tanks (each 8.5 m^3) arranged in six rows, with each row housing seven abalone raceway tanks. The effluent water from the abalone tanks in each cluster flows into the adjacent *Ulva* paddle raceway, where it is bioremediated (removal of N and P) by the *Ulva* [45] and mixed with 50% fresh seawater in the sump before being returned to the abalone raceway tanks. The continuous circulation of seawater through these tanks and *Ulva* paddle raceways ensures a steady supply of cool and aerated water for the growing abalone. Experiments in the SA IMTA lab were carried out to monitor the physical and chemical parameters of the abalone–*Ulva* IMTA system with increasing (50, 75, and 100%) recirculation rates [27].

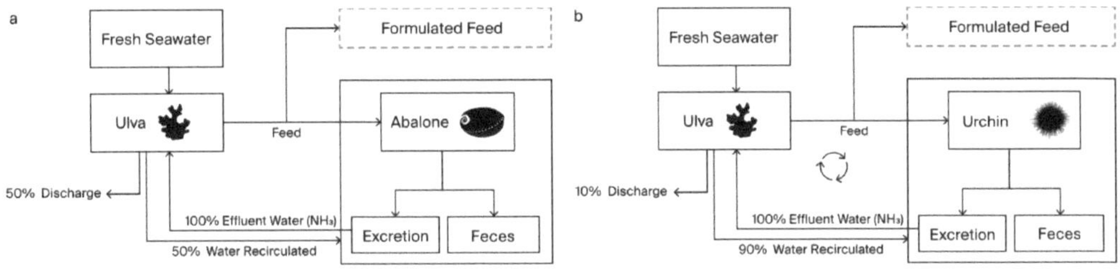

Figure 3. South African IMTA lab: (**a**) abalone/*Ulva* system; (**b**) urchin/*Ulva* system.

Operating independently but at the same farm, the second system consisted of an integrated urchin (*Tripneustes gratilla*)-*Ulva lacinulata* system that consisted of five 58,000 L glass fiber tanks linked to a 14,000 L *Ulva* paddle raceway and a 14,000 L sump. Effluent water from each of the sea urchin tanks was directed into the *Ulva* paddle raceway (where nutrients were reduced by means of the bioremediation of the seaweed), having passed through the drum filter for the removal of larger particulates. The water replacement in this system, with fresh incoming seawater from the adjacent ocean, was 10% per day.

2.2. Circularity Assessment Methodology

As part of the present study, nutrient management and the use of resources were identified as the two principal attributes through which IMTA systems contribute most to circularity. In line with Chary et al. [10], nutrients and resource use are fundamental pillars embedded in the principles applicable to aquaculture (safeguard and regenerate, avoid, prioritize, reuse and recycle, and entropy). In this sense, a set of metrics was defined for the evaluation of nutrient recycling and resource use efficiency, with the aim of providing a comparison of circularity performance between the monoculture and IMTA scenario for each laboratory.

2.2.1. Nutrient Management Metrics

Nutrient management performance was quantified as part of the evaluation of the primary function of the IMTA system, which was bioremediation. Within the circularity assessment context, bioremediation is defined as a circularity indicator that provides information on the nutrient retention efficiencies (treating waste nutrients from fed species as by-products that can be recycled by extractive species). Seaweeds absorb dissolved inorganic nutrients (DINs), shellfish filter out suspended particulate organic matter (POM), and deposit feeders such as holothuroids and polychaetes consume settled particulate organic matter.

Various approaches exist to quantify bioremediation, including the measurement of nutrient removal rates, retention capacities, nutrient balances, the utilization of tracers (e.g., stable isotopes or fatty acid composition), and modeling techniques [16]. This section describes the methodology used to calculate the bioremediation indicator through the different trials in both scenarios, monoculture, and IMTA, with the objective of quantifying this circularity pillar.

- Bioremediation in the Irish IMTA lab

Salmon nutrient emissions were calculated in accordance with the methodology for mass balance calculation developed by Wang et al. [15], which enables the estimation of N, P, and C emissions to the surrounding water from the fed fish. N and P assimilated by seaweed were estimated from elementary analysis in the laboratory, while C was calculated considering the carbon fixation and storage potential of macroalgae [46]; N and P taken up by oysters were estimated based on laboratory nutritional analysis, while C was calculated from nutrient assimilation efficiencies in integrated multi-trophic aquaculture in accordance with the literature [47]. The same reference (Nederlof et al. [16]) was used to estimate N, P and C absorption by urchins, which was based on the theoretical assimilation efficiency.

- Bioremediation in the Brazilian IMTA lab and SA IMTA lab (abalone)

For these 2 trials, bioremediation was calculated using monitored water flow data and information on nutrient concentrations across both closed and semi-closed experimental systems. Regarding the Brazil IMTA lab, one-year total solids and nutrient monitoring data were used as a basis for the evaluation of the nutrient management pillar (bioremediation indicator). Concentrations at the inlet and outlet of the shrimp, tilapia, and seaweed tanks were used to estimate the bioremediation potential for N and P.

Regarding abalone in the SA IMTA lab system, bioremediation was calculated directly from monitoring the water flows [48]. Information on total ammonia nitrogen (TAN), nitrate, nitrite, and phosphate concentrations in the inlet flow in *Ulva* raceways was compared to concentrations in the effluent water from abalone raceways to evaluate the bioremediation potential of *Ulva* [27].

- Bioremediation in the SA IMTA lab urchin trial

Urchin N emissions were calculated based on feed nutritional values; the nutrient balance was calculated based on the protein provided in feed to the N retained by urchins. The nitrogenous effluent from urchins, meaning the non-retained fraction, was used to calculate DINs in the effluent and the corresponding bio-available fraction for seaweed. Afterward, the DIN uptake efficiency of *Ulva* was estimated to finally determine the DIN balance between urchin and *Ulva* effluents.

2.2.2. Resource Use Efficiency

The use of resources was quantified through metrics that reflect the efficiency of IMTA with respect to the use of the most common resources in aquaculture, as reviewed in previous LCA studies [49]. Efficiency in the use of resources was studied under monoculture and IMTA conditions through the quantification and comparison of specific indicators related to feed, water, energy, and materials.

Regarding feed, the Feed Conversion Ratio (FCR) and the circularity of the formulations were defined as the most appropriate indicators (Equations (2) and (3) in Table 3). The former was interpreted as equivalent to the use of resources, in the sense that the higher the FCR, the more feed was required to produce one unit of fed-species and the less circular the culture. FCR considered the feed ingested when it was feasible to measure the uneaten fraction (in the case of the South African IMTA labs) and the feed delivered when ingestion was unknown (in the case of the Brazilian IMTA lab). The latter was measured based on the linearity concept adapted to aquaculture feeds [50], which reflects the percentage of ingredients that are sourced from the upcycling of products within the aquaculture systems. Whenever low-trophic species harvested from IMTA schemes were processed and incorporated as a feed ingredient in the trials under study, they were interpreted in the assessment as non-linear ingredients sourced from valorization routes. This was the case for the SA IMTA lab (urchin), where dried *Ulva* served as a feed ingredient in formulated feed for urchins.

Regarding water, the recirculation efficiency was quantified to reflect the dependency on external water sources, and it was thus an indicator applicable to semi-closed and closed systems (Equation (4) in Table 3). This indicator is particularly relevant for freshwater aquaculture systems and is therefore potentially of interest for the evaluation of RAS systems (90% of systems operating under freshwater conditions in Europe [51]). Although seawater is not considered a limited resource from a life cycle perspective [52], water recirculation in coastal areas was selected as a relevant indicator to measure not only the circularity but also the resilience capacity of farms, which is especially important when coastal sites are unable to pump seawater when toxic microalgal blooms occur.

Table 3. Circularity indicators to measure IMTA performance.

Pillar	Indicator Name	Indicator Formula	
Nutrient management	Bioremediation	$\frac{\text{C,N,P emitted by fed species}}{\text{C,N,P taken up by extractive species}}$	(1)
Resource use efficiency	Feed	$\text{FCR}\left[\frac{\text{Total feed}}{\text{biomass weight gain}}\right]$	(2)
		$\frac{\text{Total mass ingredients (g) from valorization routes}}{\text{Total mass of feed delivered}} \times 100$	(3)
	Water	$\frac{\text{Water recirculated } (m^3)}{\text{Water in the system } (m^3)} \times 100$	(4)
	Energy	$\frac{\text{kWh consumed}}{\text{biomass harvested}}$	(5)
	Infrastructure	$\frac{\text{kg infrastructure materials}}{\text{biomass harvested}}$	(6)

Likewise, the energy indicator (Equation (5) in Table 3) was measured through the estimation of energy used by each experimental IMTA lab trial in relation to the biomass harvested. Maintenance during the production phase (cleaning pump and foam fractionator) was aggregated as part of the energy indicator and interpreted in terms of kWh. Both water and energy indicators were defined to reflect the intensity of resource use but not the associated potential environmental impacts, which would be the objective of an LCA study.

Finally, the infrastructure indicator (Equation (6) in Table 3) refers to materials needed for the IMTA set ups. The manufacturing of infrastructure components involves the extraction of resource materials, especially steel- and fossil-based plastics. This indicator was defined to obtain evidence on whether integrated systems promote synergies in the use of common elements (for example, floating infrastructure) that potentially reduce the use of materials relating to the biomass harvested. In fed-species systems with low-trophic species integration, the new materials required are less than those needed in a monoculture scenario due to the shared use of infrastructure to produce more than one species. Therefore, the increase in total biomass was interpreted as a major functionality. Functionality was defined as the practicality of the infrastructure for the purpose of enabling biomass growth. The infrastructure indicator was thus calculated as the total kg of materials within each infrastructure element in relation to the total biomass produced over its lifespan. Therefore, this indicator was represented as kg of infrastructure materials/kg of biomass harvested (Equation (6) in Table 3).

Table 3 shows the pillars, aspects, and indicators that were measured for the IMTA laboratories. These indicators were defined to quantify the circular attributes that reflect the alignment of the systems with the circular principles. The evaluation of circularity performance therefore consisted of assessing the variations in these indicators between the IMTA and monoculture scenarios.

3. Results

3.1. Irish IMTA Lab

3.1.1. Nutrient Management

The bioremediation indicator was calculated through the evaluation of nutrient emissions from salmon, which were differentiated between dissolved and particulate matter. Table 4 shows the nutrient balance obtained from this experimental laboratory, in which seaweed species mitigated the emissions of the dissolved nutrient fraction, oysters absorbed particulate nutrient fractions, and urchins utilized the seaweed grown on-site. Information on the excretion and respiration of the low-trophic animals was not available and therefore excluded from the analysis.

Table 4. Nutrient balance in the Irish IMTA lab (all values refer to 1 kg of biomass—WWT).

	Emitted	Absorbed			
Nutrients (kg)	**Salmon** [(1)]	***Alaria***	***Saccharina***	**Oyster**	**Urchin**
Dissolved N	7.18×10^{-2}	1.40×10^{-4} [(2)]	8.05×10^{-5} [(2)]	-	-
Dissolved P	1.82×10^{-2}	1.97×10^{-4} [(2)]	1.97×10^{-4} [(2)]	-	-
Dissolved C	1.14×10^{-2}	1.77×10^{-3} [(3)]	1.28×10^{-3} [(3)]	-	-
Particulate N	5.57×10^{-3}	-	-	1.12×10^{-6} [(2)]	1.89×10^{-5} [(4)]
Particulate P	1.03×10^{-1}	-	-	2.99×10^{-6} [(2)]	1.89×10^{-6} [(4)]
Particulate C	1.82×10^{-2}	-	-	6.40×10^{-5} [(4)]	1.08×10^{-4} [(4)]

[(1)] Sourced from [15]; [(2)] sourced from the nutritional composition (laboratory analysis); [(3)] sourced from [47]. [(4)] sourced from [16].

3.1.2. Resource Use Efficiency

Fuel consumption refers to the petrol needed for maintenance and harvesting activities under both the monoculture and IMTA conditions. The energy within fuel consumed under the monoculture conditions was 2.53×10^{-1} kWh/kg of biomass harvested, whereas the energy within the fuel consumed under IMTA conditions was slightly lower (2.52×10^{-1} kWh/kg of biomass harvested), as derived from a previous LCA study [53].

Regarding the materials embedded in infrastructure elements, the same LCA [53] study identified the use of steel, concrete, and wood as the main materials used on the farm. The total amount of materials involved under the monoculture conditions was 0.04 kg of materials per kg of biomass harvested, while 0.05 kg of materials was used per kg of biomass harvested in IMTA.

3.2. *Brazilian IMTA Lab*

3.2.1. Nutrient Management

Bioremediation for this closed system was estimated through the monitoring of inlet and outlet nutrient concentrations. Nutrient monitoring for one year (Table 5) covered three production cycles for shrimp, two production cycles for tilapia, and four production cycles for seaweed. In the shrimp tank, the levels of ammonia, nitrate, nitrite, and phosphate increased due to the metabolic processes, while the total suspended solids (TSSs) increased due to the biofloc formation. The tilapia tank showed a similar pattern, where ammonia, nitrite, nitrate, and phosphate concentrations increased due to the biological outcomes of the nutrient assimilation of the fish, but TSSs decreased due to the capability of tilapia to eat the biofloc generated in the previous tank. Finally, the seaweed tank showed the process of bioremediation itself with nutrient retention and decreasing levels of dissolved ammonia, nitrite, nitrate, and phosphate; TSSs decreased due to the biological structure of *Ulva*, which retained part of the biofloc in the tank and was removed during the *Ulva* harvesting.

Table 5. Nutrient balance in the Brazilian IMTA lab (average values from the monitoring period).

	Shrimp Tank		Tilapia Tank		*Ulva* Tank	
Parameter	**Inlet (mg/L)**	**Outlet (mg/L)**	**Inlet (mg/L)**	**Outlet (mg/L)**	**Inlet (mg/L)**	**Outlet (mg/L)**
Ammonia	0.80	1.00	1.00	1.50	1.50	0.80
Nitrate	13.00	15.00	15.00	16.00	16.00	13.00
Nitrite	0.80	1.00	1.00	1.30	1.30	0.80
Phosphate	1.00	4.00	4.00	4.10	4.10	1.00
TSS	380.00	450.00	450.00	400.00	400.00	380.00

3.2.2. Resource Use Efficiency

For shrimp monoculture conditions, experiments carried out at the same research center indicated an FCR of 1.65 in ponds [54], which decreased to 1.34 in the BFT system integrated with tilapia [55]. Traditionally, for tilapia monoculture, the observed FCR ranged from 1.70 to 1.80 [5] and was reduced to practically half (0.85) in an integrated BFT system of marine shrimp and tilapia [55]. Therefore, the feed conversion of shrimp integrated with tilapia was improved in the BFT system compared to conventional farms [54,56], leading to increased resource use efficiency in terms of feed. Moreover, the substitution of fish meal in the conventional feed by by-products of the poultry industry in the shrimp diet implied that the linearity of the feed was reduced. Conventional fishmeal integrated 5% of fish (the source with no by-products), while FMA was totally based on by-products from the poultry industry. In particular, the system performance was not affected when 50% of the fish meal was replaced with FMA. The strategy of using poultry by-products had a positive effect on circularity performance, reducing the linearity within the fishmeal by 50%.

In terms of energy use, BFT systems imply the aeration of culture tanks to ensure the homogeneous distribution of nutrients and biofloc while avoiding biofloc deposition. BFT aeration equipment increases the energy demand compared to a pond system (the baseline of this study); therefore, energy consumption is a disadvantage of BFT [55].

To address the circularity of infrastructure materials, the kg of materials per kg of biomass harvested was compared between monoculture and IMTA conditions. To do so, components and elements identified in the LCA study [53] were taken as the basis for the indicator. For the monoculture conditions, 1.01 kg of materials per kg of biomass harvested was estimated, while the same indicator for BFT conditions resulted in 0.37 kg of materials per kg of biomass harvested.

3.3. *South African IMTA Lab (Abalone System)*

3.3.1. Nutrient Management

The bioremediation indicator was calculated based on the information sourced from Geldart, 2022 [48], who provided data on nutrient fluctuations in the system operating at 50%, 75%, and 100% recirculation (for the short term). The average concentration results from the 100% recirculation period are shown in Table 6.

Table 6. Nutrient balance in the South African IMTA lab (abalone).

	Abalone Tank		*Ulva* Tank	
Parameter	**Inlet (mg/L)**	**Outlet (mg/L)**	**Inlet (mg/L)**	**Outlet (mg/L)**
Ammonia	0.29	0.41	0.41	0.29
Nitrate	1.16	1.34	1.34	1.16
Nitrite	0.20	0.23	0.23	0.20
Phosphate	0.52	0.58	0.58	0.52

3.3.2. Resource Use Efficiency

When the abalone–*Ulva* IMTA system was implemented, the feed regime was similar to the abalone monoculture (with formulated feed for 4 days and wild harvested kelp (*Ecklonia maxima*) for 3 days per week), but the wild-harvested kelp was substituted whenever IMTA-grown *Ulva lacinulata* was available. This means that the formulated feed was provided for 4 days per week, while wild-harvested kelp was provided for 1 day, and IMTA-grown *Ulva lacinulata* was fed for 2 days per week. This feeding regime was assumed for the abalone–*Ulva* IMTA system for the purpose of this comparative analysis; however, it should be noted that in practice, the choice of seaweed depends on what is available at the time.

Combining the feeding regime with the specific FCR related to the different feed types (see Table 7), the FCR for the abalone monoculture system was 6.16, but when the abalone–*Ulva* IMTA system was implemented, the FCR was reduced to 4.01. Uneaten fractions also decreased from 1.14% to 0.71%, while the linearity of the feed decreased from 100% (no ingredient sourced from valorization routes) for the abalone monoculture system to a linearity of 71.42% for the abalone–*Ulva* IMTA system.

Table 7. Use of resources: Feed indicator, SA IMTA lab (abalone).

Production Systems	Feeding Regime	Feed Description	Ingredient (Description and Origin)	FCR
Abalone monoculture and Abalone–*Ulva*	Feed 1	Formulated feed from specialized aquatic feed (SAF)	Not available	1.4
	Feed 2	Wild-harvested kelp (*Ecklonia maxima*)	Wild-harvested	12.5
Abalone–*Ulva*	Feed 3	IMTA-grown *Ulva lacinulata*	Grown in IMTA	5

In terms of water, 50% water recirculation at the Buffeljags abalone farm played a key role in enhancing the circularity of the abalone–*Ulva* IMTA system, as it allowed for a significant reduction in the reliance on fresh incoming seawater from the adjacent ocean. This promotes resource efficiency as the system recirculates a substantial portion of the water, minimizing the demand for external resources.

Moreover, the reduced dependence on the pumping of fresh seawater contributes to energy efficiency. By recirculating 50% of the water in the system, enabled by the bioremediation (ammonia removal) capacity of *Ulva*, the system minimizes the energy expenditure associated with bringing in new seawater. The energy consumed under abalone monoculture conditions was 10.35 kWh per ton of biomass harvested, but when IMTA conditions were implemented, the energy consumption was reduced to 6.80 kWh per ton of biomass harvested. The infrastructure indicator was not calculated for the abalone–*Ulva* trial since data were not available.

3.4. *South African IMTA Lab (Urchin System)*

3.4.1. Nutrient Management

To estimate bioremediation, urchin N emissions were calculated based on the feed characteristics. Total urchin biomass (9.5 t) was fed with a pelleted formulated feed at 1.5% BW per day for 4 days a week over the entire 7-month production cycle. During IMTA, urchins were fed fresh *Ulva* at 6% BW per day for 3 days a week for 4 months and with pellets at 1.5% BW per day for 4 days a week for the remaining 3 months of the production cycle; the latter enhanced the production of the final product (the gonad). Under monoculture conditions, 17.08 t of formulated feed per production cycle was provided, with a total of 257 g of protein per kg of feed, resulting in 0.70 t DINs per production cycle with the 0.16 protein-to-N conversion factor [56]. Considering the urchins' potential N retention of 10.52%, approximately 0.63 t DINs remained in the system. The dissolved bio-available N in water was 60%, so the total N not retained by urchins was 0.38 t of N per production cycle.

The *Ulva* supplied for urchin feeding over the production cycle was 68.31 t (WW) per IMTA production cycle. The total amount of formulated feed provided was 7.32 t per IMTA production cycle. The N supplied from feed was 0.59 t of N per IMTA production cycle. The DIN uptake efficiency of *Ulva* was 80%, so the N retained by *Ulva* was 0.25 t (Figure 4).

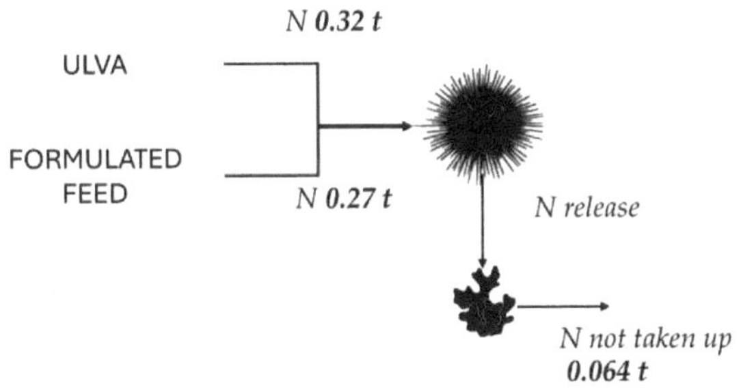

Figure 4. N mass balance in the urchin-*Ulva* system.

3.4.2. Resource Use Efficiency

In terms of feed, the FCR for formulated feed in monoculture was 0.4 (formulated feed (pellets) containing 20% DW Ulva [57]), and the FCR related to the IMTA conditions was 0.91, in accordance with the feeding regime described in Table 8 (4 months with fresh *Ulva* and 3 months with formulated feed including 20% of dried *Ulva*). The amount of uneaten feed was 2% for monoculture and 1.14% for IMTA, and the linearity of the feed was reduced from 80% under the monoculture conditions to 34.3% when IMTA was implemented.

Table 8. Use of resources: Feed indicator, SA IMTA lab (urchin).

Production System	Feeding Regime	Feed Description	Ingredient (Description and Origin)	% (Dry Weight)	FCR
Urchin monoculture	Feed 1	Formulated feed supplemented with 20% dried *Ulva lacinulata*	Wheat bran	0.321	0.4
			Maize (extruded)	0.321	
			Fish meal	0.153	
			Soybean	0.153	
			Di-Calcium phosphate	0.0184	
			Lecithin (de-oiled)	0.0138	
			Vitamin and mineral premix	0.011	
			Fish oil	0.00963	
Urchin-*Ulva*	Feed 1	Fresh *Ulva lacinulata* for the first 4 months of the production cycle	Not applicable		1.3
	Feed 2	Formulated feed supplemented with 20% dried *Ulva lacinulata* for the final 3 months of the production cycle	Wheat bran	0.321	0.4
			Maize (extruded)	0.321	
			Fish meal	0.153	
			Soybean	0.153	
			Di-Calcium phosphate	0.0184	
			Lecithin (de-oiled)	0.0138	
			Vitamin and mineral premix	0.011	
			Fish oil	0.00963	

Regarding water, implementing a high level of seawater recirculation (from 0 to 90%) in the urchin-*Ulva* IMTA system offered notable benefits in terms of the water conservation of pumped seawater resources, which is crucial for sustainable aquaculture. The provision and use of seawater are major cost components in running a land-based aquafarm. Seawater provision is a critical limiting factor in farm siting and sustainability.

With 90% seawater recirculation, the need to pump large volumes of seawater from external sources was minimized. This reduction in water exchange led to lower energy requirements for pumping, contributing to overall energy savings. The energy consumption under urchin monoculture conditions was 12.45 kWh/t of biomass harvested compared with 3.54 kWh in the urchin-*Ulva* IMTA system.

Additionally, regarding the resource use indicator, the infrastructure elements (*Ulva* D-shaped paddle raceway, pipes and fittings, pumps, and baskets, among others) were identified as part of the SA IMTA lab LCA study [53]. The monoculture and IMTA were compared considering the infrastructure material weight per kg of biomass harvested, resulting in 1.26 kg of material per kg of biomass harvested in monoculture and 0.42 kg of material per kg of biomass harvested in IMTA.

3.5. Circularity Performance of IMTA Labs

The difference between the indicators for monoculture and IMTA was calculated to evaluate the circularity performance of the experimental sites. Whenever the difference between indicators was positive, an improvement in circularity was interpreted (Table 9).

Table 9. Circularity performance interpretation.

Indicator Name	Circularity Performance When:
Bioremediation	$(\text{BIOREMEDIATION C, N, P})_{\text{IMTA}} - (\text{BIOREMEDIATION C, N, P})_{\text{Monoculture}} > 0$
Feed	$\left[\frac{FCR_{\text{monoculture}} - FCR_{\text{IMTA}}}{FCR_{\text{monoculture}}}\right] > 0$
	$\text{Linearity}_{\text{Monoculture}} - \text{Linearity}_{\text{IMTA}} > 0$
Water	$\text{Water recirculation}_{\text{IMTA}} - \text{Water recirculation}_{\text{Monoculture}} > 0$
Energy	$\left(\frac{\text{kWh consumed}}{\text{biomass harvested}}\right)_{\text{Monoc.}} - \left(\frac{\text{kWh consumed}}{\text{biomass harvested}}\right)_{\text{IMTA}} > 0$
Infrastructure	$\left(\frac{\text{kg infrastructure materials}}{\text{biomass harvested}}\right)_{\text{Monoc.}} - \left(\frac{\text{kg infrastructure materials}}{\text{biomass harvested}}\right)_{\text{IMTA}} > 0$

Regarding the IMTA lab in Ireland, the results (Table 10) revealed that the bioremediation of nutrients was the key pillar in this lab, while this system contributed comparatively less in terms of the use of resources (energy). The underlying reason was that the low increase in low-trophic biomass did not nearly offset the increase in energy demand due to the additional maintenance activities. FCR and linearity indicators were not calculated for the Irish IMTA, as no actions were taken to improve circularity in that sense. It was not applicable to calculate the water indicator in an open system.

Table 10. Irish IMTA lab circularity performance.

Pillar	Indicator	Result
Use of resources	Energy	0.38%
Nutrient management	N bioremediation	7.71%
	P bioremediation	12.08%
	C bioremediation	2.63%

The Brazilian IMTA lab demonstrated a better performance regarding the water indicator, as the system enabled a high recirculation rate compared to the baseline conditions (ponds). Secondly, the bioremediation of P in the system contributed notably to the improvement in the nutrient management pillar, followed by the improvement achieved in the circularity of shrimp feed formulation through the incorporation of by-products to replace the use of conventional (linear) ingredients (Table 11). The quantification of the improvement in FCR was calculated as the average of the two fed species, shrimp and tilapia. Nitrogen bioremediation was calculated as DINs (nitrite, nitrate, and ammonia), but only if ammonia was included in this assessment did this indicator increase (from 22.34% to 46.7%). The energy indicator is not represented, as no circularity improvement was shown, and C bioremediation was not calculated due to a lack of information.

Table 11. Brazilian IMTA lab circularity performance.

Pillar	Indicator	Result
Use of resources	Feed–FCR	37%
	Feed–Linearity	50%
	Water	99.68%
	Infrastructure	63.30%
Nutrient management	N bioremediation	22.34%
	P bioremediation	75.61%

Regarding the South African IMTA lab, results from the abalone system (Table 12) showed the potential increase in circularity due to water recirculation, thus reducing the dependency on the pumping of fresh seawater. The feasibility of operating at 75%, and at 100% recirculation for short periods, means that IMTA would increase resource efficiency not only in terms of water but also energy consumption. Under the same pillar, FCR was reduced in IMTA, and bioremediation also supported the alignment of the multi-trophic system with resource use efficiency and recycling principles. The infrastructure and C bioremediation indicators were not calculated due to a lack of information.

Table 12. South African IMTA lab (abalone) circularity performance.

Pillar	Indicator	Result
Use of resources	Feed–FCR	34.90%
	Feed–Linearity	28.58%
	Water	50.00%
	Energy	34.30%
Nutrient management	N bioremediation	18.17%
	P bioremediation	6.01%

The urchin system results (Table 13) pointed to a higher circularity regarding water recirculation, followed by N bioremediation, which was boosted by the integrated cultivation of *Ulva*. IMTA demonstrated good circularity performance in infrastructure, which translated into equipment savings compared to monoculture conditions. The integration of fresh *Ulva* as feed was quantified as a circular benefit in this system. The FCR indicator is not represented in the table since no circularity improvement was found, and the P and C bioremediation indicators were not calculated due to a lack of information.

Table 13. South African IMTA lab (urchin) circularity performance.

Pillar	Indicator	Result
Use of resources	Feed–Linearity	57.13%
	Water	90.00%
	Energy	71.57%
	Infrastructure	66.67%
Nutrient management	N bioremediation	80.00%

4. Discussion

Circular economy strategies offer a way to make better use of resources and produce less waste [2]. This study offers insights regarding the role of IMTA in the transition toward more circular aquaculture.

This work evaluates the performance of IMTA production systems, which have been shown to be suitable systems to increase circularity. In the Irish IMTA lab, Atlantic salmon release nutrients into the surrounding water, native oysters filter out particulate nutrients while the dissolved fraction is extracted by seaweeds that absorb dissolved minerals and carbon, and spiny sea urchins are fed with the seaweeds grown in this IMTA system. The

Brazilian IMTA lab cultivates the most effective biomass ratios of white shrimp, tilapia and sea lettuce in BFT systems, where organic material and nutrients are regulated. Finally the South African IMTA lab demonstrates the feasibility of increasing recirculation in the commercial abalone–*Ulva* integrated systems. The pilot commercial-scale urchin-*Ulva* system is being run in the same way but at 90% recirculation, largely due to energy cost savings as the seawater needs to be heated to 25 °C for the warm water urchin to grow. This has been run successfully in a fully grown-out trial.

Multi-trophic aquaculture provides an opportunity to not only include nutrient recovery at the production level in aquaculture farms [12] but also increase the efficiency in terms of resource use. Based on the premise that IMTA performs as a circular system, the present work reveals the need for definitions of metrics. Methodology was developed to allow for the combined assessment of different indicators that can be pertinent to the evaluation of different IMTA systems at multiple scales. The metrics reflect the two principal pillars of aquaculture impacting circularity: nutrient management and the use of resources. The natural capacity of extractive species to assimilate nutrients from the water is evidenced through the quantification of the corresponding indicators. However, the benefits achieved due to the incorporation of macroalgae contribute not only to bioremediation but also to the increased resilience of systems to harmful algal blooms and other adverse events (e.g., chemical/oil spills) [48] (e.g., abalone in the South African IMTA lab with 100% recirculation that can isolate the system from the surrounding environment for short (3–4-day) periods [58]).

The review of nutrient-retention efficiency provided by Nederlof suggested that 40–75% of nutrient emissions could be mitigated by extractive species [16]. In the present study, the interpretation of results may consider suggest that the relevance of the bioremediation potential achieved by IMTA is determined by the scale of the production of low-trophic species. In this sense, the Irish IMTA lab did not reflect the total capacity in nutrient management that would be possible by increasing extractive species biomass. On the other hand, the bioremediation indicator was focused on P and N in Brazil and South Africa, as these were the main nutrients mitigated (especially toxic compounds such as ammonia) by the extractive species in the systems, but further evaluation of carbon balance would be relevant to obtain a fuller interpretation. In this context, the potential CO_2 sequestration by mollusk shells was excluded from this study, as that is still controversial [59].

Given the relevance of bioremediation as a circular benefit of IMTA, more precise approaches are recommended to determine nutrient budgets. In mass balance approaches, there are many variations that can be incorporated for nutrient loading estimates, making them more appropriate than the use of static input values [60]. Moreover, the present assessment was not entirely performed with primary data, and secondary sources were needed (e.g., assimilation efficiency and the individual growth of urchins or the uneaten fraction of fish in Ireland). Additionally, the respiration and pseudofeces from low- trophic animals in open systems were not quantified and were thus excluded from the nutrient management indicator. Finally, the nutrient mitigation capacities of IMTA systems in open environments are strongly influenced by different factors that limit the nutrient retention capacity of low-trophic species [16], which should also be considered.

Data for bioremediation calculation were not derived from the application of advanced models, and limitations to accuracy are recognized in the calculation of this indicator. Modeling techniques provide precision and are key to optimization, but they require a lot of background and high-quality data that were not available for all the laboratories assessed in the present study. Nevertheless, a combination of experimental and modeling approaches is useful to provide further insight into refining the estimation of bioremediation efficiency. Similarly, stable isotope studies allow the differentiation of the origin of nutrients, and they help to trace fish farm waste and thus confirm the capacity of low-trophic species to contribute to bioremediation [61]. The circularity assessment in this study was applied to experimental laboratories that aimed to test and validate different species under IMTA schemes, in which the implementation of models or isotope studies was not targeted.

Regarding the use of resources, the Brazilian and South African IMTA (urchin system) labs notably improved feeding performance, as both systems incorporated circular ingredients. Brazil showed circularity attributes for the reduction in FCR, meaning a reduction in resource use. The linearity reduction due to the totally circular feed based on poultry industry by-products in the Brazilian IMTA lab is aligned with the global trend of reducing the demand for fishmeal in the aquaculture sector [62]. Further assessments would be needed to evaluate if the incorporation of circular ingredients could compensate for an increase in the use of resources due to lower nutritional functionalities of novel feeds based on non-processed ingredients (e.g., urchin fed with fresh *Ulva*). Strategies oriented to increase the digestibility [50] and palatability [57] of formulated feeds would lead to more efficient use of resources, and thus, the consideration of apparent digestibility coefficients (ADCs) of fed species within the circularity assessment would provide a more comprehensive overview of the pillar of resource use [60].

The water indicator is not interpreted as a water footprint indicator (as in the approach developed in AWARE (wulca-waterlca.org)) since the present study did not assess water use-related environmental impacts. In this regard, circular economy metrics cannot easily replace LCA approaches. Moreover, the potential for LCA to evaluate aquaculture performance is particularly recognized through indicators that reflect nutrient management performance. For example, implementing nutrient recirculation strategies in aquaculture with IMTA systems possibly contributes to reducing the impact of eutrophication, making it a complementary indicator of how well the systems work under a circular approach in the nutrient pillar.

The reduction in energy consumption enabled by water recirculation is particularly beneficial in the South African context. Electricity use for seawater pumping is a major cost of these operations. South Africa also currently has an intermittent electricity supply, and the recirculation reduces the demand for farm electricity generation (from diesel generators) for periods with daily outages. This means that the system can better deal with the frequent electricity outages if the recirculation rate is high, as less water needs to be heated to 25 °C. The Brazilian IMTA lab is an intensified system that is comparatively worse than pond systems in terms of energy consumption. The energy indicator reflects the entropy principle recognized by Chary et al. [10]. However, this study does not consider energy sources, as they would not necessarily reflect the performance of multi-trophic production.

Regarding infrastructure, the present study considered the lifespan as a fundamental parameter included in the indicator. Generally, better maintenance or the substitution of infrastructure elements with increased durability would increase the functionality and thus the circularity. The study of specific indicators, in particular the MCI indicator, would provide insight as part of studies of low- trophic aquaculture systems, in which infrastructure and equipment were identified as key elements within the environmental profile [53].

For all resource aspects, it may be relevant to highlight that the metrics suggested are oriented to evaluate the production itself (the gate-to-gate approach); thus, value chains of aquaculture products are out of the scope, except for the consideration of the origin of feed ingredients (the linearity aspect). Additionally, in the resource use pillar, the evaluation of other potential resource uses, such as fertilizer, antifouling agents, antibiotics, and other chemical agents or products, was excluded, as these were not reported during the operation of each lab.

Finally, this work demonstrates that circularity can be measured at the level of the fish farm, encouraging doing so in a simplified way while bringing the aquaculture sector closer to circular economy progress monitoring. A metrics-based circularity assessment is promoted using a broad approach that is applicable to all multi-trophic production systems. However, data availability is a major constraint, and there is still much room for improvement to increase the robustness of the results obtained, allowing comparisons between similar production systems.

5. Conclusions

The lack of specific definitions of circular aquaculture and standardized methodologies might discourage aquaculture producers from identifying and communicating strategies that contribute to increasing circular performance. The present study provided a definition of indicators that reflected in a simple but robust way the efficiency of aquaculture production from the perspective of circularity. The bioremediation, feed, water, energy and infrastructure indicators would allow not-LCA practitioners to monitor cultivation performance without significantly increasing efforts in data collection and impact evaluation. The circularity indicators expressed here were not intended to be exclusive to LCA approaches, but rather, they complement and potentially encourage the sector to evaluate its contributions to the circular economy.

Our results confirmed that multi-trophic aquaculture systems perform in line with the circular attributes embedded in the essential definition of bioremediation. Metrics for bioremediation would promote the standardization of nutrient recycling rates, from which the effectiveness of the systems could be evaluated.

In addition to bioremediation, complementary indicators applied to IMTA provide evidence for the implementation of resource efficiency strategies, which further ensures the alignment of these systems with the circular economy.

Author Contributions: Conceptualization, D.C. and I.S.; methodology, D.C. and I.S.; validation, B.M.M., J.J.B., M.B.-H., P.O., A.C. and L.H.P.; formal analysis, I.S. and D.C.; investigation, B.M.M., J.J.B., M.B.-H., P.O., A.C. and L.H.P.; resources, B.M.M., J.J.B., M.B.-H., P.O. and L.H.P.; data curation, B.M.M., J.J.B., M.B.-H., P.O., A.C. and L.H.P.; writing—original draft preparation, D.C. and I.S.; writing—review and editing, B.M.M., J.J.B., M.B.-H., P.O. and L.P; visualization, I.S.; supervision, I.S.; project administration, I.S; funding acquisition, I.S. All authors have read and agreed to the published version of the manuscript.

Funding: This research was funded by the ASTRAL Project—H2020, grant agreement No. 863034.

Institutional Review Board Statement: Ethical review and approval were waived for this study, as the circularity assessment did not imply animal experimentation. The assessment was applied to studies involving animals in each laboratory that conducted the experiments according to the ethical requirements defined in the EU ASTRAL project ASTRAL, Grant Agreement Number: 863034 (Deliverable 9.3 A-Requirement No. 3). This protocol specifies that experiments in Ireland, Brazil and South Africa follow the national and international regulations and hold the necessary permits.

Informed Consent Statement: Not applicable.

Data Availability Statement: Data are contained within the article.

Acknowledgments: The authors are grateful to the ASTRAL project that has received funding from the European Union's Horizon 2020 research and innovation program under grant agreement No. 863034. Moreover, the authors are thankful to the reviewers for their constructive comments on this work.

Conflicts of Interest: The authors declare no conflicts of interest.

References

1. FAO. *The State of World Fisheries and Aquaculture 2022. Towards Blue Transformation*; FAO: Rome, Italy, 2022; ISBN 978-92-5-136364-5.
2. Regueiro, L.; Newton, R.; Soula, M.; Méndez, D.; Kok, B.; Little, D.C.; Pastres, R.; Johansen, J.; Ferreira, M. Opportunities and Limitations for the Introduction of Circular Economy Principles in EU Aquaculture Based on the Regulatory Framework. *J. Ind. Ecol.* **2022**, *26*, 2033–2044. [CrossRef]
3. Costa-Pierce, B. The Principles and Practices of Ecological Aquaculture and the Ecosystem Approach to Aquaculture. *World Aquac.* **2021**, *52*, 25–31.
4. European Commission. *Directorate-General for Environment. A New Circular Economy Action Plan: For a Cleaner and More Competitive Europe COM/2020/98 Final*; European Commission: Brussels, Belgium, 2020.
5. European Commission. *Secretariat-General: The European Green Deal COM/2019/640 Final*; European Commission: Brussels, Belgium, 2019.
6. European Commission. *Directorate-General for Health and Food Safety: A Farm to Fork Strategy for a Fair, Healthy and Environmentally-Friendly Food System COM/2020/381 Final*; European Commission: Brussels, Belgium, 2020.

7. European Commission. *Directorate-General for Maritime Affairs and Fisheries Strategic: Guidelines for a More Sustainable and Competitive EU Aquaculture for the Period 2021 to 2030 COM/2021/236 Final*; European Commission: Brussels, Belgium, 2021.
8. Kirchherr, J.; Reike, D.; Hekkert, M. Conceptualizing the Circular Economy: An Analysis of 114 Definitions. *Resour. Conserv. Recycl.* **2017**, *127*, 221–232. [CrossRef]
9. Balsells, S.; Chary, K.; Checa, D.; Enyedi, E. Policy Recommendations For a More Circular Aquaculture. *iFishIENCi* **2022**. [CrossRef]
10. Chary, K.; Van Riel, A.; Muscat, A.; Wilfart, A.; Harchaoui, S.; Verdegem, M.; Filgueira, R.; Troell, M.; Henriksson, P.J.G.; De Boer, I.J.M.; et al. Transforming Sustainable Aquaculture by Applying Circularity Principles. *Rev. Aquac.* **2023**, *16*, 656–673. [CrossRef]
11. Campanati, C.; Willer, D.; Schubert, J.; Aldridge, D.C. Sustainable Intensification of Aquaculture through Nutrient Recycling and Circular Economies: More Fish, Less Waste, Blue Growth. *Rev. Fish. Sci. Aquac.* **2022**, *30*, 143–169. [CrossRef]
12. Cooney, R.; De Sousa, D.B.; Fernández-Ríos, A.; Mellett, S.; Rowan, N.; Morse, A.P.; Hayes, M.; Laso, J.; Regueiro, L.; Wan, A.H.; et al. A Circular Economy Framework for Seafood Waste Valorisation to Meet Challenges and Opportunities for Intensive Production and Sustainability. *J. Clean. Prod.* **2023**, *392*, 136283. [CrossRef]
13. Luo, G. Review of Waste Phosphorus from Aquaculture: Source, Removal and Recovery. *Rev. Aquac.* **2023**, *15*, 1058–1082. [CrossRef]
14. Fraga-Corral, M.; Ronza, P.; Garcia-Oliveira, P.; Pereira, A.G.; Losada, A.P.; Prieto, M.A.; Quiroga, M.I.; Simal-Gandara, J. Aquaculture as a Circular Bio-Economy Model with Galicia as a Study Case: How to Transform Waste into Revalorized by-Products. *Trends Food Sci. Technol.* **2022**, *119*, 23–35. [CrossRef]
15. Wang, X.; Olsen, L.; Reitan, K.; Olsen, Y. Discharge of Nutrient Wastes from Salmon Farms: Environmental Effects, and Potential for Integrated Multi-Trophic Aquaculture. *Aquacult. Environ. Interact.* **2012**, *2*, 267–283. [CrossRef]
16. Nederlof, M.A.J.; Verdegem, M.C.J.; Smaal, A.C.; Jansen, H.M. Nutrient Retention Efficiencies in Integrated Multi-trophic Aquaculture. *Rev. Aquac.* **2022**, *14*, 1194–1212. [CrossRef]
17. Carvalho Pereira, J.; Lemoine, A.; Neubauer, P.; Junne, S. Perspectives for Improving Circular Economy in Brackish Shrimp Aquaculture. *Aquac. Res.* **2022**, *53*, 1169–1180. [CrossRef]
18. Villar-Navarro, E.; Garrido-Pérez, C.; Perales, J.A. Recycling "Waste" Nutrients Back into RAS and FTS Marine Aquaculture Facilities from the Perspective of the Circular Economy. *Sci. Total Environ.* **2021**, *762*, 143057. [CrossRef] [PubMed]
19. Colombo, S.M.; Roy, K.; Mraz, J.; Wan, A.H.L.; Davies, S.J.; Tibbetts, S.M.; Øverland, M.; Francis, D.S.; Rocker, M.M.; Gasco, L.; et al. Towards Achieving Circularity and Sustainability in Feeds for Farmed Blue Foods. *Rev. Aquac.* **2023**, *15*, 1115–1141. [CrossRef]
20. Ahmad, A.; Hassan, S.W.; Banat, F. An Overview of Microalgae Biomass as a Sustainable Aquaculture Feed Ingredient: Food Security and Circular Economy. *Bioengineered* **2022**, *13*, 9521–9547. [CrossRef] [PubMed]
21. Sanz-Lazaro, C.; Sanchez-Jerez, P. Regional Integrated Multi-Trophic Aquaculture (RIMTA): Spatially Separated, Ecologically Linked. *J. Environ. Manag.* **2020**, *271*, 110921. [CrossRef] [PubMed]
22. Granada, L.; Sousa, N.; Lopes, S.; Lemos, M.F.L. Is Integrated Multitrophic Aquaculture the Solution to the Sectors' Major Challenges? – A Review. *Rev. Aquac.* **2016**, *8*, 283–300. [CrossRef]
23. Chopin, T. Aquaculture, Integrated Multi-Trophic (IMTA). In *Encyclopedia of Sustainability Science and Technology*; Springer: New York, NY, USA, 2013; Volume 12, pp. 542–564. ISBN 978-1-4614-5796-1.
24. The Aquaculture Advisory Council (AAC). *8.-AAC-Recommendation-Shellfish-Farming-as-a-Nitrogen-Sink_2023_8*; The Aquaculture Advisory Council (AAC): Brussels, Belgium, 2023; 11p.
25. Costa, L.C.D.O.; Carvalho, A.; Holanda, M.; Santos, J.; Borges, L.; Guterres, B.; Nam Junior, J.; Fonseca, V.; Muller, L.; Romano, L.; et al. Biological Responses of Oyster Crassostrea Gasar Exposed to Different Concentrations of Biofloc. *Fishes* **2023**, *8*, 586. [CrossRef]
26. Kang, Y.H.; Shin, J.A.; Kim, M.S.; Chung, I.K. A Preliminary Study of the Bioremediation Potential of Codium Fragile Applied to Seaweed Integrated Multi-Trophic Aquaculture (IMTA) during the Summer. *J. Appl. Phycol.* **2008**, *20*, 183–190. [CrossRef]
27. De Prisco, J.A. An Investigation of Some Key Physico-Chemical Water Quality Parameters of an Integrated Multi-Trophic Aquaculture (IMTA) System Operating Recirculation Methodology in the Western Cape of South Africa. Master's Thesis, University of Cape Town, Cape Town, South Africa, 2020.
28. Chopin, T.; Cooper, J.A.; Reid, G.; Cross, S.; Moore, C. Open-water Integrated Multi-trophic Aquaculture: Environmental Biomitigation and Economic Diversification of Fed Aquaculture by Extractive Aquaculture. *Rev. Aquac.* **2012**, *4*, 209–220. [CrossRef]
29. Neveux, N.; Bolton, J.J.; Bruhn, A.; Roberts, D.A.; Ras, M. The Bioremediation Potential of Seaweeds: Recycling Nitrogen, Phosphorus, and Other Waste Products. In *Blue Biotechnology*; Wiley: Weinheim, Germany, 2018; pp. 217–239. ISBN 978-3-527-80171-8.
30. Stabili, L.; Giangrande, A.; Arduini, D.; Borghese, J.; Petrocelli, A.; Alabiso, G.; Ricci, P.; Cavallo, R.A.; Acquaviva, M.I.; Narracci, M.; et al. Environmental Quality Improvement of a Mariculture Plant after Its Conversion into a Multi-Trophic System. *Sci. Total Environ.* **2023**, *884*, 163846. [CrossRef] [PubMed]
31. Linder, M.; Sarasini, S.; Van Loon, P. A Metric for Quantifying Product-Level Circularity. *J Ind. Ecol.* **2017**, *21*, 545–558. [CrossRef]

32. Corona, B.; Shen, L.; Reike, D.; Rosales Carreón, J.; Worrell, E. Towards Sustainable Development through the Circular Economy—A Review and Critical Assessment on Current Circularity Metrics. *Resour. Conserv. Recycl.* **2019**, *151*, 104498. [CrossRef]
33. Valenti, W.C.; Kimpara, J.M.; Preto, B.D.L.; Moraes-Valenti, P. Indicators of Sustainability to Assess Aquaculture Systems. *Ecol. Indic.* **2018**, *88*, 402–413. [CrossRef]
34. Pauliuk, S. Critical Appraisal of the Circular Economy Standard BS 8001:2017 and a Dashboard of Quantitative System Indicators for Its Implementation in Organizations. *Resour. Conserv. Recycl.* **2018**, *129*, 81–92. [CrossRef]
35. Vural Gursel, I.; Elbersen, B.; Meesters, K.P.H. Monitoring Circular Biobased Economy – Systematic Review of Circularity Indicators at the Micro Level. *Resour. Conserv. Recycl.* **2023**, *197*, 107104. [CrossRef]
36. Niero, M.; Kalbar, P.P. Coupling Material Circularity Indicators and Life Cycle Based Indicators: A Proposal to Advance the Assessment of Circular Economy Strategies at the Product Level. *Resour. Conserv. Recycl.* **2019**, *140*, 305–312. [CrossRef]
37. *Ellen MacArthur Foundation Circularity Indicators: An Approach to Measuring Circularity. Methodology*; Ellen MacArthur Foundation Isle of Wight, UK, 2019.
38. *ISO 14040:2006*; Environmental Management—Life Cycle Assessment—Principles and Framework. ISO: Geneve, Switzerland 2006; 20p.
39. *ISO 14044:2006*; Environmental Management—Life cycle Assessment—Requirements and Guidelines. ISO: Geneve, Switzerland 2006; 45p.
40. European Commission. *Understanding Product Environmental Footprint and Organisation Environmental Footprint Methods*; European Commission: Brussels, Belgium, 2021.
41. Badiola, M.; Mendiola, D.; Bostock, J. Recirculating Aquaculture Systems (RAS) Analysis: Main Issues on Management and Future Challenges. *Aquac. Eng.* **2012**, *51*, 26–35. [CrossRef]
42. Garo, Q.; Tanrattana, N.; Vincent, L. Integrated Multi-Trophic Aquaculture: Integrated Marine Recirculated Aquaculture Analysis Available online: https://keep.eu/api/project-attachment/3928/get_file/ (accessed on 1 March 2024).
43. Leandro, A.; Pereira, L.; Gonçalves, A.M.M. Diverse Applications of Marine Macroalgae. *Mar. Drugs* **2019**, *18*, 17. [CrossRef]
44. Carvalho, A.; Costa, L.C.D.O.; Holanda, M.; Gonçalves, M.; Santos, J.; Costa, C.S.B.; Turan, G.; Poersch, L.H. Growth of the Macroalgae Ulva Lactuca Cultivated at Different Depths in a Biofloc Integrated System with Shrimp and Fish. *Phycology* **2023**, *3*, 280–293. [CrossRef]
45. Uttia Alstad, S. Bioremediation Potential of the Green Algae Ulva Sp. Cultivated in Wastewater from Marine Recirculating Aquaculture Systems (RAS). Master's Thesis, Norwegian University of Science and Technology, Trondheim, Norway, 2022.
46. Marine Algae as Carbon Sinks and Allies to Combat Global Warming. In *Marine Algae*; Vaz-Pinto, F.; Arenas, F. (Eds.) CRC Press: Boca Raton, FL, USA, 2014; pp. 148–185. ISBN 978-0-429-16637-2.
47. Sjøtun, K.; Fredriksen, S.; Rueness, J. Seasonal Growth and Carbon and Nitrogen Content in Canopy and First-Year Plants of Laminaria Hyperborea (Laminariales, Phaeophyceae). *Phycologia* **1996**, *35*, 1–8. [CrossRef]
48. Geldart, M. *The Potential for Increasing the Recirculation Rate in a Commercial Integrated Abalone/Ulva Aquaculture System*; Marine Biology Honours Degree project GLDMIC023; University of Cape Town: Cape Town, South Africa, 2023.
49. Bohnes, F.A.; Hauschild, M.Z.; Schlundt, J.; Laurent, A. Life Cycle Assessments of Aquaculture Systems: A Critical Review of Reported Findings with Recommendations for Policy and System Development. *Rev. Aquac.* **2019**, *11*, 1061–1079. [CrossRef]
50. Sánchez, I.; Checa, D. Report on Circularity of the iFishIECNCi Approach; iFishIENCi Project GA No: 818036; 2023.
51. European Commission. *Directorate General for Maritime Affairs and Fisheries. y EUMOFA—2020—Recirculation Aquaculture Systems*; European Commission: Brussels, Belgium, 2020.
52. Vasquez-Mejia, C.M.; Shrivastava, S.; Gudjónsdóttir, M.; Manzardo, A.; Ögmundarson, Ó. Current Status and Future Research Needs on the Quantitative Water Use of Finfish Aquaculture Using Life Cycle Assessment: A Systematic Literature Review. *J. Clean. Prod.* **2023**, *425*, 139009. [CrossRef]
53. Surendran, P.; Sanchez, I.; Checa, D.; Poersch, L.; Macleod, A.; Michalek, K.; O'Donohoe, P.; Casserly, J.; Waters, C.; Macey, M.; et al. *Life Cycle Assessment—ASTRAL Project (GA 863034)*; ASTRAL Project GA no 863034; 2023; 63p. Available online: https://www.astral-project.eu/ (accessed on 1 March 2024).
54. Poli, M.A.; Legarda, E.C.; De Lorenzo, M.A.; Martins, M.A.; Do Nascimento Vieira, F. Pacific White Shrimp and Nile Tilapia Integrated in a Biofloc System under Different Fish-Stocking Densities. *Aquaculture* **2019**, *498*, 83–89. [CrossRef]
55. Minaz, M.; Yazıcı, İ.S.; Sevgili, H.; Aydın, İ. Biofloc Technology in Aquaculture: Advantages and Disadvantages from Social and Applicability Perspectives. *Ann. Anim. Sci.* **2023**. [CrossRef]
56. FAO. *Nitrogen and Protein Content Measurement and Nitrogen to Protein Conversion Factors for Dairy and Soy Protein-Based Foods: A Systematic Review and Modelling Analysis*; FAO: Rome, Italy, 2019.
57. Cyrus, M.; Bolton, J.; Scholtz, R.; Macey, B. The Advantages of Ulva (Chlorophyta) as an Additive in Sea Urchin Formulated Feeds: Effects on Palatability, Consumption and Digestibility. *Aquac. Nutr.* **2014**, *21*, 578–591. [CrossRef]
58. Bolton, J.J.; Robertson-Andersson, D.V.; Shuuluka, D.; Kandjengo, L. Growing Ulva (Chlorophyta) in Integrated Systems as a Commercial Crop for Abalone Feed in South Africa: A SWOT Analysis. *J. Appl. Phycol.* **2009**, *21*, 575–583. [CrossRef]
59. Alonso, A.A.; Álvarez-Salgado, X.A.; Antelo, L.T. Assessing the Impact of Bivalve Aquaculture on the Carbon Circular Economy. *J. Clean. Prod.* **2021**, *279*, 123873. [CrossRef]

50. Reid, G.K.; Chopin, T.; Robinson, S.M.C.; Azevedo, P.; Quinton, M.; Belyea, E. Weight Ratios of the Kelps, Alaria Esculenta and Saccharina Latissima, Required to Sequester Dissolved Inorganic Nutrients and Supply Oxygen Atlantic Salmon, Salmo Salar, in Integrated Multi-Trophic Aquaculture Systems. *Aquaculture* **2013**, *408–409*, 34–46. [CrossRef]
51. Sanz-Lazaro, C.; Sanchez-Jerez, P. Mussels Do Not Directly Assimilate Fish Farm Wastes: Shifting the Rationale of Integrated Multi-Trophic Aquaculture to a Broader Scale. *J. Environ. Manag.* **2017**, *201*, 82–88. [CrossRef]
52. Luthada-Raswiswi, R.; Mukaratirwa, S.; O'Brien, G. Animal Protein Sources as a Substitute for Fishmeal in Aquaculture Diets: A Systematic Review and Meta-Analysis. *Appl. Sci.* **2021**, *11*, 3854. [CrossRef]

Disclaimer/Publisher's Note: The statements, opinions and data contained in all publications are solely those of the individual author(s) and contributor(s) and not of MDPI and/or the editor(s). MDPI and/or the editor(s) disclaim responsibility for any injury to people or property resulting from any ideas, methods, instructions or products referred to in the content.

fishes

MDPI

Article

Exploring Biofouling Control by the California Sea Cucumber (*Apostichopus californicus*) in Integrated Multi-Trophic Aquaculture (IMTA) with Organic Chinook Salmon (*Oncorhynchus tshawytscha*)

Emaline M. Montgomery [1], Barb L. Cannon [2] and Christopher M. Pearce [1,*]

[1] Fisheries and Oceans Canada, Pacific Biological Station, 3190 Hammond Bay Road, Nanaimo, BC V9T 6N7, Canada
[2] Creative Salmon Co., Ltd., 612 Campbell Street, Tofino, BC V0R 2Z0, Canada
* Correspondence: chris.pearce@dfo-mpo.gc.ca

Abstract: The growth of biofouling on aquaculture infrastructure is a universal challenge. Standard industry practices to remove biofouling in finfish aquaculture typically include in situ net cleaning via power washing. Since those cleaning practices can be potentially harmful to fish-gill health and expensive, development of other non-toxic biofouling controls is an industry priority. Deposit-feeding sea cucumbers are potentially well suited for biofouling control due to their feeding mechanism, but remain relatively untested in this capacity. We examined the use of California sea cucumbers (*Apostichopus californicus*) to control biofouling on cages containing adult Chinook salmon (*Oncorhynchus tshawytscha*) at a commercial farming operation. Four cage types were established: cages with salmon and sea cucumbers, cages with salmon only, cages with sea cucumbers only, and cages without either species. Results showed that the sea cucumbers actively fed on biofouling when salmon were absent (~16% cleaner on average) but preferred to consume uneaten feed/faeces at the bottom of the cages, neglecting the biofouling, when the salmon were present. It is hypothesized that biofouling control in cages with salmon may be possible with an increased density of sea cucumbers. This is the first study to examine the use of sea cucumbers as a direct net biofouling control agent with adult fish. Our results will be beneficial for industry to develop standard operating procedures for using California sea cucumbers as a biofouling control and could contribute to the development of a management framework for sea cucumber/salmon integrated multi-tropic aquaculture.

Keywords: *Apostichopus californicus*; biofouling control; net cleaning; sea cucumber; sustainability

Key Contribution: Results showed that the sea cucumbers actively fed on net biofouling when salmon were absent (~16% cleaner nets on average) but preferred to consume uneaten feed/faeces at the bottom of the cages, neglecting the biofouling, when the salmon were present. This is the first study to examine the use of sea cucumbers as a direct net biofouling control agent with adult fish.

Citation: Montgomery, E.M.; Cannon, B.L.; Pearce, C.M. Exploring Biofouling Control by the California Sea Cucumber (*Apostichopus californicus*) in Integrated Multi-Trophic Aquaculture (IMTA) with Organic Chinook Salmon (*Oncorhynchus tshawytscha*). *Fishes* **2023**, *8*, 430. https://doi.org/10.3390/fishes8090430

Academic Editor: Bernardo Baldisserotto

Received: 29 June 2023
Revised: 1 August 2023
Accepted: 15 August 2023
Published: 23 August 2023

Copyright: © 2023 by the authors. Licensee MDPI, Basel, Switzerland. This article is an open access article distributed under the terms and conditions of the Creative Commons Attribution (CC BY) license (https://creativecommons.org/licenses/by/4.0/).

1. Introduction

The growth of biofouling material on aquaculture infrastructure is an expensive and universal challenge faced by the industry. A recent study in Norway calculated that the cost of biofouling management per production cycle could be as much as USD 500,000 per eight standard production finfish cages [1]. The blockage of net openings by biofouling organisms can negatively impact fish health by reducing water flow through the nets [2] and affect local hydrodynamic conditions around farm sites [3,4], leading to poor flushing of waste and/or hypoxic conditions. Standard industry practice to remove biofouling typically includes in situ net cleaning via power washing on a monthly or even bi-weekly frequency when fouling levels are highest in the summer. However, that cleaning practice has recently been recognized as being potentially harmful to fish gills due to the plume of biofouling particulates, particularly stinging-capable hydroids, released into the water column as a result of the pressure-washing process [5–8]. The release of biofouling material through power washing can also promote the spread of invasive species to other cages in the vicinity and into the general marine environment [5].

A number of innovations have been explored by the finfish aquaculture industry in an attempt to improve the cost effectiveness and lower the environmental impacts of power washing, including novel antifouling coatings for nets [9,10], adding bubble streams around cages [11], in situ cleaning via net-crawling robots [12,13], and in situ cleaning with organisms such as the lumpfish (*Cyclopterus lumpus* [14,15]), ballan wrasse (*Labrus bergylta* [16]), West Indian spider crab (*Maguimithrax spinosissimus* [17]), and California sea cucumbers (*Apostichopus californicus* [18]). Economic and environmental impact analyses reveal that a combined approach of biofouling prevention and active control methods will likely need to be used together in order to achieve a sustainable and permanent solution for the aquaculture industry [1]. Therefore, development of a non-impactful toolkit of cost-effective and environmentally-sustainable options for the industry to rely on should continue to be a priority for the biofouling research community.

Deposit-feeding sea cucumbers are well known for their ability to ingest and assimilate organic material from sedimentary environments and have been shown to be efficient organic extractive species in integrated multi-trophic aquaculture (IMTA; [19–22]). They are also behaviourally passive, not prone to disease, and have well-established global markets [23], making them a safe and lucrative option for IMTA with finfish and shellfish. They are lesser known, however, for their potential biofouling mitigative properties. Nearly all studies involving sea cucumbers and biofouling control have focused on using bioactive compounds from their tissues and extracts as novel antifouling coatings [24–26]. However, many sea cucumber species have large peltate feeding tentacles that brush the surface of soft and hard substrates, picking up organic-rich material for ingestion. This feeding mechanism could potentially dislodge young/small biofouling organisms on nets used in finfish farming, providing active removal and control of biofouling. Indeed, one previous study in Alaska (USA) housed California sea cucumbers in cages with pink (*Oncorhynchus gorbuscha*) and chum (*O. keta*) salmon fry with results showing that nets were 58% cleaner when sea cucumbers were present than when they were absent [18]. Despite their potential as net cleaners, no other studies have tested the biofouling control capacity of California sea cucumbers (or any other sea cucumber species) in finfish cages, leading to questions about the effectiveness of the species with other stages of salmon (e.g., juveniles, adults) or with other commercial finfish species. Indeed, the Ahlgren (1998) [18] study is the only one to examine the active biofouling mitigative properties of any sea cucumber species globally.

The objective of the present study was to examine the biofouling control capacity of California sea cucumbers in open net cages with mature Chinook salmon (*O. tshawytscha*) broodstock. This study was the first of its kind for the Canadian aquaculture industry and the first study worldwide to assess deposit-feeding sea cucumbers for active biofouling control in cages with adult finfish. The results will be beneficial for the salmon aquaculture industry in order to develop standard operating procedures for using California sea cucumbers as net cleaners and will contribute to the development of a management framework for sea cucumber/salmon IMTA.

2. Materials and Methods

2.1. Study Area, Farm Parameters, and Source of Sea Cucumbers

The study was conducted in the summer of 2021 at two organic Chinook salmon farm sites in Clayoquot Sound, British Columbia (BC), Canada, which are Canada-owned and operated by Creative Salmon Co., Ltd. (Figure 1). One site, Dawley Pass (DP), was located in Fortune Channel (49°09′56.3″ N, 125°46′10.7″ W) and the other, Warne Island (WI), was situated off the southwest shore of Warne Island (W49°07′41.6″ N, 125°44′58.6″ W). Both farm sites used for this study had mature Chinook salmon broodstock housed within designated net cages (15 × 15 × 15-m Viking-style cages with 4-cm stretch nylon netting). Fish density was dependent on normal industry practice and stock availability, with densities in the 3–4 kg m^{-3} range in this study. Fish were fed daily, to satiation, using underwater cameras to ensure adequate amounts of food were made available. Four hundred wild California sea cucumbers (>15-cm contracted length, measured after 5 s of consistent handling in air) were hand collected by SCUBA in Clayoquot Sound at a distance >1 km from the farm sites used in the study (Figure 1).

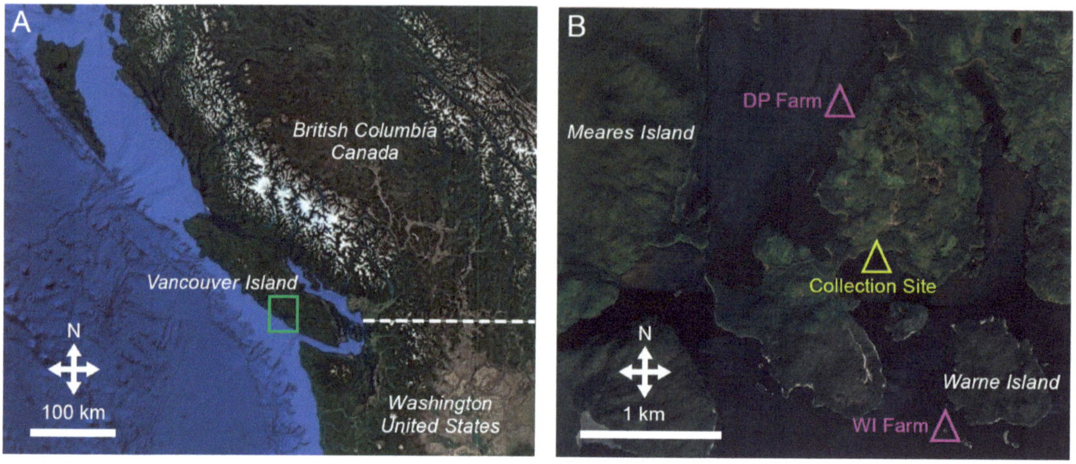

Figure 1. (**A**) Map showing Western North America and Vancouver Island, British Columbia (BC), Canada, where the study took place. The dotted white line indicates the border division between Canada and the United States. (**B**) Chinook salmon farm locations used for the project in Clayoquot Sound, BC. Farm sites are indicated by purple triangles, DP = Dawley Pass, WI = Warne Island. Both farm sites are owned and operated by Creative Salmon Co., Ltd. (Tofino, BC, Canada). The sea cucumber collection site is shown by a yellow triangle.

2.2. Experimental Design

Four cage types were established at both farm sites in order to test the capacity of sea cucumbers to control biofouling on Chinook salmon net pens: (1) fish only without sea cucumbers (F), (2) sea cucumbers only without fish (S), (3) fish and sea cucumbers (F + S), and (4) no fish or sea cucumbers, designated as an "empty" cage control (E). Each cage with sea cucumbers had 100 individuals (density = 1 ind 11.25 m^{-2} of net surface area), and was chosen based on the density used by Ahlgren (1998) [18]. Four cages were randomly chosen from the cage array (4 × 4) at both farm sites, one for each cage type, with three replicate one-month trials being conducted at both sites beginning in July, August, and September 2021 (i.e., replication was conducted temporally in three "trials"). Once assigned, cages were used for all three replicate one-month trials. During each replicate trial, sea cucumbers were allowed to actively feed on biofouling material in S and F + S cages. All nets were cleaned after each trial via sun drying and power washing, and everything was reset for subsequent temporal replicates. Sea cucumbers were transferred by divers to floating cages in the center of each fish pen during net cleaning operations in order to prevent damage or stress to them while the nets were being manipulated (Figure 2). Sea cucumbers were reused among the trials.

Figure 2. Large floating cage used to house sea cucumbers during regular net cleaning activities. Sea cucumbers were hand caught via SCUBA and transferred into the cage. It was left submerged and was attached in the middle of the cage before net cleaning began, preventing damage and stress to the sea cucumbers. This protocol and the cage design were developed by Creative Salmon Co., Ltd. personnel.

2.3. Analysis of Net Percent Cleanliness

Net percent cleanliness (PC) was defined as the percent area of net openings that was clear of biofouling growth. This was measured in each cage at the end of each one-month trial using image analysis of video footage collected by SCUBA at three different depth ranges: the top of the cage (~0–5 m), the middle of the cage (5–10 m), and the bottom of the cage (10–15 m). Three video transects per cage were taken at each sample date, with positions of the transects being chosen haphazardly. Image analysis was conducted by converting still frames from the video footage at each depth range into black/white threshold images in *ImageJ* software [27] and then calculating the percents of net openings that were clear (modified from [28]). A minimum of 25 net openings, chosen randomly, were analyzed at each net depth range per each of the three transects at each sampling point (~225 per cage per trial).

2.4. Stable Isotopes

Stable isotope ratios were used to explore feeding patterns of sea cucumbers housed with and without fish at the end of each 1-month replicate (i.e., August, September, October). Values for $d^{13}C$ and $d\ ^{15}N$ were obtained for sea cucumber muscle bands and body walls from S and F + S cages as well as wild-caught control individuals (from the original collection site, October only) using Elemental Analyzer Isotope Ratio Mass Spectrometry (EA-IRMS). Three sea cucumbers per cage type (F and F + S) were sampled at each time point. All samples were sub-sampled, run twice, and then averaged to account for sample heterogeneity. This work was completed by the Stable Isotopes Facility in the Faculty of Forestry at the University of British Columbia in Vancouver, BC.

2.5. Statistical Analysis

All statistical analyses were conducted using *R* statistical software [29] and graphics were produced using the *ggplot2* package [30]. Significance was set at $\alpha = 0.05$ for all tests. We first eliminated the possibility of seasonal variability and site variability on PC values using two-way Aligned Ranks Transformation (ART) ANOVA, since assumptions of normality were not met [31,32]. Since no significant seasonal or site effects were detected

(Table 1), PC values were pooled from each one-month replicate trial and both farm sites for all subsequent analyses. Principle component analysis (PCA) was then used to compare patterns of PC values across net depths among experimental cage types (i.e., F, S, F + S, E). Follow-up two-way and one-way ART ANOVAs were performed to further explore the influence of net depth and cage type on PC values. Post-hoc pairwise comparisons were performed using an align-and-rank procedure with a Tukey correction of the *p*-values to correct for multiple pairwise comparisons.

Table 1. Mean (±SD) percent cleanliness (%) by trial month and site and cage type and net depth. Statistical significance ($p < 0.05$) indicated by asterisks (two-factor aligned rank transform (ART) ANOVA) and pairwise groupings indicated by superscript letters (ART post-hoc multiple comparisons test).

Factor		Mean ± SD (%)	ART Two-Way ANOVA
Trial month	August	59.2 ± 25.8	*NS*
	September	70.8 ± 16.5	
	October	72.8 ± 24.3	
Site	Dawley Pass	68.8 ± 23.0	*NS*
	Warne Island	66.3 ± 24.9	
Trial month × site			*NS*
Cage type	Fish only (F)	64.6 ± 25.5	*NS*
	Sea cucumbers only (S)	79.4 ± 16.9	
	Fish + sea cucumbers (F + S)	62.1 ± 24.9	
	Empty cage (E)	64.2 ± 21.6	
Net depth *	Top (0–5 m)	46.3 ± 33.5^{A}	$F_{2,60} = 8.28, p < 0.01$
	Middle (5–10 m)	79.0 ± 18.1^{B}	
	Bottom (10–15 m)	77.5 ± 20.0^{B}	
Cage type × net depth			*NS*

A two-way ANOVA was then used to compare individual stable isotope ratios among factors (i.e., cage type, site) after each one-month replicate trial in order to assess the cumulative effects of sea cucumber exposure to biofouling at Chinook salmon farm sites. Parametric test assumptions of normality and homoscedasticity were confirmed using Shapiro tests and a plot of the residuals, respectively. Tukey post-hoc tests were used for pairwise comparisons of the treatments. No significant difference in stable isotope ratios between farm sites was detected.

3. Results

3.1. Net Percent Cleanliness

Over the course of the study, individual PC values exhibited a large range, from <10% to 100% (Figure 3). Average (±SD) PC values for the four cage types ranged from 62.1 ± 24.9% (F + S) to 79.4 ± 16.9% (S), but they did not differ significantly (Table 1). In addition, there was no significant cage-type × net-depth interaction (Table 1). There was, however, a significant net-depth effect, the top 5 m of the nets, with an average (±SD) PC value of 46.3 ± 33.5%, being significantly less clean at the end of each one-month trial than the middle (5−10 m) and bottom (10−15 m) regions of the nets, which had average PC values of 79.0 ± 18.1% and 77.5 ± 20.0%, respectively (Table 1).

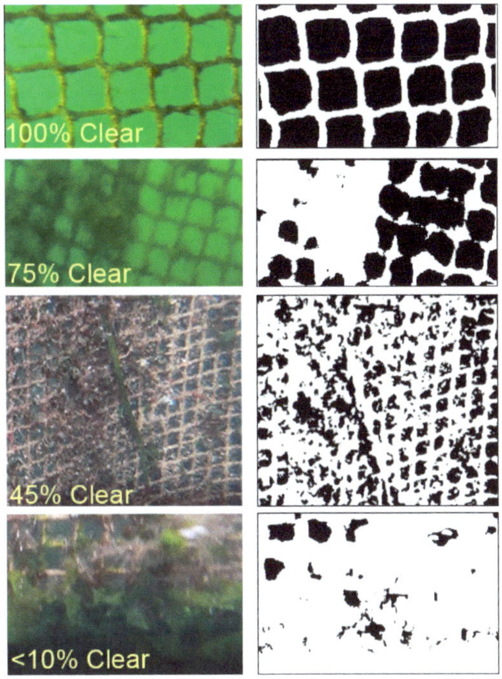

Figure 3. Sample of various net percent cleanliness values from actual photographs (**left**) and edited images (**right**). Black in the edited images (**right**) indicates a clear space for water to pass through the net. ImageJ [27] was used for the analysis.

Principle component analysis indicated two potential groupings, one containing the S and E cages and another containing the F and F + S cages (Figure 4). It also revealed that patterns of net PC values among the cage types were primarily being driven by cleanliness at the bottom of the nets only (Figure 4). Closer analysis of these bottom PC values revealed significant differences among the four cage types: F (65.3 ± 25.8%), S (94.2 ± 6.8%), F + S (67.2 ± 19.5%), and E (83.2 ± 9.2%) (Table 2). Pairwise comparisons showed significantly higher PC values in the S cages relative to the F and F + S cages, with no other significant pairwise comparisons. There were no significant differences among the cage types in the top and middle regions (Table 2). Overall, the S cages had the cleanest nets of all cage types (16% cleaner on average).

Table 2. Mean (± SD) percent cleanliness values (%) at each net depth across the four experimental cage types. Statistical significance ($p < 0.05$) indicated by asterisks (two-factor aligned rank transform (ART) ANOVA) and pairwise groupings indicated by superscript letters (ART post-hoc multiple comparisons test).

Net Depth	Cage Type, Mean ± SD (%)				ART One-Way ANOVA
	Empty Cage (E)	Fish (F)	Fish + Sea Cucumbers (F + S)	Sea Cucumbers (S)	
Top (0–5 m)	33.0 ± 37.1	49.1 ± 37.9	45.5 ± 26.4	57.5 ± 35.8	*NS*
Middle (5–10 m)	76.3 ± 18.5	79.4 ± 13.0	73.7 ± 28.6	86.7 ± 8.2	*NS*
Bottom (10–15 m) *	83.2 ± 9.2 ^{AB}	65.3 ± 25.8 ^{A}	67.2 ± 19.5 ^{A}	94.2 ± 6.8 ^{B}	$F_{3,20} = 4.32$, $p = 0.02$

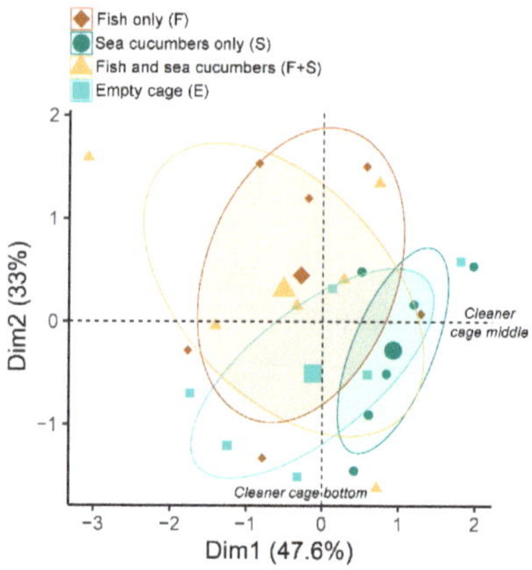

Figure 4. Principle component analysis (PCA) comparing patterns of percent net cleanliness values among the four cage types: brown diamond = fish only (F), teal circle = sea cucumbers only (S), yellow triangle = fish and sea cucumbers (F + S), and light blue square = empty cage (E). Ellipses indicate 95% confidence intervals for each cage type. The center of each ellipse is indicated by a larger icon corresponding to each cage type. Dimension 1 (*X*-axis) was primarily explained by differences in middle-cage cleanliness while Dimension 2 (*Y*-axis) was primarily explained by differences in bottom-cage cleanliness.

3.2. Stable Isotopes

Temporal patterns of d^{13}C and d^{15}N in the tissues of S and F + S sea cucumbers showed changes in diet between these two cage types after each sequential one-month replicate trial (PCA, Figure 5). These differences were significant after three months of sea cucumber integration into salmon culture for both d^{13}C (two-way ANOVA, $F_{1,15} = 7.60$, $p < 0.01$) and d^{15}N ($F_{2,15} = 3.94$, $p = 0.04$). At the three-month time point, F + S (+12.08‰) and S (+11.91‰) sea cucumbers were both enriched in ^{15}N (Tukey Test, $p = 0.04$) relative to wild-caught reference (+11.14‰) sea cucumbers, and F + S (−18.07‰) individuals were depleted in ^{13}C relative to S (−16.87‰) and wild-caught reference (−17.34‰) sea cucumbers ($p < 0.01$).

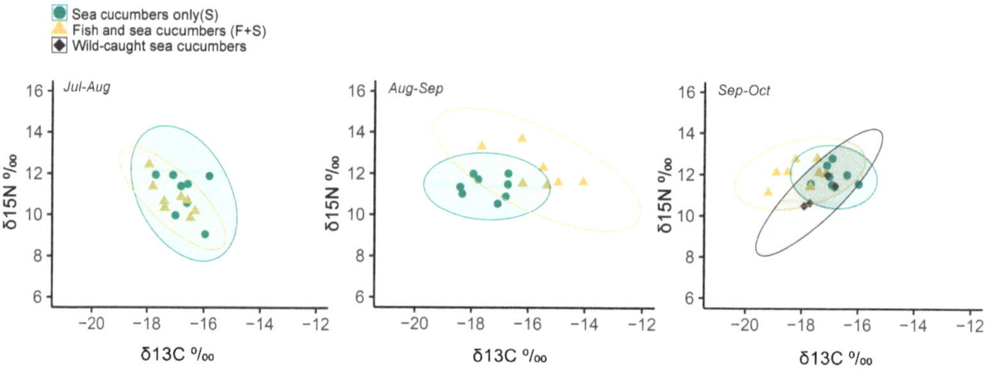

Figure 5. Scatter plots of ^{15}N versus ^{13}C stable isotope ratios in tissues of California sea cucumbers at the end of each consecutive one-month replicate trial: teal circle = sea cucumbers only (S), yellow triangle = fish and sea cucumbers (F + S), and black diamond = wild-caught sea cucumbers (October only). Ellipses indicate 95% confidence intervals for each cage type.

4. Discussion

The present study tested the biofouling control capacity of California sea cucumbers in open net cages with mature Chinook salmon broodstock. Though it appeared that sea cucumbers were feeding on biofouling material in both cage types containing them, sea cucumbers at the present density could not remove enough biofouling to have a significant effect on net cleanliness when adult Chinook were present. This result was surprising given the high biofouling removal levels (58% cleaner with sea cucumbers than without them) reported by Ahlgren (1998) [18] in the only other study to conduct this type of work using deposit-feeding sea cucumber species to directly feed on and remove biofouling from finfish cages. However, there are two key differences between the studies that could explain why the sea cucumbers were less able to control biofouling in the present work. Firstly, Ahlgren (1998) [18] tested biofouling removal in cages with salmon fry whereas we tested it with mature adults. Salmon fry are very small and were likely fed a different diet and feed rate than adults would be, in addition to producing less faeces and less dissolved wastes. The likely difference in available nutrients at the farm sites used by Ahlgren (1998) [18] versus our study could have resulted in increased growth of biofouling when adult fish were present, making it more difficult for the same density of sea cucumbers to clean [33,34]. Additionally, Ahlgren (1998) [18] conducted their study in the late winter/early spring when baseline biofouling levels are typically lower, whereas we conducted our study in late summer, when biofouling levels are at their highest in the Northern Hemisphere [8,35,36]. Taken together, these differences suggest that our density of sea cucumbers (N = 100 per cage) was too low to control biofouling in cages with adult fish in the late summer and that this density should be increased for subsequent studies.

Interestingly, sea cucumbers housed alone (S) in the present study were more efficient at removing biofouling than those housed with fish (F + S) (S cages being on average 16% cleaner than the other cage types). The presence of fish can significantly increase levels of net biofouling [33,34], but sea cucumbers in the S and F + S cages actually exhibited different feeding behaviours that could also explain differences in their vertical net cleaning efficiencies. Individuals in the S and F + S cages had different depth preferences for feeding, with the former feeding at all net depths and the latter only feeding at the bottom of cages (Montgomery, pers. obs.; Figure 6). Sea cucumbers in the F + S cages seemed to feed primarily on waste fish food and faeces that sank to the bottom of the cages, rather than on net biofouling. We observed these partially digested fish pellets and faeces in the guts of F + S sea cucumbers that were not seen in the guts of S individuals. Sea cucumbers in the F + S cages, therefore, did not need to climb the nets in order to access quality food, which led to reduced net cleaning efficiency (Figure 6). However, as stocking densities of sea cucumbers are adjusted, intraspecific competition may encourage better net cleaning in F + S cages if individuals are forced to leave the bottom of the cages to access other food resources.

The observed partitioned feeding behaviour of sea cucumbers in the F + S versus S cages was also supported by the stable isotopes results, which confirmed that sea cucumbers in those two cage types were eating different diets. Interestingly, the isotopic ratios of F + S versus S sea cucumbers shifted after each sequential one-month replicate (Figure 5). After three months of consecutive exposure to salmon culture, S sea cucumbers were more similar to wild-caught individuals, but F + S individuals were distinctly different, showing slight shifts in dietary carbon source and trophic position (higher $d^{15}N$ enrichment; Figure 5). These shifts would correspond with the observation of F + S sea cucumbers feeding on commercial salmon pellets containing organic wheat and fish extracts or their faeces. Previous studies of sea cucumbers cultured on the benthos near finfish cages have also reported shifts in $d^{13}C$ and $d^{15}N$ values corresponding to organic matter transfer from fish wastes to sea cucumber tissues, indicating a potential preference for these food sources by deposit-feeding sea cucumber species [37–39]. Though the sea cucumbers in the F + S cages were not efficient at cleaning net biofouling, their preference for salmon waste in the present study does support previous suggestions of using this species as a benthic bioremediatory in IMTA systems with finfish (see [19,40,41]).

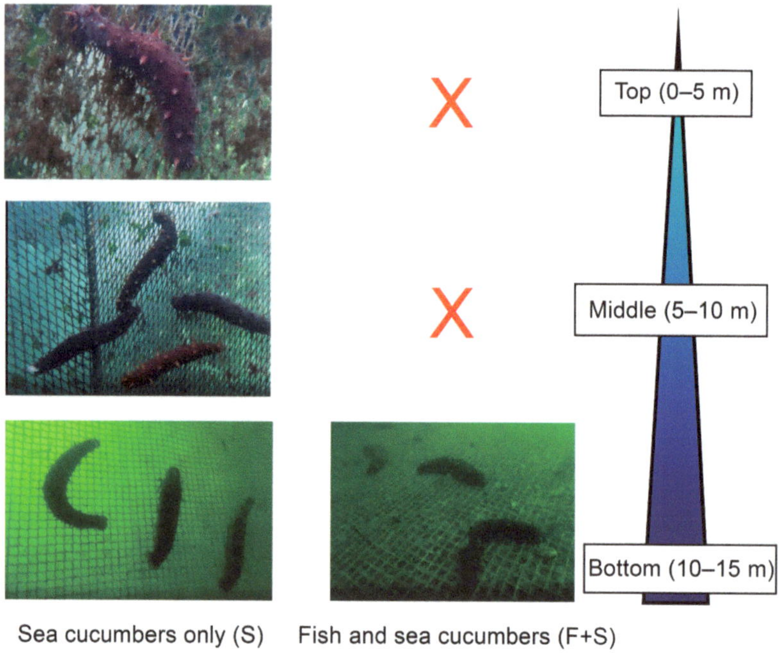

Figure 6. Typical vertical position (0−15 m) of California sea cucumbers in sea cucumber only (S) and fish and sea cucumber (F + S) cages. Red crosses represent positions that sea cucumbers were not observed in.

Previous IMTA studies of sea cucumbers and finfish have reported additional compositional changes in sea cucumber tissues when they are exposed to fish wastes, including changes in fatty acids [39,42] and amino acids/proteins [38]. Compositional changes can positively or negatively influence taste and market favourability, since fatty acids/amino acids are often the building blocks for volatile compounds such as aldehydes, alcohols, and aromatics when sea cucumber tissues are exposed to various seasoning and processing methods [43]. It will therefore be critical to assess any compositional or nutritional differences in sea cucumbers integrated in salmon culture in order to assess the viability of industry selling such sea cucumbers as a secondary revenue stream.

5. Conclusions

The present study tested the biofouling control capacity of California sea cucumbers in open net cages with mature Chinook salmon broodstock. Results showed that the sea cucumbers actively fed on net biofouling when salmon were absent (~16% cleaner on average than all other net types) but preferred to consume uneaten feed/faeces at the bottom of the cages, neglecting the biofouling, when the fish were present. It is hypothesized that biofouling control in cages with salmon may be possible with an increased density of sea cucumbers. These results will be beneficial for industry to develop standard operating procedures for using California sea cucumbers (or other deposit-feeding sea cucumbers) as a net biofouling control, and they could contribute to the development of a management framework for sea cucumber/salmon integrated multi-tropic aquaculture.

Author Contributions: Conceptualization, E.M.M. and C.M.P.; methodology, E.M.M. and C.M.P.; software, E.M.M.; validation, E.M.M. and C.M.P.; formal analysis, E.M.M.; investigation, E.M.M.; resources, B.L.C. and C.M.P.; data curation, E.M.M. and C.M.P.; writing—original draft preparation, E.M.M.; writing—review and editing, E.M.M., B.L.C. and C.M.P.; visualization, E.M.M.; supervision, C.M.P.; project administration, E.M.M., B.L.C. and C.M.P.; funding acquisition, E.M.M. and C.M.P. All authors have read and agreed to the published version of the manuscript.

Funding: This research was funded by the Aquaculture Collaborative Research and Development Program (ACRDP) of Fisheries and Oceans Canada (grant number ACRDP 20-P-05) and Creative Salmon Co., Ltd. (Tofino, British Columbia, Canada).

Institutional Review Board Statement: The study was conducted according to the guidelines of the Declaration of Helsinki and approved by the Institutional Review Board (Pacific Region Animal Care Committee) of Fisheries and Oceans Canada, Pacific Biological Station (AUP 21-005, approved 13 April 2021).

Data Availability Statement: The data are not publicly available.

Acknowledgments: The authors would like to thank the Aquaculture Collaborative Research and Development Program (ACRDP, Project: 20-P-05) of Fisheries and Oceans Canada and Creative Salmon Co., Ltd. (Tofino, British Columbia, Canada) for the funding and opportunity to conduct this study. In particular, E.M. would like to thank Rodrigo Leme, the field crews of Creative Salmon Co., Ltd., and the dive team at Mulder Marine for their enthusiasm and facilitation of the work at the farm sites. The authors would also like to thank Tim Green, Sarah Leduc, UB Diving, the Deep Bay Marine Field Station of Vancouver Island University for hosting us for preliminary laboratory tests on sea cucumber escapement, and the Stable Isotopes Lab at the University of British Columbia for helping conduct the stable isotopes work.

Conflicts of Interest: The sponsors had no role in the design, execution, or interpretation of the study. The sponsors did participate in a final review of the manuscript and provided resources during the study.

References

1. Bloecher, N.; Floerl, O. Towards Cost-Effective Biofouling Management in Salmon Aquaculture: A Strategic Outlook. *Rev. Aquac.* **2021**, *13*, 783–795. [CrossRef]
2. Cornejo, P.; Guerrero, N.M.; Montes, R.M.; Quiñones, R.A.; Sepúlveda, H.H. Hydrodynamic Effect of Biofouling in Fish Cage Aquaculture Netting. *Aquaculture* **2020**, *526*, 735367. [CrossRef]
3. Chen, Q.-P.; Bi, C.-W.; Zhang, Z.-X.; Zhao, Y.-P. Hydrodynamic Effect of Different Biofouling Types on Aquaculture Netting. *Ocean. Eng.* **2023**, *279*, 114430. [CrossRef]
4. Yu, S.; Qin, H.; Li, P.; Gong, F. Drag Force Coefficient and Flow Field Variations of Net with Different Levels of Biological Fouling under Large-Eddy Simulation. *J. Mar. Sci. Technol.* **2023**, *28*, 506–523. [CrossRef]
5. Floerl, O.; Sunde, L.M.; Bloecher, N. Potential Environmental Risks Associated with Biofouling Management in Salmon Aquaculture. *Aquac. Environ. Interact.* **2016**, *8*, 407–417. [CrossRef]
6. Åtland, Å.; Dale, T.; Bloecher, N. Environmental Considerations in Aquaculture Health Management. In *Aquaculture Health Management*; Elsevier: Amsterdam, The Netherlands, 2020; pp. 235–280.
7. Østevik, L.; Stormoen, M.; Nødtvedt, A.; Alarcón, M.; Lie, K.-I.; Skagøy, A.; Rodger, H. Assessment of Acute Effects of in situ Net Cleaning on Gill Health of Farmed Atlantic Salmon (*Salmo salar*). *Aquaculture* **2021**, *545*, 737203. [CrossRef]
8. Fletcher, L.M.; Davidson, I.C.; Bucknall, B.G.; Atalah, J. Salmon Farm Biofouling and Potential Health Impacts to Fish from Stinging Cnidarians. *Aquaculture* **2023**, *568*, 739315. [CrossRef]
9. Bloecher, N.; Floerl, O. Efficacy Testing of Novel Antifouling Coatings for Pen Nets in Aquaculture: How Good Are Alternatives to Traditional Copper Coatings? *Aquaculture* **2020**, *519*, 734936. [CrossRef]
10. Ashraf, P.M.; Lekshmi, N.M.; Chinnadurai, S.; Anjitha, S.; Archana, M.; Kumar, C.M.V.; Sandhya, K.M.; Gop, A.P. Impact Assessment of Biofouling Resistant Nano Copper Oxide–Polyaniline Coating on Aquaculture Cage Nets. *Aquac. Fish* **2023**, *8*, 538–543. [CrossRef]
11. Haberlin, D.; McAllen, R.; Doyle, T.K. Field and Flume Tank Experiments Investigating the Efficacy of a Bubble Curtain to Keep Harmful Jellyfish out of Finfish Pens. *Aquaculture* **2021**, *531*, 735915. [CrossRef]
12. Ohrem, S.J.; Kelasidi, E.; Bloecher, N. Analysis of a Novel Autonomous Underwater Robot for Biofouling Prevention and Inspection in Fish Farms. In Proceedings of the 2020 28th Mediterranean Conference on Control and Automation (MED), Saint-Raphaël, France, 15–18 September 2020; IEEE: Piscataway, NJ, USA, 2020; pp. 1002–1008.

13. Haugaløkken, B.O.A.; Kelasidi, E.; Mulelid, M.; Bloecher, N. Docking Stations for Net-Crawling Underwater Vehicles in Aquaculture Net Pens. In Proceedings of the OCEANS 2021: San Diego–Porto, San Diego, CA, USA, 20–23 September 2021; IEEE: Piscataway, NJ, USA, 2021; pp. 1–10.
14. Boissonnot, L.; Kharlova, I.; Iversen, N.S.; Staven, F.R.; Austad, M. Characteristics of Lumpfish (*Cyclopterus lumpus*) with High Cleaning Efficacy in Commercial Atlantic Salmon (*Salmo salar*) Production. *Aquaculture* **2022**, *560*, 738544. [CrossRef]
15. Imsland, A.K.D.; Reynolds, P. In Lumpfish We Trust? The Efficacy of Lumpfish *Cyclopterus lumpus* to Control *Lepeophtheirus salmonis* Infestations on Farmed Atlantic Salmon: A Review. *Fishes* **2022**, *7*, 220. [CrossRef]
16. Leclercq, E.; Zerafa, B.; Brooker, A.J.; Davie, A.; Migaud, H. Application of Passive-Acoustic Telemetry to Explore the Behaviour of Ballan Wrasse (*Labrus bergylta*) and Lumpfish (*Cyclopterus lumpus*) in Commercial Scottish Salmon Sea-Pens. *Aquaculture* **2018**, *495*, 1–12. [CrossRef]
17. Zeinert, L.R.; Brooks, A.M.L.; Couturier, C.; McGaw, I.J. Potential Use of the Caribbean Spider Crab *Maguimithrax spinosissimus* for Biofouling Removal on Marine Aquaculture Cages. *Aquaculture* **2021**, *545*, 737202. [CrossRef]
18. Ahlgren, M.O. Consumption and Assimilation of Salmon Net Pen Fouling Debris by the Red Sea Cucumber *Parastichopus californicus*: Implications for Polyculture. *J. World Aquac. Soc.* **1998**, *29*, 133–139. [CrossRef]
19. Hannah, L.; Pearce, C.M.; Cross, S.F. Growth and Survival of California Sea Cucumbers (*Parastichopus californicus*) Cultivated with Sablefish (*Anoplopoma fimbria*) at an Integrated Multi-Trophic Aquaculture Site. *Aquaculture* **2013**, *406–407*, 34–42. [CrossRef]
20. Tolon, M.T.; Emiroglu, D.; Gunay, D.; Ozgul, A. Sea Cucumber (*Holothuria tubulosa* Gmelin, 1790) Culture under Marine Fish Net Cages for Potential Use in Integrated Multi-Trophic Aquaculture (IMTA). *Indian J. Geo-Mar. Sci.* **2017**, *46*, 749–756.
21. Zamora, L.N.; Yuan, X.; Carton, A.G.; Slater, M.J.; Marine, L. Role of Deposit-Feeding Sea Cucumbers in Integrated Multitrophic Aquaculture: Progress, Problems, Potential and Future Challenges. *Rev. Aquac.* **2018**, *10*, 57–74. [CrossRef]
22. Cutajar, K.; Falconer, L.; Massa-Gallucci, A.; Cox, R.E.; Schenke, L.; Bardócz, T.; Sharman, A.; Deguara, S.; Telfer, T.C. Culturing the Sea Cucumber *Holothuria poli* in Open-Water Integrated Multi-Trophic Aquaculture at a Coastal Mediterranean Fish Farm. *Aquaculture* **2022**, *550*, 737881. [CrossRef]
23. Purcell, S.W.; Mercier, A.; Conand, C.; Hamel, J.; Toral-Granda, M.V.; Lovatelli, A.; Uthicke, S. Sea Cucumber Fisheries: Global Analysis of Stocks, Management Measures and Drivers of Overfishing. *Fish Fish.* **2013**, *14*, 34–59. [CrossRef]
24. Darya, M.; Sajjadi, M.; Yousefzadi, M.; Sourinejad, I.; Zarei, M. Bioactive Compounds of Sea Cucumber (*Holothuria leucospilota*) Added to Epoxy Resin as Environmental Friendly Antifouling Coats. *Aquat. Physiol. Biotechnol.* **2020**, *8*, 47–66.
25. Darya, M.; Abdolrasouli, M.H.; Yousefzadi, M.; Sajjadi, M.M.; Sourinejad, I.; Zarei, M. Antifouling Coating Based on Biopolymers (PCL/PLA) and Bioactive Extract from the Sea Cucumber *Stichopus herrmanni*. *AMB Express* **2022**, *12*, 24. [CrossRef] [PubMed]
26. Kamyab, E.; Goebeler, N.; Kellermann, M.Y.; Rohde, S.; Reverter, M.; Striebel, M.; Schupp, P.J. Anti-Fouling Effects of Saponin-Containing Crude Extracts from Tropical Indo-Pacific Sea Cucumbers. *Mar. Drugs* **2020**, *18*, 181. [CrossRef]
27. Abràmoff, M.D.; Magalhães, P.J.; Ram, S.J. Image Processing with ImageJ. *Biophotonics Int.* **2004**, *11*, 36–41.
28. Gansel, L.C.; Bloecher, N.; Floerl, O.; Guenther, J. Quantification of Biofouling on Nets: A Comparison of Wet Weight Measurements and Optical (Image Analysis) Methods. *Aquac. Int.* **2017**, *25*, 679–692. [CrossRef]
29. R Core Team. *R: A Language and Environment for Statistical Computing*; R Foundation for Statistical Computing: Vienna, Austria, 2020.
30. Wickham, H. *Ggplot2*; Use R! Springer International Publishing: Cham, Switzerland, 2016; ISBN 978-3-319-24275-0.
31. Wobbrock, J.O.; Findlater, L.; Gergle, D.; Higgins, J.J. The Aligned Rank Transform for Nonparametric Factorial Analyses Using Only ANOVA Procedures. In Proceedings of the SIGCHI Conference on Human Factors in Computing Systems, Yokohama, Japan, 8–13 May 2011; pp. 143–146.
32. Leys, C.; Schumann, S. A Nonparametric Method to Analyze Interactions: The Adjusted Rank Transform Test. *J. Exp. Soc. Psychol.* **2010**, *46*, 684–688. [CrossRef]
33. Strain, P.M.; Hargrave, B.T. Salmon Aquaculture, Nutrient Fluxes and Ecosystem Processes in Southwestern New Brunswick. In *Environmental Effects of Marine Finfish Aquaculture*; Springer: Berlin/Heidelberg, Germany, 2005; pp. 29–57.
34. Hodson, S.L.; Burke, C. Microfouling of Salmon-Cage Netting: A Preliminary Investigation. *Biofouling* **1994**, *8*, 93–105. [CrossRef]
35. Bloecher, N.; Olsen, Y.; Guenther, J. Variability of Biofouling Communities on Fish Cage Nets: A 1-Year Field Study at a Norwegian Salmon Farm. *Aquaculture* **2013**, *416–417*, 302–309. [CrossRef]
36. Bloecher, N.; Powell, M.; Hytterød, S.; Gjessing, M.; Wiik-Nielsen, J.; Mohammad, S.N.; Johansen, J.; Hansen, H.; Floerl, O.; Gjevre, A.G. Effects of Cnidarian Biofouling on Salmon Gill Health and Development of Amoebic Gill Disease. *PLoS ONE* **2018**, *13*, e0199842. [CrossRef]
37. Park, H.J.; Han, E.; Lee, W.C.; Kwak, J.H.; Kim, H.C.; Park, M.S.; Kang, C.K. Trophic Structure in a Pilot System for the Integrated Multi-Trophic Aquaculture off the East Coast of Korean Peninsula as Determined by Stable Isotopes. *Mar. Pollut. Bull.* **2015**, *95*, 207–214. [CrossRef]
38. Sadoul, B.; Caprioli, J.-P.; Barrier-Loiseau, C.; Cimiterra, N.; Laugier, T.; Lagarde, F.; Chary, K.; Callier, M.D.; Guillermard, M.-O.; D'Orbcastel, E.R. Is *Holothuria tubulosa* the Golden Goose of Ecological Aquaculture in the Mediterranean Sea? *Aquaculture* **2022**, *554*, 738149. [CrossRef]
39. Cutajar, K.; Falconer, L.; Massa-Gallucci, A.; Cox, R.E.; Schenke, L.; Bardócz, T.; Andolina, C.; Signa, G.; Vizzini, S.; Sprague, M.; et al. Stable Isotope and Fatty Acid Analysis Reveal the Ability of Sea Cucumbers to Use Fish Farm Waste in Integrated Multi-Trophic Aquaculture. *J. Environ. Manag.* **2022**, *318*, 115511. [CrossRef] [PubMed]

40. Orr, L.C. Co-Culture of Invertebrates with Sablefish (*Anoplopoma fimbria*) in IMTA in British Columbia: Use of Laboratory Feeding Trials to Assess the Organic Extractive Potential of Various Candidate Species. Master's Thesis, University of Victoria, Victoria, BC, Canada, 2012.
41. Hudson, B.; Hauser, L.; Vadopalas, B.; Hetrick, J.; Pride, A.; Hatchery, S.; Seward, H.; Carson, W. *Development of Red Sea Cucumber (Parastichopus californicus) Poly-Aquaculture for Nutrient Uptake and Seafood Export*; Final Report of the Saltonstall-Kennedy Program #NA15NMF4270322; 2019. Available online: https://www.pacshell.org/pdf/Final%20Report%20SK%20Cucumber%202019.pdf (accessed on 28 June 2023).
42. Sun, J.; Hamel, J.-F.; Gianasi, B.L.; Graham, M.; Mercier, A. Growth, Health and Biochemical Composition of the Sea Cucumber *Cucumaria frondosa* after Multi-Year Holding in Effluent Waters of Land-Based Salmon Culture. *Aquac. Environ. Interact.* **2020**, *12*, 139–151. [CrossRef]
43. Li, X.; Dong, Y.; Jiang, P.; Qi, L.; Lin, S. Identification of Changes in Volatile Compounds in Sea Cucumber *Apostichopus japonicus* during Seasonings Soaking Using HS-GC-IMS. *LWT* **2022**, *154*, 112695. [CrossRef]

Disclaimer/Publisher's Note: The statements, opinions and data contained in all publications are solely those of the individual author(s) and contributor(s) and not of MDPI and/or the editor(s). MDPI and/or the editor(s) disclaim responsibility for any injury to people or property resulting from any ideas, methods, instructions or products referred to in the content.

fishes

MDPI

Article

Effects of Snail *Bellamya purificata* Farming at Different Stocking Densities on the Algal and Fungal Communities in Sediment

Yiran Hou [1,2], Mengmeng Zhou [1,2], Rui Jia [1,2], Wei Sun [2], Yanhong Yang [3], Xiongjian Huang [3], Bing Li [1,2,*] and Jian Zhu [1,2,*]

1 Wuxi Fisheries College, Nanjing Agricultural University, Wuxi 214081, China; houyr@ffrc.cn (Y.H.); zhoumengmeng1006@gmail.cn (M.Z.); jiar@ffrc.cn (R.J.)
2 Key Laboratory of Integrated Rice-Fish Farming Ecology, Ministry of Agriculture and Rural Affairs, Freshwater Fisheries Research Center, Chinese Academy of Fishery Sciences, Wuxi 214081, China; sunwei_ffrc@hotmail.com
3 Honghe Hani and Yi Autonomous Prefecture Fisheries Management Station, Mengzi 661100, China; yangyh_hani@hotmail.com (Y.Y.); huangxiongjian@hotmail.com (X.H.)
* Correspondence: lib@ffrc.cn (B.L.); zhuj@ffrc.cn (J.Z.); Tel.: +86-510-85550535 (B.L.); +86-510-85550414 (J.Z.)

Abstract: The snail *Bellamya purificata* is recognized as a potential bio-remediation species, and is commonly employed in polyculture to enhance resource utilization efficiency and realize culture environment regulation. In order to enrich the microbiome studies on elucidating the ecological effects of snail *B. purificata* farming, we assessed the effect of *B. purificata* farming activities, at varying stocking densities, on the algal and fungal communities in sediment. Four experimental groups were established in our study, each corresponding to a different stocking density: 0, 234.38, 468.75, and 937.5 g/m^2, represented as CON, LD, MD, and HD, respectively. High-throughput sequencing based on ITS and 23S ribosomal RNA (rRNA) genes was employed to analyze the variations in algal and fungal communities under *B. purificata* farming activities at different stocking densities. *B. purificata* farming activities had no significant effect on the alpha diversities of fungal and algal communities, but significantly altered the compositions of fungal and algal communities in sediments, especially *B. purificata* farming activity at low stocking density. *B. purificata* farming activities at low stocking density could significantly increase the relative abundances of fungal genera *Paraconiothyrium* and *Penicillium* compared with the CON group. The promoting effect diminished with increasing density. *B. purificata* farming activities at low or medium stocking density also could enhance the relative abundances of algal genera *Microchloropsis*, *Scenedesmus*, and *Auxenochlorella*. Hence, *B. purificata* farming activity at low stocking density might be the optimum density to enhance resource utilization efficiency and minimize environmental pollution.

Keywords: *Bellamya purificata* cultivation; aquaculture; algal community; fungal community; sediment

Key Contribution: The findings of this study provide valuable insights into the impacts of *B. purificata* farming activities at different stocking densities on the algal and fungal communities in sediment and contribute a broader understanding for the snail *B. purificata*'s ecological effects and serve as a valuable theoretical reference for the rational application and development of *B. purificata* farming.

Citation: Hou, Y.; Zhou, M.; Jia, R.; Sun, W.; Yang, Y.; Huang, X.; Li, B.; Zhu, J. Effects of Snail *Bellamya purificata* Farming at Different Stocking Densities on the Algal and Fungal Communities in Sediment. *Fishes* **2023**, *8*, 488. https://doi.org/10.3390/fishes8100488

Academic Editors: Wagner C. Valenti and Eduardo Luis Ballester

Received: 7 August 2023
Revised: 25 September 2023
Accepted: 27 September 2023
Published: 28 September 2023

Copyright: © 2023 by the authors. Licensee MDPI, Basel, Switzerland. This article is an open access article distributed under the terms and conditions of the Creative Commons Attribution (CC BY) license (https://creativecommons.org/licenses/by/4.0/).

1. Introduction

Global aquaculture production approached a record high of 122.6 million tons valued at USD281.5 billion in 2020, with China contributing nearly 70% to the total world aquaculture output [1]. The development of aquaculture has contributed to ensuring global food security and meeting the increasing demand of the growing world population for high-quality proteins [1,2]. Along with rapid development, the aquaculture industry is facing several pressing challenges. Aquaculture ecosystems, including ponds, lakes, reservoirs, and rivers, play a vital role in maintaining the quality of aquatic products and ensuring food security supply. However, traditional aquaculture always pursues maximizing yield and benefits through blindly increasing stocking densities and feeding amounts, which in turn lead to a low utilization of resource, large amounts of residual organic waste, and various environment issues in its own and surrounding aquatic ecosystems [3–8].

To address this issue for sustainable aquaculture, integrated multi-trophic aquaculture (IMTA) has been studied and developed in recent years, which represents an innovative approach to aquaculture that aims to optimize resource utilization and minimize negative impacts on the environment [9–13]. IMTA refers to the polyculture of multiple species from different trophic levels in a mutually beneficial relationship, where waste products from one species serve as nutrients for others [9]. The snail *Bellamya purificata* is recognized as a potential bio-remediation species, and commonly employed in polyculture to enhance the resource utilization efficiency and realize the culture environment regulation [14–16]. The snail *B. purificata* is a highly representative freshwater snail and is widely distributed across ponds, lakes, reservoirs, rivers, and other aquatic ecosystems in China, which has a natural preference for inhabiting silt and consuming organic debris and algae in its surrounding environment [17,18]. There have been some studies on the ecological effects of *B. purificata* in aquaculture or water purification processes [14,19,20]. The snail *B. purificata* can enhance organic matter degradation within sediment and promote material circulation at the sediment–water interface, in addition to purifying the culture water [14,19,20]. We have attempted to explore the mechanisms or pathways of *B. purificata*'s ecological effects from a microbiological perspective by determining the bacterial communities in sediments [21]. However, fungi and algae are also important components of microorganisms, although the biomass of bacteria may be ten times that of fungi in the sediment [22].

Fungi and algae play a vital role in material circulation and biogeochemical processes in the aquatic ecosystem. Algae serve as crucial primary producer and food chain driver in aquatic ecosystems [23]. Fungi and algae can provide insight into dynamic variations in ponds, lakes, reservoirs, rivers, and other aquatic ecosystems through the algae's and fungi's structural, functional, and physiological features [23–26]. A previous study has revealed that different farming practices and farming species markedly affect the fungal communities in sediments [27]. Xu [25] first reported the dynamic variations of fungal community and diversity in the integrated rice–crab farming system to better understand and optimize the farming ecosystem. Even so, there is still a lack of studies focusing on the fungal and algal communities In aquaculture ecosystems, particularly the influences of *B. purificata* farming activities on the fungal and algal communities. We believe that enriching the content of microbiome studies to demonstrate the snail *B. purificata*'s ecological effects is crucial for the rational application and development of *B. purificata* farming.

Hence, we performed high-throughput sequencing based on ITS and 23S ribosomal RNA (rRNA) genes in sediment to assess the effects the *B. purificata* farming activities with different stocking densities on the fungal and algal communities. The findings of this study would lead to a broader understanding of the snail *B. purificata*'s ecological effects and serve as a valuable theoretical reference for the rational application and development of *B. purificata* farming.

2. Materials and Methods

2.1. Experiment Design

The experiments were conducted at the Freshwater Fisheries Research Center of the Chinese Academy of Fishery Sciences (120.250479° E, 31.51581° N; Wuxi, China). The experimental snail and sediment were collected from aquaculture ponds located at the Dapu aquaculture facility (119.939129° E, 31.316981° N; Wuxi, China). Before commencing the experiment, a period of 14 days was allotted for the *B. purificata* snails to acclimatize to the controlled laboratory environment by placing them in a glass tank. To ensure homogeneity and consistency, the sediment used in the experiment underwent drying, grinding, sieving, and mixing according to the pre-processing steps implemented in the previous studies [28–30]. Twelve glass tanks (80 × 40 × 45 cm) were employed in the experiment. All the glass tanks were covered with sediment to a depth of 7 cm on the bottom, filled with aerated and filtered tap water, and then left to be precipitated and stabilized for 14 d before the experiment. For the experiment, four separate groups, including one control group and three treatment groups, were established according to four different stocking densities and each group with three replicates. The four different stocking densities were 0, 234.38, 468.75, and 937.50 g/m^2, respectively. The corresponding groups were abbreviated as CON, LD, MD, and HD, respectively. After acclimation, healthy snails with an average wet weight of 2.53 ± 0.01 g were collected and randomly distributed between glass tanks. During the experiment period, the commercial feed (Zhejiang Haida Feed Co., Ltd., Shaoxing, China) was utilized as the experimental diet. The snails were fed every day at 4:00 pm, with the amount approximating 2% of their individual body weights. The experimental conditions were maintained at a constant water temperature of 26.5 ± 0.5 °C and a dissolved oxygen (DO) level of about 6.5 mg/L. One-third of the water in each glass tank was changed every two days. The experimental lighting condition was a natural light/dark cycle. The experimental period lasted for 80 days, during which all snails were observed to be in good health with no mortalities recorded.

2.2. Sample Collection

At the end of the experiments, sediment samples were collected using plastic tubes with a diameter of 2 cm from ten randomly chosen sampling points in each glass tank. The sediment samples were taken from the surface sediment layer, which was between 0–1 cm deep. To maintain consistency, all the sediment samples from each identical glass tank were mixed thoroughly. Sediment samples designated for the analysis of fungal and algal communities were promptly stored at −80 °C to preserve their integrity for further analysis.

2.3. PCR Amplification and Sequencing

The DNA of fungi and algae in the sediment samples was extracted using the E.Z.N.A.® soil DNA Kit (Omega Bio-tek, Norcross, GA, USA) according to the manufacturer's protocol. The quality and concentration of DNA were evaluated using 1.0% agarose gel electrophoresis and a NanoDrop® ND-2000 spectrophotometer (Thermo Scientific Inc., Waltham, MA, USA), and subsequently, the DNA was stored at −80 °C until further use. Specific primers were designed and synthesized to amplify the ITS1-ITS2 and 23S rDNA regions using an ABI GeneAmp® 9700 PCR thermocycler (ABI, CA, USA). The ITS1-ITS2 region was amplified with primer pairs ITS1 (5′–CTTGGTCATTTAGAGTAAGTAA–3′) and ITS2 (5′–GCTGTGTTCATCGATGC–3′). The 23S rDNA region was amplified with two forward primers (A23SrVF1: 5′–AGACARAAAGACCCTATG–3′ and A23SrVF2: 5′–CARAAAGACCTATTGMAGCT–3′) and two reverse primers (A23SrVR1: 5′–AGATCAGC CTTTATCC–3′ and A23SrVR2: 5′–TCAGCCTGTTATCCTAG–3′) [31].

PCR amplification for ITS1-ITS2 region was carried out three times in 20 μL reaction mixtures consisting of 2 μL of 10× Buffer, 2 μL of 2.5 mM dNTPs, 0.8 μL of forward primer (5 μM), 0.8 μL of reverse primer (5 μM), 0.2 μL of rTaq Polymerase, 0.2 μL of BAS, 10 ng of template DNA, and double-distilled H_2O to the final volume. PCR amplification for the 23S rDNA region was carried out three times in 20 μL reaction mixtures including 4 μL of 5× FastPfu Buffer, 2 μL of 2.5 mM dNTPs, 0.8 μL of forward primer (5 μM), 0.8 μL of reverse primer (5 μM), 0.4 μL of FastPfu Polymerase, 0.2 μL of BAS, 10 ng of template DNA, and double-distilled H_2O to the final volume. The PCR amplification was performed using the following cycling conditions: initial denaturation at 95 °C for 3 min, followed by 30 cycles of denaturing at 95 °C for 30 s, annealing at 55 °C for 30 s and extension at 72 °C for 45 s, and single extension at 72 °C for 10 min, and end at 10 °C. The PCR product was extracted with a 2% agarose gel. The extracted PCR product was then purified and quantified by the AxyPrep DNA Gel Extraction Kit (Axygen Biosciences, Union City,

CA, USA) and Quantus™ Fluorometer (Promega, WI, USA) following the manufacturer's protocol, respectively.

Purified amplicons were pooled in equimolar amounts and subjected to paired-end sequencing on an Illumina MiSeq PE300 platform/NovaSeq PE250 platform (Illumina, CA, USA) at Majorbio Bio-Pharm Technology Co., Ltd. (Shanghai, China). Raw sequencing reads were deposited into the NCBI Sequence Read Archive (SRA) database (Accession Number: PRJNA993837).

2.4. Data Processing

The raw FASTQ files underwent de-multiplexing through an in-house perl script. Subsequently, raw FASTQ files were subjected to quality filtering using fastp version 0.19.6 and merged using FLASH version 1.2.7 [32,33]. The optimized sequences were clustered into operational taxonomic units (OTUs) using UPARSE 7.1 with a 97% sequence similarity level [34]. The most abundant sequence for each OTU was chosen as a representative sequence. To ensure accuracy, the OTU table was manually filtered, and chloroplast sequences were eliminated from all samples. In order to mitigate the effects of sequencing depth on the alpha and beta diversity measure, the number of ITS DNA and 23S rDNA sequences from each sample were rarefied to 44,639 and 13,331, respectively. The taxonomy of each OTU representative sequence was analyzed using RDP Classifier version 2.2 against the ITS DNA and 23S rDNA database (Unite ITS 8.0 and NT v20210917) with a confidence threshold of 0.7, respectively [31,35].

2.5. Statistical Analysis

Bioinformatic analysis for the sediment samples was conducted by the Majorbio Cloud platform (https://cloud.majorbio.com (accessed on 1 December 2022)). Alpha diversity indices including the observed richness (Sobs), Shannon, Simpson, Chao1, and ACE were calculated using Mothur v1.30.1 [36]. The Sobs, Chao1, and ACE were used for accessing the richness of the fungal and algal communities, while Shannon and Simpson were employed to evaluate the diversity. Higher values of these indices indicate higher richness or diversity. One-way ANOVA followed by the Tukey–Kramer post hoc test was used to confirm differences in the alpha diversity indices of fungal and algal communities within sediment between CON, LD, MD, and HD groups. Subsequently, for the beta diversity, the Bray–Curtis distances among different samples were calculated and the principal coordinate analysis (PCoA) based on Bray–Curtis distances was conducted to reveal the differences in the fungal and algal communities between different groups. The Adonis test accompanying the PCoA analysis was performed to further determine the differences in the fungal and algal communities between different groups. The community bar plot was conducted to demonstrate the relative abundances of the dominate phyla and genera in the fungal and algal communities. The phylum or genus in the fungal and algal communities with a relative abundance greater than 1% would be defined as a dominant phylum or genus. One-way ANOVA followed by the Tukey–Kramer post hoc test was used to confirm differences in the relative abundances of all the phyla and genera within fungal and algal communities. All analyses and related figures were completed using the vegan and ggplot2 packages in R v. 4.0.3 (R Core Team, Vienna, Austria).

3. Results

3.1. Overview of Fungal and Algal Communities in Sediment

In the present study, 2295 distinct OTUs were obtained from the sediment samples through Illumina sequencing technology based on the fungal ITS gene, and subsequentially assigned into 12 phyla and 259 genera. As shown in Figure 1, there were 124 OTUs and 11 genera unique to the CON group, 515 OTUs and 108 genera unique to the LD group, 262 OTUs and 20 genera unique to the MD group, and 336 OTUs and 17 genera unique to the HD group. Moreover, the LD group exhibited 834 OTUs and 143 genera not included in the CON group, which were far more than those exhibited by MD and HD groups.

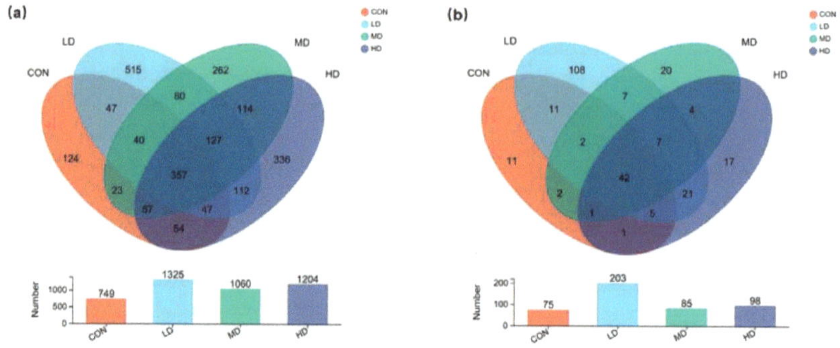

Figure 1. Venn diagrams of the fungal communities in CON, LD, MD, and HD groups at operational taxonomic unit (**a**) and genus (**b**) levels.

We also obtained 1391 algae-related OTUs based on the 23S rRNA gene, which were assigned into 44 phyla and 333 genera. As shown in Figure 2, there were 205 OTUs and 36 genera unique to the CON group, 156 OTUs and 21 genera unique to the LD group, 140 OTUs and 19 genera unique to the MD group, and 119 OTUs and 22 genera unique to the HD group. Compared with the CON group, the LD group owned 286 unique OTUs and 53 unique genera, the MD group owned 268 unique OTUs and 48 unique genera, and the HD group owned 248 unique OTUs and 53 unique genera.

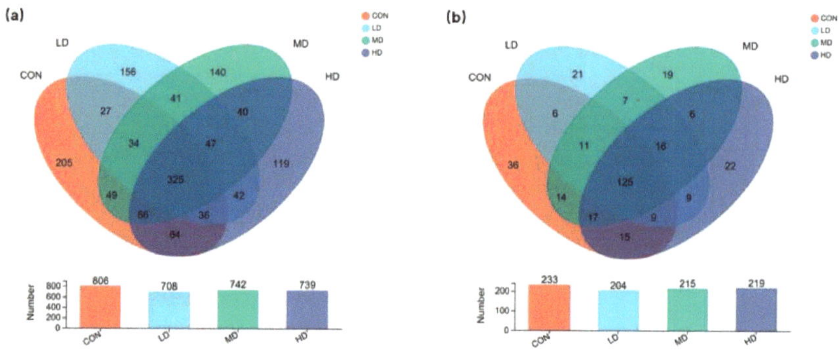

Figure 2. Venn diagrams of the algal communities in CON, LD, MD, and HD groups at operational taxonomic unit (**a**) and genus (**b**) levels.

3.2. *Alpha and Beta Diversities of Fungal Community in Sediment*

Alpha diversity indices including Sobs, Shannon, Simpson, ACE, and Chao1 were calculated to evaluate the diversity and richness of the fungal community in sediment. As shown in Figure 3, no significant differences in the Sobs, Shannon, Simpson, ACE, and Chao1 between the CON, LD, MD, and HD groups were observed according to the results of one-way ANOVA followed by the Tukey–Kramer post hoc test ($p > 0.05$). For the beta diversity, PCoA analysis was conducted to investigate the differences in fungal community based on Bray–Curtis distances. As shown in Figure 4a, PC1 and PC2 explained 25.39% and 18.94% of the total variation in the fungal community within sediment, respectively. The CON, LD, MD, and HD groups were not obviously separated, as they all have overlapping areas. The result of the Adonis test also indicated no significant differences between the CON, LD, MD, and HD groups ($p > 0.05$).

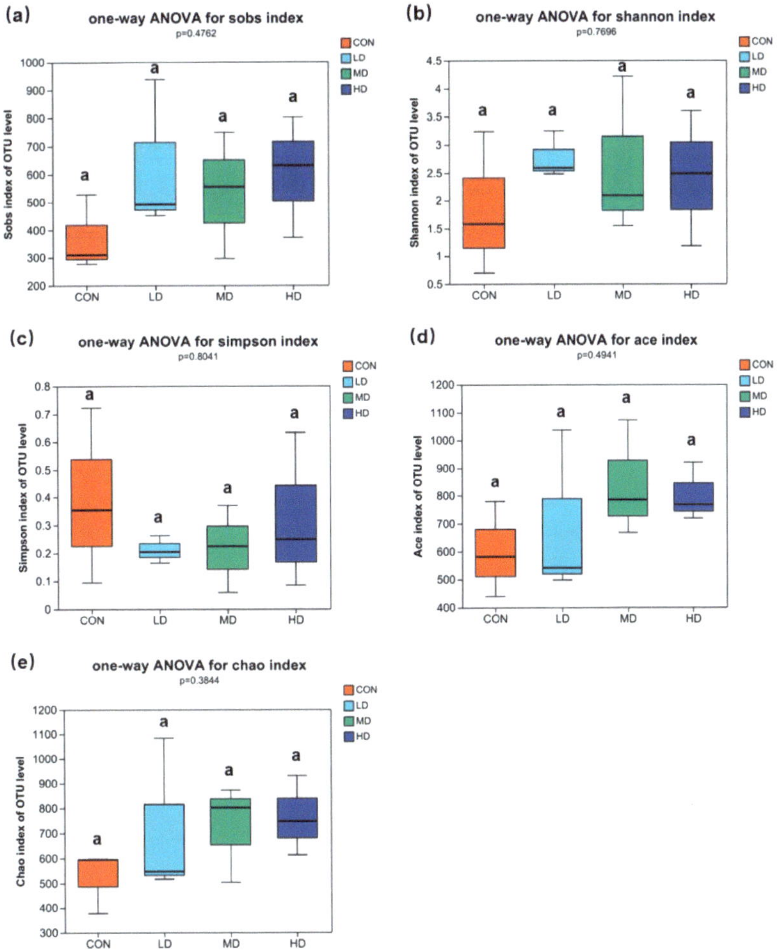

Figure 3. Differences in the alpha diversity indices of fungal communities including Sobs (**a**), Shannon (**b**), Simpson (**c**), ACE (**d**), and Chao1 (**e**) between CON, LD, MD, and HD groups. Different letters indicated significant differences between different groups.

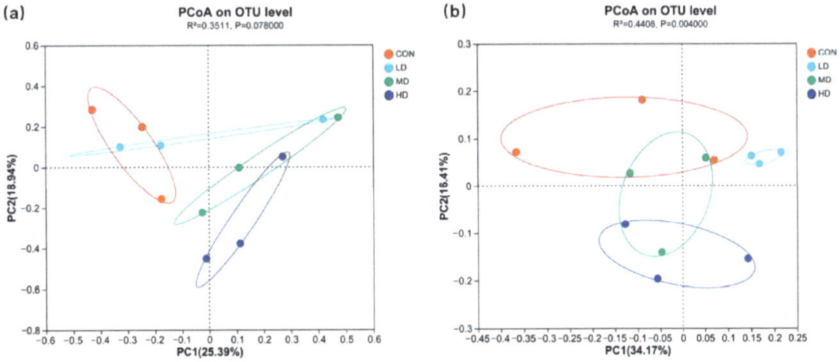

Figure 4. Principal coordinate analysis (PCoA) of fungal (**a**) and algal (**b**) communities in sediment based on Bray–Curtis distances.

3.3. Alpha and Beta Diversities of Algal Community in Sediment

According to the results of one-way ANOVA followed by the Tukey–Kramer post hoc test, there were no significant differences in Sobs, Shannon, Simpson, ACE, and Chao1 between the CON, LD, MD, and HD groups as shown in Figure 5 ($p > 0.05$). For the beta diversity, as shown in Figure 4b, the first two PCs in the PCoA analysis explained 34.17% and 16.41% of the total variation in the algal community within sediment, respectively. The LD group was obviously separated from the other three groups as revealed by the PCoA results. The result of the Adonis test also confirmed the PCoA results and indicated a significant difference between the CON, LD, MD, and HD groups ($p < 0.05$).

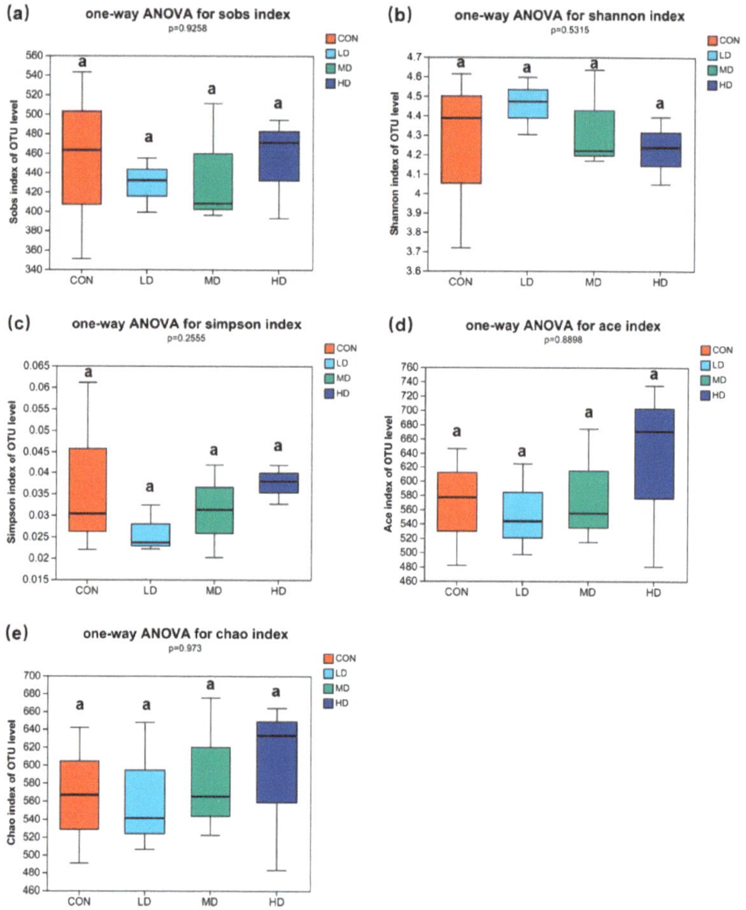

Figure 5. Differences in the alpha diversity indices of algal communities including Sobs (**a**), Shannon (**b**), Simpson (**c**), ACE (**d**), and Chao1 (**e**) between CON, LD, MD, and HD groups. Different letters indicated significant differences between different groups.

3.4. Composition of Fungal Community in Sediment

There were in total 12 phyla and 259 genera assigned from 2295 distinct OTUs in the sedimentary fungal community in the present study. The dominate phyla and genera (relative abundance > 1%) in the fungal community are shown in Figure 6. Similar with previous studies, most fungi could not be effectively annotated at either the phylum or genus level and has been represented as unclassified_k_Fungi in Figure 6 [27,37]. Excluding the unclassified fungal taxa, the dominate phyla were Ascomycota, Rozellomycota,

Basidiomycota, and Chytridiomycota, and the dominate genera were *Scutellinia*, *unclassified_p_Rozellomycota*, *unclassified_p_Chytridiomycota*, and *Apiotrichum* in the present study.

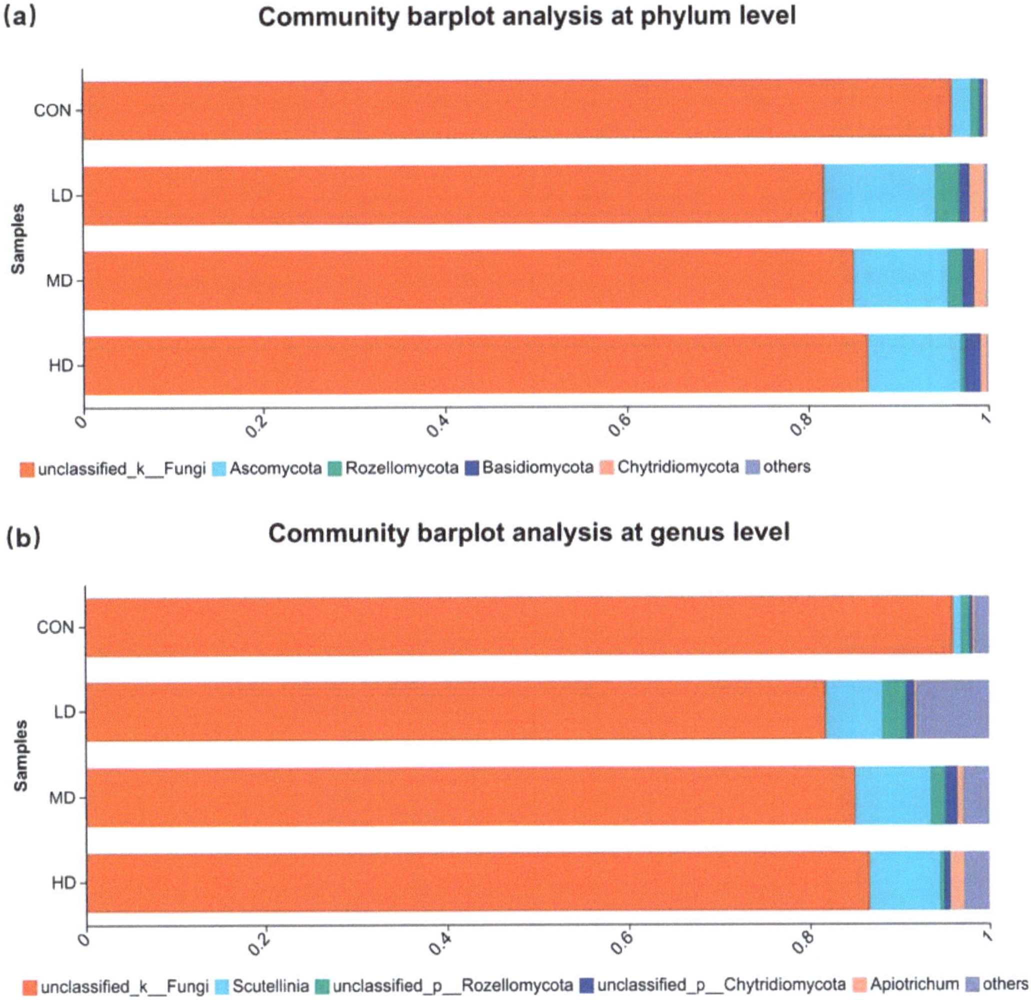

Figure 6. Fungal community composition at phylum (**a**) and genus (**b**) levels within sediment in CON, LD, MD, and HD groups.

As shown in Figure 7, the relative abundances of genera *Paraconiothyrium* and *Penicillium* were significantly different among the CON, LD, MD, and HD groups ($p < 0.05$). The relative abundance of *Paraconiothyrium* in sediment of LD groups was significantly higher than that of CON group ($p < 0.05$). However, no significant differences in the relative abundance of *Paraconiothyrium* between the LD, MD, and HD groups were observed ($p > 0.05$). The relative abundance of *Penicillium* in the LD groups was significantly higher than in the CON, MD, and HD groups ($p < 0.05$).

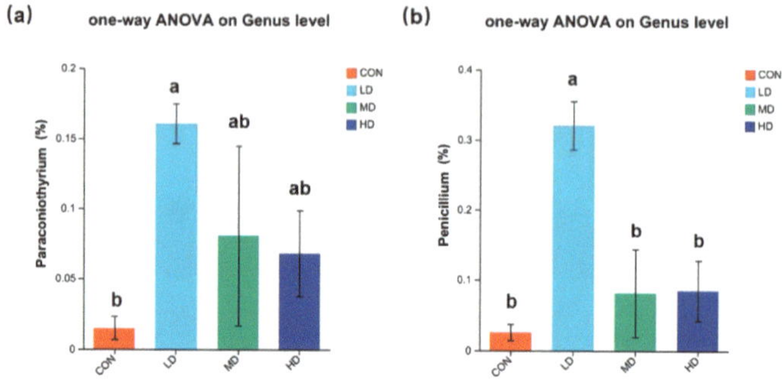

Figure 7. The significantly different genera of fungal communities in the sediment between CON, LD, MD, and HD groups, including *Paraconiothyrium* (**a**) and *Penicillium* (**b**). Different letters indicated significant differences between different groups.

3.5. Composition of Algal Community in Sediment

There were a total of 44 phyla and 333 genera in the algal community in the present study. The dominate phyla and genera (relative abundance > 1%) in the algal community were shown in Figure 8. The dominate phyla ranked in descending order of relative abundance were Verrucomicrobia, unclassified_d_unclassified, Chlorophyta, Cyanobacteria, Ignavibacteriae, Firmicutes, unclassified_d_Bacteria, unclassified_d_Eukaryota, Bacillariophyta, Candidatus_Woesebacteria, unclassified, Proteobacteria, and Oomycota. There were 23 dominate genera. Excluding the unclassified algal genus, the top 5 dominate genera were *Pedosphaera*, *Prosthecobacter*, *Desmodesmus*, *Synechococcus*, and *Opitutus*.

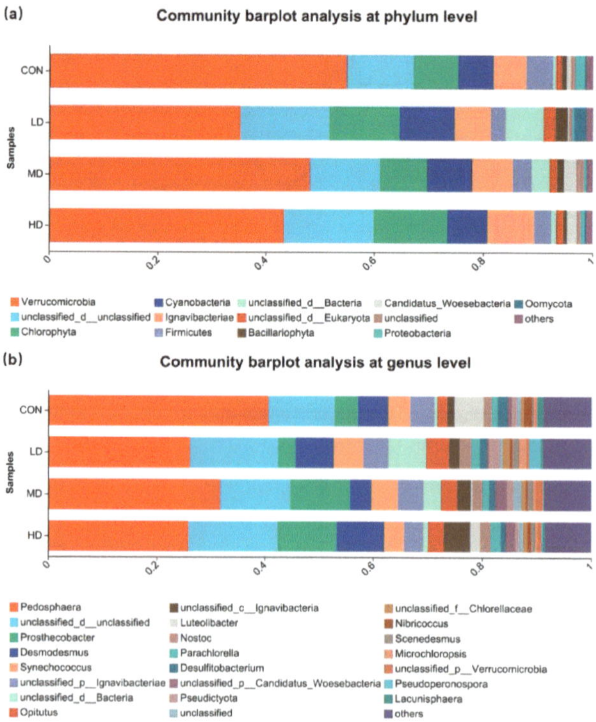

Figure 8. Algal community composition at phylum (**a**) and genus (**b**) levels within sediment in CON, LD, MD, and HD groups.

Through One-way ANOVA, there were 1 phylum and 6 genera that were significantly different among the CON, LD, MD, and HD groups. As shown in Figure 9, the relative abundance of phylum Streptophyta in the sediment of the LD group was significantly higher than that of MD group ($p < 0.05$). However, the CON, LD, and HD groups showed similar relative abundances of Streptophyta ($p > 0.05$). The relative abundance of genus *Nibricoccus* in the HD group was significantly higher than in the MD group ($p < 0.05$). The LD and MD groups exhibited significantly higher relative abundances of *Scenedesmus* compared with the CON group ($p < 0.05$), but the relative abundance of *Scenedesmus* in the HD group was similar with that in the CON group ($p > 0.05$). The relative abundance of *Microchloropsis* in the LD group was significantly higher than that in the CON, MD, and HD groups ($p < 0.05$). The relative abundance of *Auxenochlorella* in the CON group was significantly lower than that in the MD group ($p < 0.05$), and no significant differences among the LD, MD, and HD groups were observed ($p > 0.05$). The relative abundance of *Choricystis* in the LD group showed a significant decreasing trend with increasing stocking density, while the relative abundance of *Chthoniobacter* showed a significant increasing trend with increasing stocking density and reached a maximum in the MD group ($p < 0.05$).

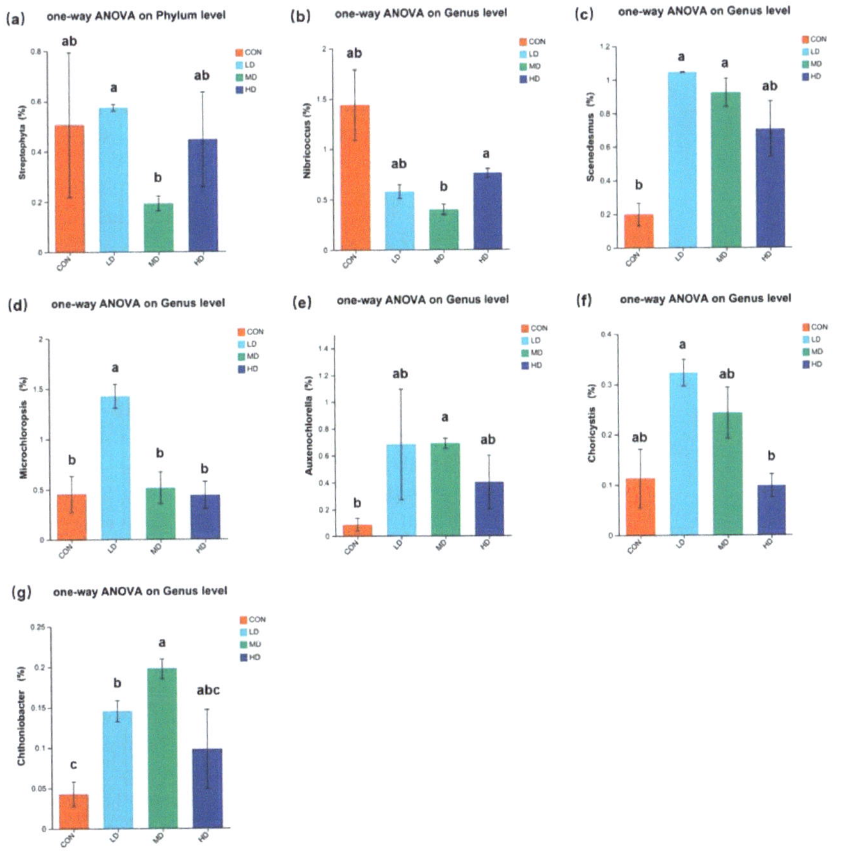

Figure 9. The significantly different phylum and genera of algal communities in the sediment between CON, LD, MD, and HD groups, including Streptophyta (**a**), *Nibricoccus* (**b**), *Scenedesmus* (**c**), *Microchloropsis* (**d**), *Auxenochlorella* (**e**), *Choricystis* (**f**), and *Chthoniobacter* (**g**). Different letters indicated significant differences between different groups.

4. Discussion

4.1. Fungal Community in Sediment Affect by *B. purificata* Farming Activities

Fungal community, as well as algal and bacterial communities, play a significant role in the biogeochemical cycle of aquatic ecosystems. Fungi have diverse morphological structures, complex community structures, and strong metabolic capabilities, the physiological and biochemical characteristics of which are affected by the surrounding environment [38]. In the present study, most fungi could not be well annotated, which could be attributed to the relatively limited DNA sequences in the existing fungal databases compared to the total amount of fungal DNA sequences [39,40]. The phyla Ascomycota, Rozellomycota, Basidiomycota, and Chytridiomycota were the dominate phyla in the present study, ranking among the top 5 in relative abundance, respectively. Fan [39] analyzed the fungal community in sediment from tilapia (*Oreochromis niloticus*) cultural ponds and reported the Basidiomycota, Ascomycetes, and Chytridiomycota as the dominate phyla. Zhang et al. [27] investigated dozens of fish, crab, and crayfish ponds and the dominant fungal phyla observed in these ponds were predominantly Ascomycota, Chytridiomycota, Rozellomycota, and Basidiomycota. Wang et al. [41] and Zhao et al. [42] also revealed the Ascomycetes and Basidiomycota as the dominant phyla in Poyang Lake and Hongze Lake. These similar results indicate that the present indoor simulation experiment could well simulate the real state of the fungal community in the cultural ecosystem.

B. purificata farming activities at different stocking densities had no obvious impacts on the fungal community diversity in the sediment, which was different with the variations in fungal community diversity under tilapia farming activities reported in a previous study [39]. Fan [39] pointed out that the fungal community in the cultural pond was not only sensitive to temperature and climate, but also to the external nutrient inputs. The accumulation of organic waste, including feed residues and faces, in sediment significantly affect the fungal community [39]. However, the snail *B. purificata* is a typical species which has been proven to play an important role in organic matter degradation in sediment [19]. Snails promote the degradation of organic matter, reduce the accumulation of organic matter in sediment, and enhance material cycling at the sediment–water interface [19]. Hence, the snail *B. purificata* might have the potential to maintain the consistency in sedimentary organic matter content between different groups by ingestion and promoting the degradation and recycling of organic matter, thereby avoiding the obvious impact derived from organic matter accumulation on the fungal community. Meanwhile, the unaffected fungal community also revealed that bioturbation by the snail *B. purificata* could not directly alter the fungal community in sediment, although *B. purificata* bioturbation could significantly change the physico-chemical properties of the sediment [19]. As the relevant physicochemical properties within the sediment were not measured in our study, this inference needs to be explored in further studies.

Although the overall impact of *B. purificata* farming activities on the fungal community in sediment is not significant according to the results of alpha and beta diversity, *B. purificata* cultivation affected several specific genera. *B. purificata* cultivation in the low stocking density (LD) group significantly increased the genera *Paraconiothyrium* and *Penicillium*. The genus *Paraconiothyrium* is widely distributed worldwide with diverse host habitats and has potential applications as a producer of antibiotics [43]. *Paraconiothyrium* can suppress the activity of harmful fungi such as *Sclerotinia sclerotiorum* in the soil, thus reducing the infection caused by pathogenic microorganisms [43]. The genus *Penicillium*, as a saprophytic fungus, is widely distributed in soil and sediment [44,45]. The secondary metabolites of the *Penicillium* are various compounds with antibacterial and antioxidant activities, which can inhibit the growth of pathogenic bacteria as well as plant-pathogenic fungi [44–50]. Hence, the significantly increased *Paraconiothyrium* and *Penicillium* in the LD group indicated that *B. purificata* farming activity at a low stocking density may effectively promote the enrichment of *Paraconiothyrium* and *Penicillium* in sediment, thereby inhibiting pathogenic microbe activity, improving the performance of cultured aquatic animals, and establishing more sustainable aquatic food production [51]. However, this promoting effect diminished with the increasing density.

4.2. Algal Community in Sediment Affect by B. purificata Farming Activities

Algae play a significant role in autochthonous primary production, providing the basis for littoral secondary production, and act as crucial regulators of nutrient dynamics within aquatic ecosystems [52,53]. Despite their importance, variations in the algal community within the aquaculture ecosystem has received relatively little attention [52]. In the present study, *B. purificata* farming activities imposed a more pronounced effect on the algal community in sediment relative to the fungal community. In particular, *B. purificata* farming activity at a low stocking density had a stronger overall effect on the algal community than that at medium and high stocking densities according to the results of the PCoA. As reported in a previous study, phyla such as Chlorophyta and Cyanobacteria, were dominated in the algal community of the East Sea [54]. Similarly, the relative abundance of phyla Chlorophyta and Cyanobacteria were also dominated in the present study, both of which ranked among the top 5 in relative abundance without unclassified taxa.

Similar to the effects of *B. purificata* farming activities at different stocking densities on the fungal community's diversity, *B. purificata* farming activities also had no obvious impacts on the algal community's diversity in the sediment. However, there has been limited research on the interactions and effects of culture species with algal communities in sediments. The dynamics of algal communities are associated with various physical and chemical factors in the aquaculture environment, among which inorganic nutrients such as nitrogen and phosphorus are fundamental substances required for the growth and reproduction of algae [55,56]. *B. purificata* can enhance the organic matter degradation within sediment and promote material circulation at the sediment–water interface, which may facilitate the growth of algae [14,19,20]. On the other hand, *B. purificata* prefers to ingest organic debris and algae in its surroundings [17,18]. Hence, we hypothesized that the non-significant impacts were likely to be attributed to a combined effect of promoting algal growth by bioturbation and inhibiting the algal community by ingestion.

Moreover, in the present study, the algal genera *Microchloropsis*, *Scenedesmus*, and *Auxenochlorella* were significantly enhanced in the LD or MD group relative to those in the CON group. The algal genus *Microchloropsis* adapts to different nutritional conditions and can effectively utilize nitrate and organic nitrogen [57,58]. The genus *Scenedesmus* is capable of removing nitrate and phosphate from the surrounding environment [59]. In addition, the genus *Auxenochlorella* is an early-appearing single-celled eukaryotic green algae, which is an efficient primary producer in ecosystems and capable of removing ammonium [60,61]. The significantly increased relative abundances of *Microchloropsis*, *Scenedesmus*, and *Auxenochlorella* in the LD or MD groups might result from the enhanced degradation of organic matter in sediment and the improved cycling of nutrients at the sediment–water interface caused by *B. purificata* snail farming. Talib et al. [62] have investigated mitigating eutrophication in lakes through nutritional control and biological manipulation and discovered that the significant increase in the abundance of the harmless genus *Scenedesmus* is an important phenomenon during the process of reducing eutrophication levels. Meanwhile, the importance of *Scenedesmus* and *Auxenochlorella* has been confirmed and they are widely employed in wastewater treatment and environmental regulation [59–61]. This revealed that the enhanced degradation of organic matter in sediment and the improved migration of nutrients from sediment to the overlying water caused by *B. purificata* bioturbation would not induce an increase in the relative abundance of harmful algae. Instead, it increased the relative abundance of harmless algae that have an environmental regulatory significance.

5. Conclusions

B. purificata farming activities had no significant effect on the alpha diversities of fungal and algal communities, but significantly altered the compositions of fungal and algal communities in sediments, especially *B. purificata* farming activity at a low stocking density. *B. purificata* farming activities at a low stocking density could significantly increase the relative abundances of fungal genera *Paraconiothyrium* and *Penicillium*, thereby inhibiting pathogenic microbe activity and improving the performance of cultured aquatic animals. *B. purificata* farming activities at low or medium stocking densities could also enhance the relative abundances of harmless algal genera *Scenedesmus* and *Auxenochlorella*, which are widely employed in wastewater treatment and environmental regulation. Therefore, the low stocking density (234.38 g/m^2) in the present study might be the most optimum density from the perspective of fungal and algal communities. Implementing *B. purificata* cultivation at a low stocking density might lead to more sustainable aquatic food production. The findings of this study might contribute to better understanding for the ecological effects of snail *B. purificata* farming and serve as a valuable theoretical reference for the rational application and development of *B. purificata* farming.

Author Contributions: Y.H.: conceptualization, methodology, formal analysis, data curation, writing — original draft. M.Z.: formal analysis, data curation, writing—original draft. R.J.: visualization, writing—review and editing. W.S.: formal analysis, data curation. Y.Y.: investigation. X.H.: investigation. B.L.: resources, investigation. J.Z.: resources, funding acquisition, supervision. All authors have read and agreed to the published version of the manuscript.

Funding: This research was funded by the Central Public-interest Scientific Institution Basal Research Fund, CAFS (grant No. 2022XT0504), the National Natural Science Foundation of China (grant No. 31802302), and the China Agriculture Research System of MOF and MARA (grant No. CARS-45).

Data Availability Statement: The datasets presented in this study can be found in online repositories. The names of the repository/repositories and accession number(s) are PRJNA993837.

Conflicts of Interest: The authors declare no conflict of interest.

References

1. Food and Agriculture Organization of the United Nations. *The State of World Fisheries and Aquaculture 2022*; Food and Agriculture Organization of the United Nations: Rome, Italy, 2022.
2. Costello, C.; Cao, L.; Gelcich, S.; Cisneros-Mata, M.A.; Free, C.M.; Froehlich, H.E.; Golden, C.D.; Ishimura, G.; Maier, J.; Macadam-Somer, I.; et al. The future of food from the sea. *Nature* **2020**, *588*, 95–100. [CrossRef]
3. Cao, L.; Wang, W. Wastewater management in freshwater pond aquaculture in China. In *Sustainability in Food and Water. Alliance for Global Sustainability Bookseries (Science and Technology: Tools for Sustainable Development)*; Sumi, A., Fukushi, K., Honda, R.K.H., Eds.; Springer: Dordrecht, The Netherlands, 2010; Volume 18.
4. Deng, L.F.; Jiang, X.L. The application and research progress of bioremediation in pond aquaculture. *Oceanol. Limnol. Sin.* **2013**, *44*, 1270–1275.
5. Liu, H.L.; Cao, X.Y.; Song, C.L.; Zhou, Y.Y. Environmental effects of organic matter enrichment and restoration strategy in sediment of aquaculture ponds. *J. Hydroecol.* **2011**, *32*, 130–134.
6. Liu, X.G. *Study on the Pond Aquaculture Pollution and Ecological Engineering Regulation Techniques*; Nanjing Agricultural University: Nanjing, China, 2011.
7. Ji, X.; Lin, X.; Xu, Z.; Lin, Y. Machenism of mariculture self-pollution and its effects on environment. *Mar. Environ. Sci.* **2000**, *19*, 66–71.
8. Zhou, H.; Jiang, C.; Zhu, L.; Wang, X.; Hu, X.; Cheng, J.; Xie, M. Impact of pond and fence aquaculture on reservoir environment. *Water Sci. Eng.* **2011**, *4*, 92–100.
9. Knowler, D.; Chopin, T.; Martinez-Espineira, R.; Neori, A.; Nobre, A.; Noce, A.; Reid, G. The economics of Integrated Multi-Trophic Aquaculture: Where are we now and where do we need to go? *Rev. Aquac.* **2020**, *12*, 1579–1594. [CrossRef]
10. Neori, A.; Chopin, T.; Troell, M.; Buschmann, A.H.; Kraemer, G.P.; Halling, C.; Shpigel, M.; Yarish, C. Integrated aquaculture: Rationale, evolution and state of the art emphasizing seaweed biofiltration in modem mariculture. *Aquaculture* **2004**, *231*, 361–391. [CrossRef]
11. Shi, H.H.; Zheng, W.; Zhang, X.L.; Zhu, M.Y.; Ding, D.W. Ecological-economic assessment of monoculture and integrated multi-trophic aquaculture in Sanggou Bay of China. *Aquaculture* **2013**, *410*, 172–178. [CrossRef]
12. Troell, M.; Joyce, A.; Chopin, T.; Neori, A.; Buschmann, A.H.; Fang, J.G. Ecological engineering in aquaculture—Potential for integrated multi-trophic aquaculture (IMTA) in marine offshore systems. *Aquaculture* **2009**, *297*, 1–9. [CrossRef]

13. Yu, L.Q.J.; Mu, Y.; Zhao, Z.; Lamc, V.W.Y.; Sumaila, U.R. Economic challenges to the generalization of integrated multi-trophic aquaculture: An empirical comparative study on kelp monoculture and kelp-mollusk polyculture in Weihai, China. *Aquaculture* **2017**, *471*, 130–139. [CrossRef]
14. Chen, J.; Song, G.; Wang, X.; She, L.; Wu, S.; Wu, M.; Ding, F. The efficiency of water purification by different densities of *Bellamya aeruginosa* and *Bellamya purificata*. *J. Anhui Agric. Sci.* **2012**, *40*, 11708–11709.
15. Ma, B.; Wang, H.; Jin, W.; Huang, B.; Li, Y.; Li, Z.; Wang, M.; Li, C. Study on the water purification effect of the algae *Elodea nuttallii* and snail *Bellamya purificata*. *Jiangxi Fish. Sci. Technol.* **2022**, *2*, 3–6.
16. Duan, X.; Xie, C.; Lv, Y.; Zhang, N.; ZHao, F.; Li, R. Feeding habits of Bellamya purificata and its function in water purification system of ecological ditch. *Fish. Mod.* **2013**, *40*, 17–21.
17. Cao, Z.; Jiang, X. The influence of environmental factors on *Bellamya purificate*. *J. Shanghai Fish. Univ.* **1998**, *7*, 200–205.
18. Yan, Y.; Jin, W.; Wen, H.; Ma, X.; Xue, T.; Sun, C.; He, Y.; Bing, X. Estimation of genetic parameters for growth traits of *Bellamya purificata* in 60 days. *Freshw. Fish.* **2018**, *48*, 108–111.
19. Hou, Y.R.; Li, B.; Luo, J.W.; Zhang, C.F.; He, J.; Zhu, J. Effect of Bellamya purificata on organic matter degradation in surface sediment as revealed by amino acids. *Aquac. Environ. Interact.* **2021**, *13*, 1–12. [CrossRef]
20. Zhao, F.; Xie, C.; Zhang, N.; Lv, Y.; Li, R. Efficiency of water purification and the nitrogen and phosphorous release of the sediment by different densities of *Bellamya purificata*. *J. Hydroecol.* **2014**, *35*, 32–38.
21. Zhou, M.; Hou, Y.; Jia, R.; Li, B.; Zhu, J. Effects of Bellamya purificata Cultivation at Different Stocking Densities on the Dynamics and Assembly of Bacterial Communities in Sediment. *Biomolecules* **2023**, *13*, 254. [CrossRef]
22. Liu, L.; Xu, M. Microbial biomass in sediments affects greenhouse gas effluxes in Poyang Lake in China. *J. Freshw. Ecol.* **2016**, *31*, 109–121. [CrossRef]
23. Sabater, S.; Buchaca, T.; Cambra, J.; Catalan, J.; Guasch, H.; Ivorra, N.; Muñoz, I.; Navarro, E.; Real, M.; Romaní, A. Structure and function of benthic algal communities in an extremely acid river. *J. Phycol.* **2003**, *39*, 481–489. [CrossRef]
24. Marino, J.A.; Denef, V.J.; Dick, G.J.; Duhaime, M.B.; James, T.Y. Fungal community dynamics associated with harmful cyanobacterial blooms in two Great Lakes. *J. Great Lakes Res.* **2022**, *48*, 1021–1031. [CrossRef]
25. Xu, S. *Diversity of Microbial Communities in Mixed Crab Breeding Rice Plantations*; Tianjing University: Tianjing, China, 2021.
26. Zhang, H.; Jia, J.; Chen, S.; Huang, T.; Wang, Y.; Zhao, Z.; Feng, J.; Hao, H.; Li, S.; Ma, X. Dynamics of Bacterial and Fungal Communities during the Outbreak and Decline of an Algal Bloom in a Drinking Water Reservoir. *Int. J. Environ. Res. Public Health* **2018**, *15*, 361. [CrossRef]
27. Zhang, Z.; Deng, Q.; Cao, X.; Zhou, Y.; Song, C. Patterns of sediment fungal community dependent on farming practices in aquaculture ponds. *Front. Microbiol.* **2021**, *12*, 542064. [CrossRef] [PubMed]
28. Hou, Y.; Li, B.; Feng, G.; Zhang, C.; He, J.; Li, H.; Zhu, J. Responses of bacterial communities and organic matter degradation in surface sediment to Macrobrachium nipponense bioturbation. *Sci. Total Environ.* **2021**, *759*, 143534. [CrossRef]
29. Nie, X.; Wu, S.; Wu, F.; Sun, S.; Jiang, C.; Tan, W.; Wan, J. Effects of the bioturbation activity of *Tubifex tubifex* on nitrogen release from sediments. *Acta Sci. Circumstantiae* **2011**, *31*, 107–113.
30. Wang, Y.; Song, J.; Weiwei, J.; Cheng, D.; Xue, J.; Yang, X.; Zhang, B.; Zhang, J. Effect of bioturbation of *Chironomid larvae* and *Limno drilus hoffmeiteri* on the release of nitrogen, oxygen and phosphate in the sediments from a river. *Acta Sci. Circumstantiae* **2015**, *35*, 2504–2511.
31. Yoon, T.-H.; Kang, H.-E.; Kang, C.-K.; Lee, S.H.; Ahn, D.-H.; Park, H.; Kim, H.-W. Development of a cost-effective metabarcoding strategy for analysis of the marine phytoplankton community. *PeerJ* **2016**, *4*, e2115. [CrossRef]
32. Chen, S.F.; Zhou, Y.Q.; Chen, Y.R.; Gu, J. fastp: An ultra-fast all-in-one FASTQ preprocessor. *Bioinformatics* **2018**, *34*, 884–890. [CrossRef]
33. Magoc, T.; Salzberg, S.L. FLASH: Fast length adjustment of short reads to improve genome assemblies. *Bioinformatics* **2011**, *27*, 2957–2963. [CrossRef]
34. Edgar, R.C. UPARSE: Highly accurate OTU sequences from microbial amplicon reads. *Nat. Methods* **2013**, *10*, 996–998. [CrossRef]
35. Nilsson, R.H.; Larsson, K.-H.; Taylor, A.F.S.; Bengtsson-Palme, J.; Jeppesen, T.S.; Schigel, D.; Kennedy, P.; Picard, K.; Gloeckner, F.O.; Tedersoo, L.; et al. The UNITE database for molecular identification of fungi: Handling dark taxa and parallel taxonomic classifications. *Nucleic Acids Res.* **2019**, *47*, D259–D264. [CrossRef] [PubMed]
36. Schloss, P.D.; Westcott, S.L.; Ryabin, T.; Hall, J.R.; Hartmann, M.; Hollister, E.B.; Lesniewski, R.A.; Oakley, B.B.; Parks, D.H.; Robinson, C.J.; et al. Introducing mothur: Open-Source, Platform-Independent, Community-Supported Software for Describing and Comparing Microbial Communities. *Appl. Environ. Microbiol.* **2009**, *75*, 7537–7541. [CrossRef] [PubMed]
37. Xu, F.; Zhu, L.; Wang, J.; Xue, Y.; Liu, K.; Zhang, F.; Zhang, T. Nonpoint Source Pollution (NPSP) Induces Structural and Functional Variation in the Fungal Community of Sediments in the Jialing River, China. *Microb. Ecol.* **2023**, *85*, 1308–1322. [CrossRef] [PubMed]
38. Jiang, G.; Yao, R.; Lin, Y. A survey of study on secondary metabolite s from marine imperfect fungi. *J. Microbiol.* **2001**, *21*, 41–44.
39. Fan, L. *Microbial Community in Tilapia (Oreochromis niloticus) Cultural Ponds*; Nanjing Agricultural University: Nanjing, China, 2015.
40. Zhang, Y.; Guo, L. Progress of fungal DNA barcode. *Mycosystema* **2012**, *31*, 809–820.
41. Wang, P.; Xiao, H.; Yuan, R.; Zhao, J.; Yu, X. Fungal community in the estuarine sediment of Poyang Lake. *Acta Sci. Circumstantiae* **2018**, *38*, 1949–1956.

42. Zhao, Y.; Zheng, S.; Fang, Z.; Yu, Z. Vertical distribution of fungal communities in Honghu Lake sediment. *Mycosystema* **2011**, *30*, 721–728.
43. Wang, J.F.; Shao, S.C.; Liu, C.S.; Song, Z.Q.; Liu, S.S.; Wu, S.H. The genus Paraconiothyrium: Species concepts, biological functions, and secondary metabolites. *Crit. Rev. Microbiol.* **2021**, *47*, 781–810. [CrossRef]
44. Du, L.; Li, D.H.; Zhu, T.J.; Cai, S.X.; Wang, F.P.; Xiao, X.; Gu, Q.Q. New alkaloids and diterpenes from a deep ocean sediment derived fungus Penicillium sp. *Tetrahedron* **2009**, *65*, 1033–1039. [CrossRef]
45. Xu, T.; Wu, Y.; Jia, Y.; Liu, B.; Ye, B. A *Penicillium* sp. XGH2321 isolated from the rhizospheric soil of *Rhizophora stylosa* griff and its antibacterial activity. *Microbiol. China* **2009**, *36*, 1682–1687.
46. Corte, A.M.; Liotta, M.; Venturi, C.B.; Calegari, L. Antibacterial activity of *Penicillium* spp. strains isolated in extreme environments. *Polar Biol.* **2000**, *23*, 294–297. [CrossRef]
47. Bringmann, G.; Lang, G.; Gulder, T.A.M.; Tsuruta, H.; Muhlbacher, J.; Maksimenka, K.; Steffens, S.; Schaumann, K.; Stohr, R.; Wiese, J.; et al. The first sorbicillinoid alkaloids, the antileukemic sorbicillactones A and B, from a sponge-derived Penicillium chrysogenum strain. *Tetrahedron* **2005**, *61*, 7252–7265. [CrossRef]
48. Petit, K.E.; Mondeguer, F.; Roquebert, M.F.; Biard, J.F.; Pouchus, Y.F. Detection of griseofulvin in a marine strain of Penicillium waksmanii by ion trap mass spectrometry. *J. Microbiol. Methods* **2004**, *58*, 59–65. [CrossRef] [PubMed]
49. Silva, M.G.; Aracari, N.; Furtado, J.C.; Pupo, M.T.; Fonseca, M.J.V.; Said, S.; da Silva, A.A.; Bastos, J.K. Antibacterial activity from Penicillium corylophilum Dierckx. *Microbiol. Res.* **2004**, *159*, 317–322. [CrossRef]
50. Xu, M.J.; Gessner, G.; Groth, I.; Lange, C.; Christner, A.; Bruhn, T.; Deng, Z.W.; Li, X.; Heinemann, S.H.; Grabley, S.; et al. Shearinines D-K, new indole triterpenoids from an endophytic Penicillium sp (strain HKI0459) with blocking activity on large-conductance calcium-activated potassium channels. *Tetrahedron* **2007**, *63*, 435–444. [CrossRef]
51. De Schryver, P.; Vadstein, O. Ecological theory as a foundation to control pathogenic invasion in aquaculture. *Isme J.* **2014**, *8*, 2360–2368. [CrossRef]
52. Francoeur, S.N.; Pinowska, A.; Clason, T.A.; Makosky, S.; Lowe, R.L. Unionid bivalve influence on benthic algal community composition in a Michigan Lake. *J. Freshw. Ecol.* **2002**, *17*, 489–500. [CrossRef]
53. James, M.R.; Hawes, I.; Weatherhead, M.; Stanger, C.; Gibbs, M. Carbon flow in the littoral food web of an oligotrophic lake. *Hydrobiologia* **2000**, *441*, 93–106. [CrossRef]
54. Xu, Z.; Chen, Y.; Meng, X.; Wang, F.; Zheng, Z. Phytoplankton community diversity is influenced by environmental factors in the coastal East China Sea. *Eur. J. Phycol.* **2016**, *51*, 107–118. [CrossRef]
55. Liang, Y.; Liu, C.; Tian, C.; Sun, M. Effects of nutrients sources and input mode on the community succession of six marine microalgal species. *Trans. Oceanol. Limnol.* **2014**, *2*, 23–30. [CrossRef]
56. Zhao, B.; Hu, W.; Li, C.; Zhou, H.; Zhao, H.; Sun, Y.; Ju, W.; Han, S. Community structure of phytoplankton and benthophyte in *Apostichopus japonicus* aquaculture ponds and water quality characteristics. *J. Guangxi Acad. Sci.* **2020**, *36*, 399–405.
57. Dal Bo, D.; Magneschi, L.; Bedhomme, M.; Billey, E.; Deragon, E.; Storti, M.; Menneteau, M.; Richard, C.; Rak, C.; Lapeyre, M.; et al. Consequences of Mixotrophy on Cell Energetic Metabolism in Microchloropsis gaditana Revealed by Genetic Engineering and Metabolic Approaches. *Front. Plant Sci.* **2021**, *12*, 628684. [CrossRef]
58. El-Sheekh, M.M.; Mansour, H.M.; Bedaiwy, M.Y.; El-shenody, R.A. Influence of nutrient supplementation and stress conditions on the biomass and lipid production of Microchloropsis salina for biodiesel production. *Biomass Convers. Biorefin.* **2022**. Available online: https://link.springer.com/article/10.1007/s13399-022-03434-9#article-info (accessed on 1 December 2022). [CrossRef]
59. Fierro, S.; Sanchez-Saavedra, M.D.P.; Copalcua, C. Nitrate and phosphate removal by chitosan immobilized Scenedesmus. *Bioresour. Technol.* **2008**, *99*, 1274–1279. [CrossRef] [PubMed]
60. Wang, Q.K.; Yu, Z.Y.; Wei, D. High-yield production of biomass, protein and pigments by mixotrophic Chlorella pyrenoidosa through the bioconversion of high ammonium in wastewater. *Bioresour. Technol.* **2020**, *313*, 123499. [CrossRef] [PubMed]
61. Zhao, X.C.; Tan, X.B.; Yang, L.B.; Liao, J.Y.; Li, X.Y. Cultivation of Chlorella pyrenoidosa in anaerobic wastewater: The coupled effects of ammonium, temperature and pH conditions on lipids compositions. *Bioresour. Technol.* **2019**, *284*, 90–97. [CrossRef] [PubMed]
62. Talib, A.; Recknagel, F.; Cao, H.; van der Molen, D.T. Forecasting and explanation of algal dynamics in two shallow lakes by recurrent artificial neural network and hybrid evolutionary algorithm. *Math. Comput. Simul.* **2008**, *78*, 424–434. [CrossRef]

Disclaimer/Publisher's Note: The statements, opinions and data contained in all publications are solely those of the individual author(s) and contributor(s) and not of MDPI and/or the editor(s). MDPI and/or the editor(s) disclaim responsibility for any injury to people or property resulting from any ideas, methods, instructions or products referred to in the content.

fishes

Article

Natural Food Intake and Its Contribution to Tambaqui Growth in Fertilized and Unfertilized Ponds

Adriana Ferreira Lima [1,2,*], Anderson Guilherme Pereira dos Reis [1], Vladimir Eliodoro Costa [3] and Wagner Cotroni Valenti [2]

1 Fisheries and Aquaculture Center, Brazilian Agricultural Research Corporation (EMBRAPA), Palmas 77008-900, TO, Brazil; andersonguilhermepereira3@gmail.com
2 Aquaculture Center of São Paulo State University—CAUNESP, Jaboticabal 14884-900, SP, Brazil; w.valenti@unesp.br
3 Stable Isotope Center, São Paulo State University—UNESP, Botucatu 18618-689, SP, Brazil; vladimir.costa@unesp.br
* Correspondence: adriana.lima@embrapa.br

Abstract: Natural food available in ponds can complement formulated feed in fed aquaculture. This study elucidated the natural food intake and its contribution to tambaqui (*Colossoma macropomum*) growth in fertilized and unfertilized ponds, using stable isotope and stomach contents analyses. Additionally, it described the impact of fertilization management on natural food availability, fish performance, and production costs. Tambaqui juveniles (93.8 ± 15.0 g) were stocked (0.55 fish/m^2) in fertilized (Fert) and unfertilized (NoFert) ponds (600 m^2) for a culture period of 10 months in quadruplicate. A lower food conversion ratio was observed in Fert ponds. The main natural food items ingested by tambaqui were insects, vegetables, and cladocerans. Plankton contributed 39.4% and 10.7% of muscle formation in Fert and NoFert ponds, respectively. Pond fertilization (2.45 g of nitrogen and 0.80 g of phosphorus per square meter every two weeks) did not significantly affect fish growth, survival, or productivity but had a slightly influence on water quality parameters. However, fertilization increased the zooplankton density (through phytoplankton) in the water, thereby increasing autochthonous food availability for tambaqui consumption. This species demonstrates the ability to alternate between natural food sources and commercial feed without compromising its development. Consequently, tambaqui exhibits suitability for farming within restorative and integrated aquaculture systems as well as intensive systems reliant on commercial feed.

Keywords: *Colossoma*; cost-effectiveness; pond fertilization; pond natural food; stable isotope; trophic ecology; plankton

Key Contribution: We described the first register of plankton consumption and their contribution to tambaqui muscle formation in the grow-out ponds. Results indicated that tambaqui can interchange between natural food and commercial feed without compromising its development.

Citation: Lima, A.F.; dos Reis, A.G.P.; Costa, V.E.; Valenti, W.C. Natural Food Intake and Its Contribution to Tambaqui Growth in Fertilized and Unfertilized Ponds. *Fishes* **2024**, *9*, 139. https://doi.org/10.3390/fishes9040139

Academic Editor: Kyeong-Jun Lee

Received: 16 March 2024
Revised: 8 April 2024
Accepted: 13 April 2024
Published: 16 April 2024

Copyright: © 2024 by the authors. Licensee MDPI, Basel, Switzerland. This article is an open access article distributed under the terms and conditions of the Creative Commons Attribution (CC BY) license (https://creativecommons.org/licenses/by/4.0/).

1. Introduction

Many aquaculture systems harness natural biological productivity to support the production of edible aquatic organisms. The extractive species inside the culture ponds can convert particulate and dissolved materials not ingestible by animals of higher trophic levels into biomass, effectively recovering inaccessible energy and nutrients. This process should be explored to restore natural resources from waste and pollution inside ponds, according to a circular economy approach [1]. Nevertheless, nutrient limitation may hinder biological productivity in aquaculture ponds. To compensate, organic and inorganic fertilization may be performed to stimulate natural productivity [2,3]. Fertilization is frequently used in ponds recently stocked with fingerlings or during early grow-out cultures to boost the development of natural food because most fishes have limitations in ingesting or digesting manufactured feed in the early life stages [4]. During the grow-out phase, fertilization has been observed to enhance fish productivity in various species including *Cirrhinus migrala* [5], different carps in polyculture [6–8], *Coptodon rendalli* [9] and *Labeo rohita* [10]. This effect is generally attributed to a rise in natural food, mainly phytoplankton, zooplankton, periphyton, and benthic invertebrates.

Most studies have evaluated the effect of fertilization on pond productivity without the supply of formulated feed. This practice results in limited productivity because the enhancement of natural food is insufficient to feed fish stocked in high densities. Productivity is low in fertilized ponds without an allochthonous diet, while intensively fed ponds bear high productivity. An allochthonous diet is the primary source of nutrients in semi-intensive or intensive fish monocultures [3,11]. Despite this, natural food can supplement the formulated feed in ponds or partially replace it [4,12]. They are even essential in farming certain species for which no suitable and well-balanced diets are available [13,14] Therefore, understanding the role of fertilization associated with an allochthonous diet may provide information for aquaculture efficiency and align with restoration approaches (sensus Alleway [1]).

The incorporation of ingested food into fish biomass and its contribution to growth is variable. The intake of allochthonous diets or natural biota organisms has been largely studied in fish production by stomach content analysis [10,15–25]. However, this technique does not reveal the contribution of each kind of feed to fish growth because part of the eaten feed is not assimilated. Stable isotope analysis is a tool for quantifying dietary support of fish growth among the ingested feed items with distinct isotope content, since the isotopic compositions of animal tissues resemble their diets [26–31]. Since formulated diet is the main cost of production in fish farms [3], understanding its utilization by fish and how it can be replaced or complemented by natural food would help to improve feeding strategies and decrease production costs.

The tambaqui (*Colossoma macropomum*) is an indigenous fish species from the Amazon River Basin and has been framed in Central and South American countries, including Panama, Honduras, Brazil, Colombia, Peru, and Venezuela. This species has also been farmed in China [32]. This Tambaqui is a low-trophic-level fish that exhibits a remarkable ability to consume natural food sources both in its natural habitat [33,34] and in rearing ponds during all life stages [20,22]. Tambaqui also shows easy captivity adaptation, a low feed conversion ratio (FCR), and fast growth [12,22]. Brazil is the primary producer [35], where tambaqui is farmed in earthen ponds [36]. The farming practices include a direct stocking of juveniles weighing between 2 and 5 g or a preliminary culture phase lasting 2 to 3 months until the fish reach a weight of 70 to 80 g, and then, they are transferred to grow-out ponds where they reach 1.5 to 3 kg within a span of 10 months [12,37].

Tambaqui culture in ponds may be an interesting model for studying the contribution of natural food in the grow-out phase to low-trophic-level fish growth. Combining the determination of stomach content and stable isotopic analyses enables us to determine both the ingested food by fish and its actual assimilation into their body mass. Therefore, the present study aims to describe the intake of natural food and its contribution to tambaqui growth in fertilized and unfertilized ponds. Furthermore, it aims to understand the impact of fertilization on natural food availability, fish performance, and production costs.

2. Material and Methods

A batch of Tambaqui fingerlings (1.18 $\pm$ 0.55 g) was purchased from a commercial hatchery (Brejinho de Nazaré, TO, Brazil; 48°35′17.93″ S, 11°1′52.95″ W) and kept in 1000 m^2 ponds at a density of 9 fish m^{-2} [12] for two months until they reached approximately 90 g. During this period, fish were fed three to four times a day with commercial feed (45% crude protein, 1–2.6 mm pellet size).

2.1. Experimental Design and Conditions

Tambaqui juveniles (93.8 ± 15.0 g and 17.6 ± 0.8 cm) were stocked in 600 m^2 ponds at 0.55 fish/m^2 and reared for 10 months. This density is within the range used in commercial farms, i.e., 0.4–0.7 fish/m^2 [37]. Fish were submitted to two experimental conditions (treatments), fertilized ponds (Fert) and unfertilized ponds (NoFert), with four replicates each, in a completely randomized design. Initially, all ponds were drained, disinfected with quicklime (100 g m^{-2}), and limed 24 h later (100 g m^{-2}). The first and biweekly fertilizations were applied only in Fert ponds using 5 g of urea, 3 g of triple superphosphate, and 10 g of rice bran per square meter. This management results in an input of 2.45 g of nitrogen (N) m^{-2} and 0.80 g of phosphorus (P) m^{-2} with N:P ratio by mass of 3:1 each two weeks. The combination of chemical and organic fertilization (rice bran) usually results in greater production [4]. The fertilization management in Fert ponds was interrupted in the ninth and tenth months because of a pronounced decrease in water transparency. The inlet water came from a local dam. Ponds received water to seepage reposition and water renovation of a maximum of 3% of the total water volume per day.

Fish were manually fed with a commercial extruded feed twice a day (9:00 h and 15:00 h), six days a week. Feeding rates were adjusted during the culture (Table 1) based on the regular practice in commercial farms. Feeding was over when the fish ceased eating or when the daily dose (Table 1) computed by feed rate was reached. The actual quantity of feed provided was recorded and used to calculate the apparent feed conversion and feed intake. Fortnightly, 30 fish were randomly sampled from each pond, weighed, and then, returned to their respective ponds. Fish mean weight and feeding rate were used to recalculate the quantity of daily feed supplied to each pond. An aliquot of each feed batch was sampled, and they were pooled for proximate analyses (Table 1).

Table 1. Feed composition, pellet size, and feeding rates used in feed management for tambaqui (*Colossoma macropomum*) grow-out in ponds for 10 months. Feeding was ceased if the fish stopped eating before reaching the daily feed rate.

Fish Weight (g)	Crude Protein (%)	Gross Energy (cal 100 g^{-1})	Ether Extract (%)	Crude Fiber (%)	Pellet Size (mm)	Feeding Rate (% of Body Weight/Day)
60–200	34.4	336.5	4.1	7.1	4	4.5
200–500	36.1	337.7	4.1	6.8	6	3.5
500–700	36.1	337.7	4.1	6.8	6	2.5
700<	30.3	337.4	3.4	6.7	10	2.5

At the end of the experiment, ponds were drained, and all fish were harvested and weighed in batches to obtain the total yield. Then, fish were counted, and 10 individuals from each pond were weighed individually. The following variables were calculated: (1) specific growth rate (% SGR day^{-1}) = 100 × [(ln final weight − ln initial weight)/days of experiment]; (2) daily feed intake (% live weight dia^{-1}) = [total feed supplied/(final biomass + initial biomass/2)]/days of experiment × 100; (3) apparent feed conversion ratio (FCR) = total feed supplied/total weight gain; (4) survival = 100 × (final number of fish/initial number of fish). Uneaten feed floated on the pond surface and the amount observed was negligible. Thus, the amount of feed supplied was used as a proxy for the actual feed intake.

2.2. Plankton Availability

Plankton was sampled monthly by dragging a 20 μm mesh net (for phytoplankton) and a 68 μm mesh net (for zooplankton) along each pond, at approximately 30 cm below the surface, over a distance of 10 m. Samples were concentrated in 500 mL and fixed in 4% formalin. Phytoplankton were counted using Neubauer chambers and results were expressed in individuals/L. Zooplankton was quantitatively analyzed using Sedwick–Rafter chambers, and results were expressed in individuals/L for each taxonomic group.

2.3. Stomach Content Analysis

Five fish from each pond were sampled monthly, euthanized with eugenol bath (35 mg/L) [38], individually weighted, and dissected for stomach removal. Fish did not receive formulated feed for 24 h before sampling to enable better observation and identification of the ingested natural food. Stomachs were dissected and fixed in 4% formalin. Then, they were weighed with and without their content. Stomach contents were analyzed qualitatively and quantitatively. Food items were grouped according to the following categories: (a) insects, (b)terrestrial plants, (c) phytoplankton, (d) sediment, (e) copepods, (f) cladocerans, (g) rotifers, (h) ostracods and decapods, and (i) formulated diet. Frequency of occurrence (FO) was calculated as the ratio between the number of stomachs containing a specific food item and the total number of stomachs analyzed [39]. The relative abundance (RA) of each item was calculated considering the relative-fullness method, in which the contribution of each prey category is visually scored as the percentage of the total food content in the stomach [40]. Food selectivity was measured using the Ivlev Electivity Index (Ei) [41] calculated using the equation Ei = (r − p)/(r + p), in which r is the percentage of each item in the fish stomach and p is the percentage of each item in the pond environment. The resulting index values range from −1 to +1. Positive values indicate preferred items, whereas negative values indicate avoided items; h zero indicates no selectivity.

2.4. Isotopic Analysis

Muscle samples were collected from eight fish at the beginning of the study and eight fish per treatment at the end. Feed with different granulometry was sampled and pooled for isotopic analyses. The plankton sample was obtained from a pool of sub-samples collected monthly in Fert and NoFert ponds. The plankton sample included insects, net plankton, ostracods, and decapods. All samples were dried in an oven at 65 °C for 72 h and kept at 4 °C until analysis. The isotopic analyses were carried out at the Center for Stable Isotopes of the Universidade Estadual Paulista (UNESP, Botucatu, Brazil). Each sample was homogenized individually in a cryogenic mill at −196 °C and then weighed in tin capsules (~0.5 mg). Homogenization was necessary to reduce the sample internal variation of feed, because of its granulation, and for plankton because of the diversity of organisms. The capsules were subjected to the simultaneous analysis of isotopic ratios $R(^{15}N/^{14}N)$ sample and $R(^{13}C/^{12}C)$ sample in a CF-IRMS continuous flow isotopic ratio mass spectrometry system, using an IRMS (Delta V, Thermo Scientific, Albuquerque, New Mexico) coupled to an elemental analyzer EA (Flash 2000, Thermo Scientific) through a gas interface (ConFlo IV, Thermo Scientific). The isotopic ratios were expressed as a relative difference from the isotopic ratio $\delta^{15}N$ and $\delta^{13}C$ according to Equations (1) and (2) below [42] reported on mUr [43], where $R(^{15}N/^{14}N)$ air and $R(^{13}C/^{12}C)$ VPDB are the international standard isotope ratios for N and C, respectively. The standard uncertainty of the simultaneous measure was estimated at ±0.15 mUr and ±0.20 mUr for $\delta^{15}N$ and $\delta^{13}C$, respectively. Values of $\delta^{15}N$ e $\delta^{13}C$ were normalized according isotopic reference materials USGS61, USGS62 and USGS63, respectively [44,45].

$$\delta^{15}N = \left[\frac{R\left(\frac{15N}{14N}\right)_{sample}}{R\left(\frac{15N}{14N}\right)_{Air}} - 1 \right] \quad (1)$$

$$\delta^{13}C = \left[\frac{R\left(\frac{13C}{12C}\right)_{sample}}{R\left(\frac{13C}{12C}\right)_{VPDB}} - 1 \right] \quad (2)$$

2.5. Assimilation of Feed into Muscle Using Isotopic Analysis

First, we estimated feed-specific trophic discrimination factors (Δ) that correct differences in isotope $\delta^{13}C$ values between consumers and feeds due to the differential assimilation of feed components and metabolic fractionation [46]. To do so, we used the difference between the steady-state isotope $\delta^{13}C$ values for the fish tissue at the beginning of the experiment and the average isotope $\delta^{13}C$ values for the formulated feed that fish were consuming before this experiment (Equation (3)).

$$\Delta^{13}C = \delta^{13}C_{tissue} - \delta^{13}C_{feed} \quad (3)$$

To obtain the fish isotopic values without the trophic discrimination factors ($\delta_{fish\text{-}\Delta}$), we calculate the difference between the average fish isotopic values at the end of the experiment and the $\Delta^{13}C$. The proportion of plankton (P) in the muscle (expressed in %) of the fish at the end of the experiment in each treatment was assessed using an isotopic balance with $\delta_{fish\text{-}\Delta}$ and the isotopic values of the two feeds available ($\delta_{zooplankton}$, δ_{feed}) (Equation (4)) [42].

$$P = \frac{\delta_{fish-\Delta} - \delta_{feed}}{\delta_{plankton} - \delta_{feed}} \quad (4)$$

2.6. Water Quality Analyses

Temperature, dissolved oxygen, pH, and total ammonia nitrogen (TAN) were measured in situ three times a week with a specific probe (YSI Professional Plus, Yellow Springs, OH, USA). Transparency was also measured three times a week with a Secchi disc. Alkalinity and hardness, were measured biweekly with commercial colorimetric kits (Alfakit®, Florianópolis, SC, Brazil). Phosphorus was analyzed monthly according to the method described in APHA [47].

2.7. Economic Analysis

A partial budget analysis [48] was used to assess the economic impact of fertilization management on tambaqui grow-out. Only the production costs associated with formulated feed, fertilization inputs, and relevant labor were considered in this analysis since these factors varied between treatments, consistent with previous studies performed by Veverica et al. [49] and Liranço et al. [50]. The fertilization management in each pond was estimated to be completed in 20 min, which was considered to estimate labor costs. The values of each input are described in Table 2.

Table 2. Values of the items considered in the partial budget economic analysis to introduce fertilization in tambaqui (*Colossoma macropomum*) grow-out in ponds. Prices were obtained in Palmas-TO, Brazil, in January 2020. (US$ 1.00 = R$ 5.26).

Items	Value (US$)	Unit
Formulated feed (4 mm)	0.49	US$/kg
Formulated feed (6 mm)	0.49	US$/kg
Formulated feed (10 mm)	0.43	US$/kg
Urea	0.73	US$/kg
Triple superphosphate	0.30	US$/kg
Rice bran	0.18	US$/kg
Labor	1.27	h

2.8. Statistical Analysis

Data normality (Shapiro–Wilk) and homogeneity (Bartlett) of the residues were evaluated and the transformation of Box & Cox [51] was used when these premises were not met. To assess the differences between treatments for final weight, SGR, daily feed ingested, feed conversion rate, fish yield, survival, alkalinity, hardness, and Zooplankton density, data were subjected to Student's *t*-test. Phytolankton density, dissolved oxygen concentration, transparency, pH and TAN data were analyzed considering the periods of evaluation and using a mixed model as a technique of measures repeated in time. Ponds were considered experimental units and evaluated as aleatory effects. Treatments, months, and their interactions were evaluated as fixed effects. When significant ANOVA was obtained, the means were compared using the Tukey test. The other variables did not meet normality and/or homoscedasticity even with transformed data, and thus, they were analyzed using a non-parametric Kruskal–Wallis test. The data were presented as the mean ± standard deviation, and statistical significance was assumed at a level of p close to 0.05. All statistical analyses were performed using the 4.2.3 R software [52].

2.9. Legal and Ethical Aspects

The study complied with official Brazilian guidelines for the care and use of animals for scientific and educational purposes (Concea—CEUA protocol 42/2018).

3. Results

3.1. Tambaqui Performance in Ponds

The growth of fish during the culture period, final mean weight, survival, and yield showed no significant difference between the Fert and NoFert treatments (Figure 1A, Table 3). The daily feed intake and feed conversion ratio (FCR) were observed to be lower in Fert ponds compared to NoFert ponds (Table 3).

Table 3. Performance of tambaqui (*Colossoma macropomum*) produced in fertilized (Fert) and unfertilized (No Fert) ponds for 10 months. SGR = specific growth rate.

Parameters	Treatment		*p* Value
	Fert	NoFert	
Final mean weight (g)	1697.5 ± 88.8 a	1628.4 ± 63.8 a	0.2525
SGR (% day^{-1})	0.95 ± 0.02 a	0.94 ± 0.01 a	0.3903
Daily feed intake (% day^{-1})	1.16 ± 0.02 b	1.25 ± 0.05 a	0.0077
Feed conversion rate	2.04 ± 0.04 b	2.22 ± 0.08 a	0.0090
Fish yield (t/ha/cycle)	6.2 ± 0.3 a	6.1 ± 0.2 a	0.7552
Survival (%)	91.6 ± 5.6 a	94.4 ± 1.6 a	0.3701

Different letters in the same row indicate significant differences by *t*-test.

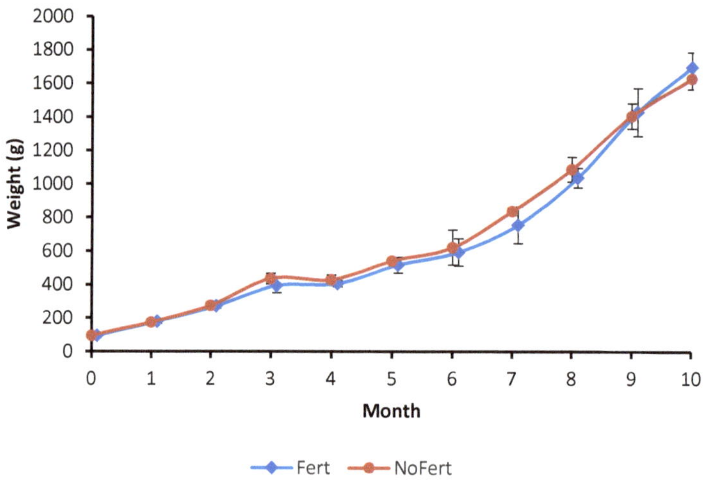

(A)

Figure 1. *Cont.*

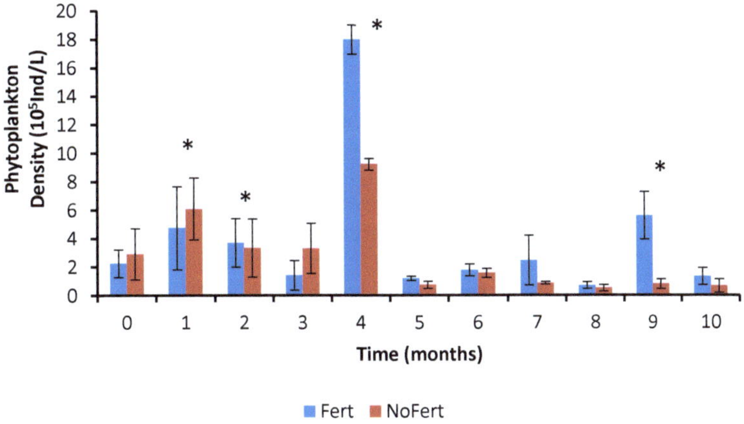

(B)

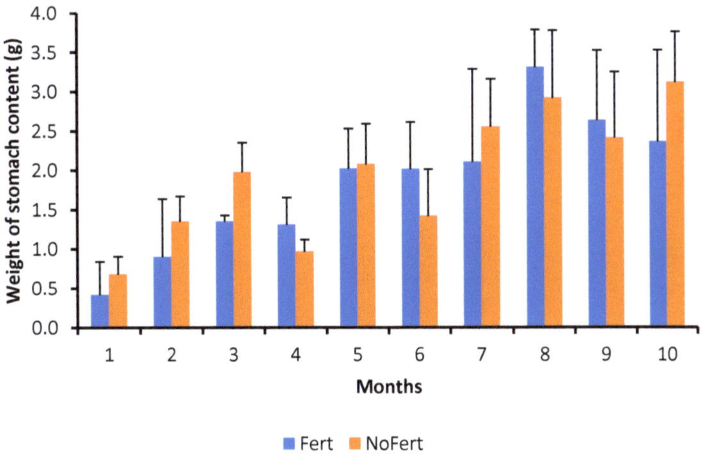

(C)

Figure 1. Growth of tambaqui (*Colossoma macropomum*) (**A**), the variation of phytoplankton in water pond (**B**), and weight of stomach content of tambaqui (**C**) during grow-out in fertilized (Fert) and unfertilized (NoFert) ponds. No significant difference in fish growth was observed between treatments ($p = 0.2525$). The weight of stomach content ($p = 0.0019$) was higher in fertilized ponds. Whiskers indicate SD. Ind. = individuals; * indicates significant differences between treatments each time.

3.2. Plankton Community

The interaction between fertilization management and culture time was significant for phytoplankton density ($p < 0.0001$). Phytoplankton abundance was initially lower in fertilized ponds during the first month, but significantly higher during the second, fourth, and ninth months (Figure 1B). The peak of phytoplankton in the fourth month coincided with an abnormally high concentration of phosphorous in the inlet water. The average zooplankton density in the water of fertilized ponds (1079 ± 261 individuals/L) was higher than in unfertilized ponds (716 ± 161 individuals/L) ($p = 0.0561$).

Fertilization increased the densities of cladocerans, copepods, and rotifers inside ponds by approximately twofold (Table 4). However, the great variability in the samples obtained in the different ponds during culture decreased the power of the *t*-test [51]. Thus, no significant statistical differences were observed. The frequency of occurrence and relative abundance of food items in stomach contents had significant variations over time, but it did not exhibit any pattern. Insects, followed by vegetable fragments, cladocerans, and copepods showed the highest frequency of occurrence inside stomachs (Table 4). There were interactions between treatments and culture duration regarding the frequency of occurrence for copepods; higher frequencies were observed in fertilized ponds during the fifth month, while lower frequencies occurred during the seventh.

Table 4. Density of zooplankton groups in pond water and frequency of occurrence (FO) and relative abundance (RA) of food items in the stomach content of tambaqui (*Colossoma macropomum*) produced in fertilized (Fert) and unfertilized (NoFert) ponds. FM = fertilization management. DIP = density inside ponds. NQ = not quantified. Different letters in the same column indicate significant differences between treatments.

Food Items	FM	DIP (Individuals/L)	FO (%)	RA (%)
Cladocerans	Fert	77 ± 129	42 ± 35	10 ± 13 a
	NoFert	39 ± 34	29 ± 35	6 ± 12 b
Copepods	Fert	458 ± 493	25 ± 29	4 ± 7
	NoFert	246 ± 224	17 ± 27	4 ± 9
Rotifers	Fert	641 ± 1199	2 ± 7	0 ± 0
	NoFert	361 ± 336	0 ± 0	0 ± 0
Phytoplakton	Fert	360 ± 376	6 ± 19	1 ± 3
	NoFert	310 ± 367	6 ± 22	0 ± 2
Terrestrial Plants	Fert	NQ	40 ± 32	12 ± 16
	NoFert	NQ	41 ± 35	14 ± 21
Insects	Fert	1 ± 1	62 ± 33	26 ± 24
	NoFert	1 ± 1	65 ± 30	26 ± 21
Digested food	Fert	NQ	58 ± 36	15 ± 13
	NoFert	NQ	57 ± 33	16 ± 16
Sediment	Fert	NQ	7 ± 15	1 ± 2
	NoFert	NQ	11 ± 23	1 ± 3
Feed	Fert		10 ± 23	1 ± 4
	NoFert	-	9 ± 24	1 ± 5
Ostracods and decapods	Fert	NQ	10 ± 22	2 ± 4
	NoFert	NQ	5 ± 11	0 ± 1

3.3. Stomach Content

The interaction between fertilization management and culture time was not significant for the weight of tambaqui stomach content, but an independent effect was observed. The weight of stomach content showed a tendency to increase with time in both treatments ($p < 0.0001$) and was higher in unfertilized than in fertilized ponds ($p = 0.0019$) (Figure 1C).

The most abundant food items inside the stomachs were insects, vegetable fragments, and cladocerans (Table 4). Cladocerans were more abundant in fertilized ponds, while all the others showed similar abundance in both treatments (Table 4). Formulated feed showed a low abundance and occurred in the fish stomach content from the seventh month onward (Table 4). However, we have observed that tambaqui ingested commercial feed since the first month. The adults and larvae of insects that live in the water column or are associated with the water surface were observed. No typically benthic insects were identified. Electivity indexes were very similar in both treatments. Thus, data were pooled and showed together. Electivity indexes for cladocerans and copepods did not show a pattern during the culture. However, they showed that tambaqui had a slight preference for cladocerans and avoidance of copepods (Figure 2). Tambaqui showed great preference for ingesting insects and avoided rotifers during all growth phases (Figure 2).

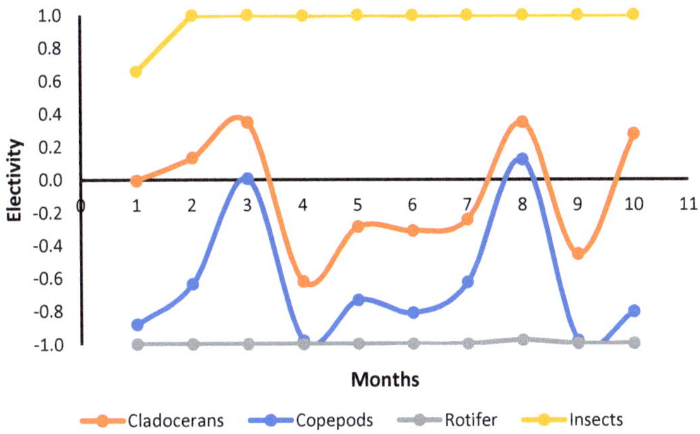

Figure 2. Variation in Ivlev electivity index (Ei) for tambaqui, *Colossoma macropomum*, produced in ponds in each month of culture.

3.4. Stable Carbon and Nitrogen Isotopes as Indicators of Food Assimilation

At the beginning of the experiment, the fish showed $\delta^{15}N$ = +7.37 ± 0.64 mUr and $\delta^{13}C$ = −20.86 ± 0.32 mUr. At the end, fish produced in fertilized ponds showed $\delta^{15}N$ = +8.78 ± 0.74 mUr and $\delta^{13}C$ = −19.07 ± 0.57 mUr; similar values were found for fish produced in unfertilized ponds, which showed $\delta^{15}N$ = +8.27 ± 0.46 mUr and $\delta^{13}C$ = −18.59 ± 0.40 mUr. Trophic discrimination factors were $\Delta^{15}N$ = +4.16 mUr and $\Delta^{13}C$ = +1.72 mUr between fish muscle and diet. The dispersion of isotopic values is shown in Figure 3. The commercial feed showed $\delta^{15}N$ = +1.69 mUr and $\delta^{13}C$ = −19.40 mUr, the plankton of fertilized ponds $\delta^{15}N$ = +4.51 mUr and $\delta^{13}C$ = −22.92 mUr, and the plankton of unfertilized pond showed $\delta^{15}N$ = +10.47 mUr and $\delta^{13}C$ = −27.96 mUr. At the end of the experiment, the proportions of plankton and commercial feed in fish muscle were 39.4% and 60.6% for fish produced in fertilized ponds and 10.67% and 89.33% for unfertilized ponds, respectively. These proportions were calculated from $\delta^{13}C$ values. Vegetables were not considered for isotopic analysis and only insects caught in zooplankton samples were included.

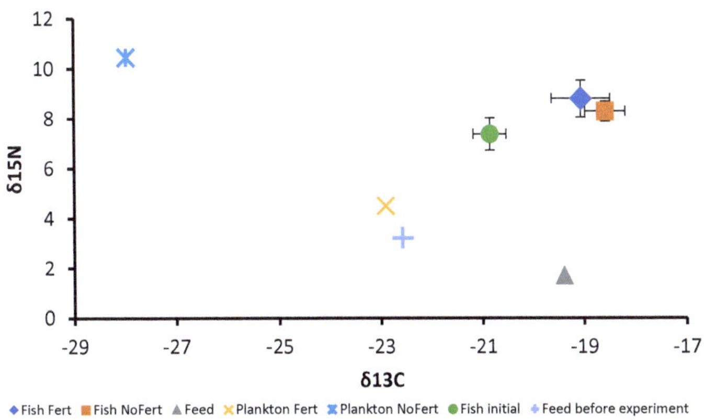

Figure 3. Dispersion of the mean and standard deviation (Whiskers) values of $\delta^{15}N$ (mUr) and $\delta^{13}C$ (mUr) of plankton, formulated feed, and tambaqui *Colossoma macropomum* before and after the grow-out phase subjected to fertilization (Fert) and no fertilization (NoFert) management.

3.5. Water Quality

The temperature was very similar in all ponds during the entire culture. Only water transparency and dissolved oxygen showed a significant interaction between fertilization management and time of culture (Table 5). Lower transparency occurred in the third and ninth months, while higher oxygen values were observed in Fert ponds in the seventh and ninth months. The pH was affected by the fertilization management and the time of culture without interactions between these factors. However, the variation among ponds was very low. TAN was influenced only by the time of culture, without differences among treatments. The concentration of phosphorous in the water was higher in Fert ponds and varied over time without a pattern. Alkalinity and hardness showed no temporal pattern and did not differ between the treatments.

Table 5. Mean values (±SD) of water quality parameters obtained in ponds of tambaqui (*Colossoma macropomum*) produced in fertilized (Fert) and unfertilized (NoFert) ponds for 10 months. Alkalinity and hardness were not subjected to ANOVA with measures repeated in time because they did not meet normality or homoscedasticity even after transformations. TAN = total ammonia nitrogen.

Water Parameters	Fertilization Management (FM)		*p* Values		
	Fert	NoFert	FM	Time	F × Time
Temperature (°C)	28.58 ± 0.06	28.56 ± 0.05	0.5289	<0.0001	0.0001
Transparency (cm)	66.26 ± 7.86	69.3 ± 12.26	0.3679	<0.0001	0.0029
Dissolved oxygen (mg L^{-1})	4.09 ± 0.31	3.61 ± 0.36	0.0006	<0.0001	0.0038
pH	7.26 ± 0.04 a	7.27 ± 0.09 b	0.0042	<0.0001	0.8512
TAN (mg L^{-1})	0.2725 ± 0.04	0.2675 ± 0.02	0.8682	<0.0001	0.2613
Phosphorus (mg/L)	0.41 ± 0.26	0.33 ± 0.22	0.0424	<0.0001	0.7673
Alkalinity (mg de $CaCO_3$/L)	20.69 ± 2.52	21.12 ± 1.59	0.7860	-	-
Hardness (mg de $CaCO_3$/L)	15.34 ± 1.76	16.04 ± 2.02	0.6723	-	-

Different letters in the same row indicate significant differences with Tukey test.

3.6. Economic Analysis

As no difference was observed for final weight and survival among treatments, the average biomass of 370.6 kg was used to perform the economic analysis. The feed conversion rate (FCR) was used to calculate the feed needed to produce the fish biomass in each treatment. In general, it was necessary to have 8% of 4 mm, 35% of 6 mm, and 57% of 10 mm formulated feed for tambaqui grow-out. These percentages were used to estimate the total feed cost. Fourteen fertilization events were computed for each pond during culturing. These values were used to estimate fertilization and the relevant labor costs. Despite the fertilization decrease FCR by 9%, this procedure increases production costs by 8% (Table 6).

Table 6. Economic costs of feed, fertilization, and labor to supply fertilizers in tambaqui (*Colossoma macropomum*) produced in fertilized (Fert) and unfertilized (NoFert) ponds for 10 months. Prices were obtained in Palmas-TO, Brazil, in January 2020. Monetary values are in US Dollars (US$ 1.00 = R$ 5.26).

Items	Fert	NoFert
Fish biomass (kg/600 m^2 pond)	370.61	370.61
Feed conversion ratio	2.04	2.22
Total formulated feed (kg)	756.04	822.75
Formulated feed (4 mm) (US$)	28.47	30.98
Formulated feed (6 mm) (US$)	127.16	138.38
Formulated feed (8 mm) (US$)	180.35	196.27
Total feed costs (US$)	335.99	365.63
Total fertilizer costs [1] (R$)	53.66	0
Fertilization management labor costs [1,2] (US$)	5.85	0
Total cost (US$)	395.49	365.63
Cost increment in Fert management (%)	8.2%	

[1] Fourteen applications of fertilizer were computed for each pond; [2] operator spent 20 min for each fertilization.

4. Discussion

Tambaqui consumed zooplankton, fragments of vegetables, and formulated feed during the entire grow-out phase in both fertilized and unfertilized ponds. Results indicated that pond fertilization increased zooplankton density in the pond water, through phytoplankton, thereby enhancing natural food availability. This allowed a greater exploration of natural biota as food, characterized by a selective increase in zooplankton ingestion by the tambaqui. Consequently, zooplankton accounted for 39.4% of all assimilated food, whereas it constituted only 10.7% in unfertilized ponds. On the other hand, this greater assimilation of natural food did not lead to higher growth or productivity. This suggests that this species can interchange between natural food and a commercial diet without compromising its development.

The limited availability of natural food may have constrained the contribution of this food to muscle tissue growth, as evidenced by zooplankton density in unfertilized ponds. The lack of natural food was compensated by the increase in formulated feed ingestion, showed by a higher feed conversion rate (computed only using the feed supplied) of tambaqui in unfertilized ponds. A high contribution of natural food to fish growth was also observed in juvenile channel catfish (*Ictalurus punctatus*) reared in nursery ponds receiving eutrophic water [53]. In this culture, natural food accounted for 57–58% of muscle tissue growth when fish were fed commercial feeds at varying rates, suggesting that the availability of formulated feed did not determine the consumption of natural food, and rather, it was primarily influenced by its own abundance [53].

The stable isotope analysis has been utilized in food assimilation studies to quantify the contribution of natural food to fish growth in farm ponds [27–30,54]. However, this is the first study of tambaqui in ponds, despite the importance of this species for aquaculture. As previously observed for other species, alterations in the isotopic signal were detected in both treatments due to intake and nutrient assimilation from natural food and commercial feed [53–55]. Fish reared in unfertilized ponds exhibited higher $\delta^{13}C$ enrichment, indicating greater formulated feed assimilation as confirmed by food-type assimilation analysis and lower cladocerans ingestion resulting in a higher FCR. On the other hand, zooplankton impacts more intensively in $\delta^{14}N$ enrichment [56], which explains the higher $\delta^{14}N$ enrichment in tambaqui produced in fertilized ponds, corroborating the higher contribution of natural food in this pond management. The high assimilation of plankton by tambaqui demonstrates that these low-trophic fish may play an important role in recovering energy and nutrients in pond systems even when receiving allochthonous feed. This characteristic places the species as a promising candidate for restorative aquaculture in fed or unfed cultures in which aquaculture provides ecological benefits to the environment leading to improved environmental sustainability and ecosystem services, in addition to the supply of food [1].

Interestingly, differences in plankton isotopic signals from fertilized and unfertilized ponds were observed, which can be related to fertilization management. Narimbi et al. [30] observed that an inorganic nitrogen source was the main isotopic signal responsible for phytoplankton in ponds of tilapia culture even when organic fertilizer was available. Similarly, in the present study, lower $\delta^{14}N$ values were observed in plankton from fertilized ponds. This observation can be explained by the utilization of urea as a nitrogen source, which presents a $\delta^{14}N$ signal close to zero [57].

The density of phytoplankton and concentration of TAN in the pond water did not differ between fertilized and unfertilized ponds during most of the culture period. Probably, in fertilized ponds, phytoplankton rapidly assimilated the added nitrogen and phosphorous, and the populations were regulated by the zooplankton, which increased with the cascade effect. Zooplankton supports the nutrient flow between phytoplankton and fish, contributing to restoring nutrients and energy in the pond system [58]. However, Brucet et al. [59] and Vakkilainen et al. [60] highlighted a more significant influence of fish predation than nutrient enrichment (via phytoplankton) in regulating zooplankton populations.

We observed a high presence of insects and terrestrial plants in tambaqui stomachs. The vegetables were seeds and leaf fragments, possibly from the vegetation that surrounded ponds. Conversely, insects and terrestrial plants have low importance in tambaqui food in Amazonian rivers [33,56]. The differences may be related to the availability of these food items in the water or a change in the food preference of tambaqui in culture ponds. In the present study, isotopic analyses neglected vegetables and insects because we chose the principal items observed in the natural field. Although insects were present in the zooplankton sample, it may not have covered all the diversity and abundance of insects present in the ponds. Thus, the contribution of these natural food items to tambaqui muscle tissue growth was subestimated.

The density of cladocerans, copepods, and rotifers was quite variable during the culture; however, on average, it was approximately double in fertilized ponds. Nevertheless, only cladocerans were more abundant in the stomachs of fish raised in fertilized ponds, indicating a feed preference for this group. These results may be due to the capacity to catch prey or selectivity pre- or post-capture. Larvae of tambaqui showed substantial ingestion of rotifers [61], while juveniles [62] and adults (present results) showed low consumption of it. Thus, it is likely that a gill raker limitation to capture this small prey is developed with the growth of tambaqui. On the other hand, copepods have similar or higher sizes than cladocerans, and thus lower consumption of them indicates an avoidance and a real preference for cladorerans. Therefore, further research should be conducted on fertilization management to boost cladocerans and insect populations. An additional benefit of increased zooplankton density in fishponds is the action of this community to improve the effectiveness of the biological carbon pump that regulates the atmospheric carbon dioxide levels, which is an ecosystem service of aquaculture [58,63]. Techniques promoting the intensification of natural food such as fertilization, the addition of substrates for growing periphyton, and bottom nutrients suspension may contribute to reducing tambaqui's reliance on formulated feed and enable ecological intensification of production through ecological processes and functions [64].

The natural biota plays a crucial role in nutrient cycling within aquaculture ponds, contributing to biogeochemical cycles and solar energy fixation. These fundamental processes form the basis for a restorative aquaculture practice [1,58]. The observed 8% decrease in the feed conversion ratio (FCR) in fertilized ponds indicates that the natural biota can partially substitute commercial diet in tambaqui production without compromising productivity, thereby reducing reliance on formulated feed. Therefore, an increase in natural food is desirable in this production system. Isotopic analyses showed that about 40% of tambaqui biomass comes from the aquatic biota. Therefore, tambaqui may be a candidate for restorative aquaculture practices. This characteristic may bring a new perspective to the production of this Amazon species.

The lower weight of tambaqui stomach content in fish produced in fertilized ponds could be because of the fastest movement of natural food in the gut [65] and its greater digestibility [65,66]. On the other hand, higher ingestion of formulated feed and its high participation in muscle building in unfertilized ponds justify the greater weight of stomach content in this treatment because it is less digestible. The stomachs were fixed in formalin solution until the analysis, and natural food may have had higher dehydration than formulated feed due to its humidity (~90%), increasing the difference observed [67]. Formulated feed was observed inside tambaqui stomachs despite the feeding stopped a day before gut samples. This presence is due to the long time elapsed from ingestion to gastric evacuation. In tambaqui, the gut is empty after 24 h of the last feeding only in 50% of fish [68]. In pacu, *Piaractus mesopotamicus*, a species of the same family of tambaqui, gastric emptying can occur in 56 h, depending on water temperature [69]. The ingestion rate of tambaqui decreased during the development, as it was demonstrated by the ratio of stomach weight and total fish weight.

In general, fertilization promotes an increase in natural food productivity in fishponds, which supplement formulated feed and result in higher fish growth and productivity [4].

The positive effect of fertilization on fish production was described for many fish species, such as carps in monoculture and polyculture [2,5,8,70], red-breasted tilapia (*Coptodon rendalli*, former *Tilapia rendalli*) [9], Nile tilapia (*Oreochromis niloticus*), North African catfish (*Clarias gariepinus*) [49], and pirarucu (*Arapaima gigas*) [71]. However, in the present study, fertilization did not improve tambaqui productivity, similar to what was observed by Gomes and Silva [22] in tambaqui and Duodu et al. [72] in Nile tilapia (*Oreochromis niloticus*) in combining feed with fertilization (N:P ratio of 1:1 and 1.6:1, respectively). In the present study, the commercial feed supplied in unfertilized ponds compensated for the lower natural food ingestion and did not affect fish growth. Thus, commercial feed available for tambaqui is probably well-balanced and allows fish to grow with a low dependence on natural food. As tambaqui is not a territorial or aggressive fish, it may be produced in intensive systems where natural food is negligible.

The water quality parameters meet those recommended for fish production by Boyd [73] in both fertilized and unfertilized ponds. Therefore, they did not affect the growth and survival of tambaqui. The temperature was similar in ponds of both treatments and fluctuated according to weather variations over the study (*cf.* INMET data [74]). Transparency was lower, and phytoplankton density was higher in some months in fertilized ponds, corroborating the impact of fertilization procedure in ponds described by Boyd [4]. The dissolved oxygen increased up to the fifth month, when it decreased, possibly because of an elevation in fish biomass that demands more oxygen. TAN decreased until the middle of the production cycle when it increased owing to an elevation in tambaqui biomass, as observed previously by Gomes and Silva [22] and indicated by Boyd [75]. Boyd [73] reports that feed and fertilizers are the primary sources of TAN in aquaculture systems. The similarity among treatments suggests that fertilizers had a low impact on TAN levels or that the ammonia was rapidly assimilated by phytoplankton in fertilized ponds. However, phosphorous concentration was significantly higher in fertilized ponds, suggesting that the ratio N:P in fertilizer was unbalanced for fulfilling the phytoplankton requests. Alkalinity and hardness were similar in all ponds of both treatments, probably because they were equally limed. Generally, the results indicated that the fertilization of earthen ponds using 5 g of urea, 3 g of triple superphosphate, and 10 g of rice bran per square meter produces a low effect on water quality in tambaqui ponds.

Fertilization management increased production costs despite reduced feed use. The costs of fertilizer and labor to handle it surpassed the expenditure reduction in commercial feed by 8%. Duodu et al. [72] also observed a slight increment in the production costs of Nile tilapia (*Oreochromis niloticus*) in Ghana when fish were produced by combining inorganic fertilization (1.6:1, N:P ratio) with commercial feed. As inorganic fertilizers are expensive, organic fertilizers could be an alternative to decrease the costs [76]. Further studies should be conducted to define the cost-effective composition and quantity supplied of fertilizers to lead to the best composition of zooplankton to feed tambaqui.

Generally, a large amount of nutrients is accumulated in fishpond bottoms [77]. Thus, techniques to suspend the sediments, such as aeration, may contribute to making nutrients laid in sediment available to the phytoplankton community [60]. The use of benthic species integrated with the culture of tambaqui may resuspend nutrients in the sediment by bioturbation with no cost [77–79]. Therefore, the culture of tambaqui combined with benthic feeders or iliophagus species may be advantageous to recovering sequestered nutrients, saving fertilizers and commercial diet. This practice increases the circularity and restorative capacity of the production system. In addition, it increases the productivity and diversity of products, transforming pollution into biomass of high economic value. Further studies should be performed to determine the best species and stocking densities and ratios for improving productivity in integrated systems.

5. Conclusions

Results indicated that the fertilization of tambaqui farming ponds with 2.45 g of nitrogen and 0.80 g of phosphorus per square meter (N:P ratio by mass of 3:1) every two weeks combined with a commercial diet does not affect fish growth, survival, and productivity and has a limited effect on water quality and production costs. However fertilization can increase zooplankton density in the water column (via phytoplankton) which raises natural biota and autochthonous food availability. Natural food may represent about 40% (at least) of all assimilated food in the tambaqui muscles. However, this species can interchange between natural food and commercial feed without compromising its development. Therefore, tambaqui may be farmed in restorative and integrated aquaculture systems, following the principles of the circular economy, but also in intensive systems that rely on commercial feed. Further studies should be conducted increasing the levels of nitrogen and phosphorus and varying their ratio (N:P) in fertilizers, as well as the frequency of fertilization to determine the optimal fertilization management to improve tambaqui production in semi-intensive aquaculture.

Author Contributions: Conceptualization, A.F.L.; data curation, A.F.L.; formal analysis, A.F.L. A.G.P.d.R., V.E.C. and W.C.V.; investigation, A.F.L. and A.G.P.d.R.; methodology, A.F.L. and V.E.C. project administration, A.F.L.; writing—original draft, A.F.L. and V.E.C.; writing—review and editing W.C.V. All authors have read and agreed to the published version of the manuscript.

Funding: This research was part of BRS Aqua Program, supported by the Brazilian Development Bank—Technologic Fund (BNDES-Funtec), grant number 16.2.0225.1, Ministry of Agriculture, Livestock And Food Supply (MAPA), grant number 00350.000111/2014-50, Brazilian Agricultural Research Corporation (EMBRAPA), grant number 21.17.02.001.05.00, and the Brazilian National Council For Scientific and Technological Development (CNPq), grant number 421502/2017-7.

Institutional Review Board Statement: The study was conducted in accordance with the Brazilian guidelines for the care and use of animals for scientific and educational purposes and approved by the Animal Ethics Committee of EMBRAPA FISHERIES AND AQUACULTURE (protocol code 42/2018).

Informed Consent Statement: Not applicable.

Data Availability Statement: Data are contained within the article.

Conflicts of Interest: Author Adriana F. Lima and Anderson G. P. Reis were employed by the company Brazilian Agricultural Research Corporation (EMBRAPA). The remaining authors declare that the research was conducted in the absence of any commercial or financial relationships that could be construed as a potential conflict of interest.

References

1. Alleway, H.K.; Waters, T.J.; Brummett, R.; Cai, J.; Cao, L.; Cayten, M.R.; Costa-Pierce, B.A.; Dong, Y.; Hansen, S.C.B.; Liu, S.; et al. Global principles for restorative aquaculture to foster aquaculture practices that benefit the environment. *Conserv. Sci. Pract.* **2023**, *5*, e12982. [CrossRef]
2. Bhakta, J.N.; Bandyopadhyayl, P.K.; Jana, B.B. Effect of different doses of mixed fertilizer on some biogeochemical cycling bacterial population in carp culture pond. *Turk. J. Fish. Aquat. Sci.* **2006**, *6*, 165–171. Available online: http://www.trjfas.org/pdf/issue_6_2/165_171.pdf (accessed on 11 August 2021).
3. Boyd, C.E.; D'Abramo, L.R.; Glencross, B.D.; Huyben, D.C.; Juarez, L.M.; Lockwood, G.S.; McNevin, A.A.; Tacon, A.G.J.; Teletchea, F.; Tomasso, J.R.; et al. Achieving sustainable aquaculture: Historical and current perspectives and future needs and challenges. *J. World Aquac. Soc.* **2020**, *51*, 578–633. [CrossRef]
4. Boyd, C.E. Aquaculture pond fertilization. *CAB Rev.* **2018**, *13*, 1–12. [CrossRef]
5. Garg, S.K.; Bhatnagar, A. Effect of fertilization frequency on pond productivity and fish biomass in still water ponds stocked with *Cirrhinus mrigala* (Ham.). *Aquac. Res.* **2000**, *31*, 409–414. [CrossRef]
6. Garg, S.K.; Bhatnagar, A. Effect of different doses of organic fertilizer (cow dung) on pond productivity and fish biomass in stillwater ponds. *J. Appl. Ichthyol.* **1999**, *15*, 10–18. [CrossRef]
7. Bhakta, J.N.; Sarkar, D.; Jana, S.; Jana, B.B. Optimizing fertilizer dose for rearing stage production of carps under polyculture. *Aquaculture* **2004**, *239*, 125–139. [CrossRef]
8. Tabinda, A.B.; Ayub, M. Effect of high phosphate fertilization rate on pond phosphate concentrations, chlorophyll a, and fish growth in carp polyculture. *Aquac. Int.* **2010**, *18*, 285–301. [CrossRef]
9. Kang'Ombe, J.; Brown, J.A.; Halfyard, L.C. Effect of using different types of organic animal manure on plankton abundance, and on growth and survival of *Tilapia rendalli* (Boulenger) in ponds. *Aquac. Res.* **2006**, *37*, 1360–1371. [CrossRef]

10. Rahman, M.M.; Nagelkerke, L.A.J.; Verdegem, M.C.J.; Wahab, M.A.; Verreth, J.A.J. Relationships among water quality, food resources, fish diet and fish growth in polyculture ponds: A multivariate approach. *Aquaculture* **2008**, *275*, 108–115. [CrossRef]
11. Milstein, A.; Azim, M.E.; Wahab, M.A.; Verdegem, M.C.J. The effects of periphyton, fish and fertilizer dose on biological processes affecting water quality in earthen fish ponds. *Environ. Biol. Fishes* **2003**, *68*, 247–260. [CrossRef]
12. Woynárovich, A.; Van Anrooy, R. Field Guide to the Culture of Tambaqui *Colossoma macropomum*, Cuvier, 1816. In *FAO Fisheries and Aquaculture Technical Paper*; FAO: Rome, Italy, 2019; Volume 624. Available online: https://books.google.com.br/books?id=WbKIDwAAQBAJ&printsec=frontcover&source=gbs_ge_summary_r&cad=0#v=onepage&q&f=false (accessed on 11 November 2022).
13. Azim, M.E.; Verdegem, M.C.J.; Khatoon, H.; Wahab, M.A.; Van Dam, A.A.; Beveridge, M.C.M. A comparison of fertilization, feeding and three periphyton substrates for increasing fish production in freshwater pond aquaculture in Bangladesh. *Aquaculture* **2002**, *212*, 227–243. [CrossRef]
14. Azim, M.E.; Little, D.C. Intensifying aquaculture production through new approaches to manipulating natural food. *CAB Rev.* **2007**, *1*, 23. [CrossRef]
15. Cremer, M.C.; Smitherman, R.O. Food habits and growth of silver and bighead carp in cages and ponds. *Aquaculture* **1980**, *20*, 57–64. [CrossRef]
16. Spataru, P.; Wohlfarth, G.W.; Hulata, G. Studies on the natural food of different fish species in intensively manured polyculture ponds. *Aquaculture* **1983**, *35*, 283–298. [CrossRef]
17. Chakrabarti, R.; Jana, B.B. Plankton intake as a function of body weight by common carp fry in different feeding conditions. *Aquaculture* **1991**, *93*, 21–34. [CrossRef]
18. Sipauba-Tavares, L.H.; Braga, F.M.D.S. Study on feeding habits of *Piaractus mesopotamicus* (Pacu) larvae in fish ponds. *Naga ICLARM Q.* **1995**, *22*, 24–29.
19. Mai, M.G.; Zaniboni-Filho, E. The effect of storage age in external tanks in the larvicultura performance of *Salminus brasiliensis* (Osteichthyes, Characidae). *Acta Sci. Anim. Sci.* **2005**, *27*, 287–296.
20. Sipaúba-Tavares, L.H.; Braga, F.M.S. The feeding activity of Colossoma macropomum larvae (tambaqui) in fishponds with water hyacinth (*Eichhornia crassipes*) fertilizer. *Braz. J. Biol.* **2007**, *67*, 459–466. [CrossRef]
21. Ribeiro, D.F.O.; Oliveira Nuñer, A.P. Feed preferences of *Salminus brasiliensis* (Pisces, Characidae) larvae in fish ponds. *Aquaculture* **2008**, *274*, 65–71. [CrossRef]
22. Gomes, L.C.; Silva, C.R. Impact of pond management on tambaqui, *Colossoma macropomum* (Cuvier), production during growth-out phase. *Aquac. Res.* **2009**, *40*, 825–832. [CrossRef]
23. Castelo Branco, M.A.; Arruda, M.A.; Gamito, S. Feeding habits of Solea senegalensis in earthen ponds in Sado estuary. *J. Sea Res.* **2010**, *64*, 446–450. [CrossRef]
24. Aya, F.A.; Corpuz, M.N.C.; Garcia, L.M.B. Diet composition, feed preferences and mouth morphology of early stage silver therapon (*Leiopotherapon plumbeus*, Kner 1864) larvae reared in outdoor tanks. *J. Appl. Ichthyol.* **2015**, *31*, 77–82. [CrossRef]
25. Lima, A.F.; Tavares-Filho, A.; Moro, G.V. Natural food intake by juvenile *Arapaima gigas* during the grow-out phase in earthen ponds. *Aquac. Res.* **2018**, *49*, 2051–2058. [CrossRef]
26. Cyrus, M.D.; Bolton, J.J.; Macey, B.M. The use of stable isotope ratios $\delta^{13}C$ and $\delta^{15}N$ to track the incorporation of *Ulva* and other important dietary ingredients into the gonads of the sea urchin *Tripneustes gratilla*. *Aquac. Nutr.* **2020**, *26*, 174–185. [CrossRef]
27. Asano, Y.; Hayashizaki, K.I.; Eda, H.; Khonglaliang, T.; Kurokura, H. Natural foods utilized by Nile tilapia, Oreochromis niloticus, in fertilizer-based fish ponds in Lao PDR identified through stable isotope analysis. *Fish. Sci.* **2010**, *76*, 811–817. [CrossRef]
28. Filbrun Jesse, E.; Reynolds, C.A.; Culver, D.A. Effects of feeding rate on habitat quality in fish rearing ponds. *J. World Aquac. Soc.* **2013**, *44*, 198–209. [CrossRef]
29. Abimorad, E.G.; Ducatti, C.; Castellani, D.; Jomori, R.K.; Portella, M.C.; Sartori, M.M.P.; Carneiro, D.J. Dietary lysine deficiency increases the incorporation rate of nitrogen-free compounds in muscle of pacu (*Piaractus mesopotamicus*). *Aquaculture* **2017**, *479*, 522–527. [CrossRef]
30. Narimbi, J.; Mazumder, D.; Sammut, J. Stable isotope analysis to quantify contributions of supplementary feed in Nile Tilapia *Oreochromis niloticus* (GIFT strain) aquaculture. *Aquac. Res.* **2018**, *49*, 1866–1874. [CrossRef]
31. Zhou, H.; Gu, B. Using stable isotope analysis to assess the relationship among dietary protein sources, growth, nutrient turnover and incorporation in Nile tilapia (*Oreochromis niloticus*). *Aquac. Nutr.* **2020**, *26*, 1443–1452. [CrossRef]
32. Hilsdorf, A.W.S.; Hallerman, E.; Valladão, G.M.R.; Zaminhan-Hassemer, M.; Hashimoto, D.T.; Dairiki, J.K.; Takahashi, L.S.; Corrêa Albergaria, F.C.; Gomes, M.E.S.; Venturieri, R.L.L.; et al. The farming and husbandry of *Colossoma macropomum*: From Amazonian waters to sustainable production. *Rev. Aquac.* **2022**, *14*, 993–1027. [CrossRef]
33. Goulding, M.; Carvalho, M. Life history and management of the tambaqui (*Colossoma macropomum*, Characidae): An important Amazonian food fish. *Rev. Bras. Zool.* **1982**, *1*, 107–133. [CrossRef]
34. Da Silva, J.A.M.; Pereira-Filho, M.; de Oliveira-Pereira, M.I. Seasonal variation of nutrients and energy in tambaqui's (*Colossoma macropomum* Cuvier, 1818) natural food. *Braz. J. Biol.* **2000**, *60*, 599–605. [CrossRef]
35. FAO. Fishery and Aquaculture Statistics—Yearbook 2020. In *FAO Yearbook of Fishery and Aquaculture Statistics*; FAO: Rome, Italy, 2023. [CrossRef]
36. Valladão, G.M.R.; Gallani, S.U.; Pilarski, F. South American fish for continental aquaculture. *Rev. Aquac.* **2018**, *10*, 351–369. [CrossRef]

37. Valenti, W.C.; Barros, H.P.; Moraes-Valenti, P.; Bueno, G.W.; Cavalli, R.O. Aquaculture in Brazil: Past, present and future. *Aquac. Rep.* **2021**, *19*, 100611. [CrossRef]
38. Roubach, R.; Gomes, L.C.; Leão Fonseca, F.A.; Val, A.L. Eugenol as an efficacious anaesthetic for tambaqui, *Colossoma macropomum* (Cuvier). *Aquac. Res.* **2005**, *36*, 1056–1061. [CrossRef]
39. Hyslop, E.J. Stomach contents analysis—A review of methods and their application. *J. Fish Biol.* **1980**, *17*, 411–429. [CrossRef]
40. Amundsen, P.-A.; Sánchez-Hernández, J. Feeding studies take guts—Critical review and recommendations of methods for stomach contents analysis in fish. *J. Fish Biol.* **2019**, *95*, 1364–1373. [CrossRef]
41. Ivlev, V.W. *Experimental Ecology of the Feeding of Fishes*; Yale University Press: New Haven, CT, USA, 1961; p. 302.
42. Coplen, T.B. Guidelines and recommended terms for expression of stable-isotope-ratio and gas-ratio measurement results. *Rapid Commun. Mass Spectrom.* **2011**, *25*, 2538–2560. [CrossRef]
43. Brand, W.A.; Coplen, T.B. Stable isotope deltas: Tiny, yet robust signatures in nature. *Isot. Environ. Health Stud.* **2012**, *48*, 393–409. [CrossRef]
44. Paul, D.; Skrzypek, G.; Fórizs, I. Normalization of measured stable isotopic compositions to isotope reference scales—A review. *Rapid Commun. Mass Spectrom* **2007**, *21*, 3006–3014. [CrossRef]
45. Schimmelmann, A.; Qi, H.; Coplen, T.B.; Brand, W.A.; Fong, J.; Meier-Augenstein, W.; Kemp, H.F.; Toman, B.; Ackermann, A.; Assonov, S.; et al. Organic reference materials for hydrogen, carbon, and nitrogen stable isotope-ratio measurements: Caffeines, n-alkanes, fatty acid methyl esters, glycines, l-valines, polyethylenes, and oils. *Anal. Chem.* **2016**, *88*, 4294–4302. [CrossRef]
46. Britton, J.R.; Busst, G.M.A. Stable isotope discrimination factors of omnivorous fishes: Influence of tissue type, temperature, diet composition and formulated feeds. *Hydrobiologia* **2018**, *808*, 219–234. [CrossRef]
47. APHA. *Standard Methods for the Examination of Water and Waste Water*, 21st ed.; American Public Health Association: Washington, DC, USA, 2017.
48. Engle, C.R. *Aquaculture Economics and Financing: Management and Analysis*; John Wiley & Sons: Hoboken, NJ, USA, 2010.
49. Veverica, K.; Bowman, J.; Gichuri, W.; Izaru, P.; Mwau, P.; Popma, T. Relative Contribution of supplemental feed and inorganic fertilizers in semi-intensive tilapia production. In *Seventeenth Annual Technical Report*; McElwee, K., Burke, D., Niles, M., Cummings, X., Egna, H., Eds.; Pond Dynamics/Aquaculture CRSP, Oregon State University: Corvallis, OR, USA, 2000; pp. 23–28.
50. Liranço, A.D.D.S.; Romagosa, E.; Scorvo-Filho, J.D. Growth performance of *Pseudoplatystoma corruscans* stocked in rearing systems: Semi-intensive (ponds) and intensive (cages). *Ciência Rural.* **2011**, *41*, 524–530. Available online: http://www.scielo.br/scielo.php?pid=S0103-84782011000300027&script=sci_arttext&tlng=es (accessed on 9 March 2018). [CrossRef]
51. Box, G.E.P.; Cox, D.R. An analysis of transformation. *J. R. Stat. Soc.* **1964**, *26*, 211–243. [CrossRef]
52. R Core Team, R. A Language and Environment for Statistical Computing; R Foundation for Statistical Computing: 2021, Viena, Austria. Available online: https://www.r-project.org/ (accessed on 15 June 2023).
53. Filbrun, J.E.; Culver, D.A. Stable isotopes reveal live prey support growth of juvenile channel catfish reared under intensive feeding regimens in ponds. *Aquaculture* **2014**, *433*, 125–132. [CrossRef]
54. Jomori, R.K.; Ducatti, C.; Carneiro, D.J.; Portella, M.C. Stable carbon ($\delta^{13}C$) and nitrogen ($\delta^{15}N$) isotopes as natural indicators of live and dry food in *Piaractus mesopotamicus* (Holmberg, 1887) larval tissue. *Aquac. Res.* **2008**, *39*, 370–381. [CrossRef]
55. Lima, A.F.; Rodrigues, A.P.O.; Costa, V.E. Frozen zooplankton is efficient as natural food during pirarucu *Arapaima gigas* weaning. *Aquac. Res.* **2021**, *52*, 4227–4236. [CrossRef]
56. Oliveira, A.C.B.; Martinelli, L.A.; Moreira, M.Z.; Soares, M.G.M.; Cyrino, J.E.P. Seasonality of energy sources of *Colossoma macropomum* in a floodplain lake in the Amazon—Lake Camaleão, Amazonas, Brazil. *Fish. Manag. Ecol.* **2006**, *13*, 135–142. [CrossRef]
57. Bateman, A.S.; Kelly, S.D. Fertilizer nitrogen isotope signatures. *Isot. Environ. Health Stud.* **2007**, *43*, 237–247. [CrossRef]
58. Lomartire, S.; Marques, J.C.; Gonçalves, A.M. The key role of zooplankton in ecosystem services: A perspective of interaction between zooplankton and fish recruitment. *Ecol. Indic.* **2021**, *129*, 107867. [CrossRef]
59. Brucet, S.; Boix, D.; Quintana, X.D.; Jensen, E.; Nathansen, L.W.; Trochine, C.; Meerhoff, M.; Gascón, S.; Jeppesena, E. Factors influencing zooplankton size structure at contrasting temperatures in coastal shallow lakes: Implications for effects of climate change. *Limnol. Oceanogr.* **2010**, *55*, 1697–1711. [CrossRef]
60. Vakkilainen, K.; Kairesalo, T.; Hietala, J.; Balayla, D.M.; Becares, E.; Van De BUnd, W.J.; Van Donk, E.; Fernandez-Alaez, M.; Gyllstrom, M.; Hansson, L.-A.; et al. Response of zooplankton to nutrient enrichment and fish in shallow lakes: A pan-European mesocosm experiment. *Freshw. Biol.* **2004**, *49*, 1619–1632. [CrossRef]
61. Sipaúba-Tavares, L.H. Análise da seletividade alimentar em larvas de tambaqui (*Colossoma macropomum*) e tambacu (híbrido, pacu *Piaractus mesopotamicus* e tambaqui *Colossoma macropomum*) sobre os organismos zooplanctônicos. *Acta Limnol. Bras.* **1993**, *6*, 114–132.
62. Lima, A.F.; Pereira, A.S.; de Oliveira Costa-Fernandes, T.; Rodrigues, A.P.O.; Costa, V.E.; Maciel-Honda, P.O. The effect of nursery production system (in cage and pond) on performance, health status, and plankton ingestion of the low trophic level fish tambaqui, *Colossoma macropomum*. *Aquaculture* **2024**, *586*, 740721. [CrossRef]
63. Chary, K.; van Riel, A.; Muscat, A.; Wilfart, A.; Harchaoui, S.; Verdegem, M.; Filgueira, R.; Troell, M.; Henriksson, P.J.G.; de Boer, I.J.M.; et al. Transforming sustainable aquaculture by applying circularity principles. *Rev. Aquac.* **2023**, *16*, 656–673. [CrossRef]

64. Aubin, J.; Callier, M.; Rey-Valette, H.; Mathé, S.; Wilfart, A.; Legendre, M.; Slembrouck, J.; Caruso, D.; Chia, E.; Masson, G.; et al. Implementing ecological intensification in fish farming: Definition and principles from contrasting experiences. *Rev. Aquac.* **2019**, *11*, 149–167. [CrossRef]
65. Watanabe, T.; Kiron, V. Prospects in larval fish dietetics. *Aquaculture* **1994**, *124*, 223–251. [CrossRef]
66. Bakke, A.M.; Glover, C.; Krogdahl, Å. Feeding, digestion and absorption of nutrients. *Fish Physiol.* **2010**, *30*, 57–110. [CrossRef]
67. Mitra, G.; Mukhopadhyay, P.K.; Ayyappan, S. Biochemical composition of zooplankton community grown in freshwater earthen ponds: Nutritional implication in nursery rearing of fish larvae and early juveniles. *Aquaculture* **2007**, *272*, 346–360. [CrossRef]
68. da Silva Reis, Y.; Leite, J.L.R.; de Almeida, C.A.L.; Pereira, D.S.P.; Vidal, L.V.O.; de Araujo, F.G.; Fortes-Silva, R. New insights into tambaqui (*Colossoma macropomum*) feeding behavior and digestive physiology by the self-feeding approach: Effects on growth, dial patterns of food digestibility, amylase activity and gastrointestinal transit time. *Aquaculture* **2019**, *498*, 116–122. [CrossRef]
69. Dias-Koberstein, T.C.R.; Carneiro, D.J.; Urbinati, E.C. Gastrointestinal passage time length and gastric emptying of pacu (*Piaractus mesopotamicus*) in different temperatures. *Acta Scientiarum Anim. Sci.* **2005**, *27*, 413–417.
70. Garg, S.K.; Bhatnagar, A. Effect of varying closes of organic and inorganic fertilizers on plankton production and fish biomass in brackish water fish ponds. *Aquac. Res.* **1996**, *27*, 157–166. [CrossRef]
71. Rezende, F.P.; Lima, A.F. Effect of pond fertilization on growth performance of pirarucu during grow-out phase. *Lat. Am. J. Aquat. Res.* **2022**, *50*, 22–30. [CrossRef]
72. Duodu, C.; Boateng, D.; Edziyie, R. Effect of pond fertilization on productivity of tilapia pond culture in Ghana. *J. Fish. Coast. Manag.* **2020**, *2*, 56–64. [CrossRef]
73. Boyd, C.E. *Water Quality: An Introduction*; Springer: Cham, Switzerland, 2020; 440p. [CrossRef]
74. INMET. INMET Tempo. 2021. Available online: https://tempo.inmet.gov.br/ (accessed on 19 October 2021).
75. Boyd, C.E.; Ammonia Nitrogen Dynamics in Aquaculture. Global Aquaculture Advocate. Available online: https://www.aquaculturealliance.org/advocate/ammonia-nitrogen-dynamics-in-aquaculture/ (accessed on 8 October 2018).
76. Das, S.K.; Jana, B.B. Pond fertilization regimen: State-of-the-art. *J. Appl. Aquac.* **2003**, *13*, 35–66. [CrossRef]
77. Flickinger, D.L.; Costa, G.A.; Dantas, D.P.; Moraes-Valenti, P.; Valenti, W.C. The budget of nitrogen in the grow-out of the Amazon River prawn (*Macrobrachium amazonicum* Heller) and tambaqui (*Colossoma macropomum* Cuvier) farmed in monoculture and in integrated multitrophic aquaculture systems. *Aquac. Res.* **2019**, *50*, 3444–3461. [CrossRef]
78. Flickinger, D.L.; Costa, G.A.; Dantas, D.P.; Proença, D.C.; David, F.S.; Durborow, R.M.; Moraes-Valenti, P.; Valenti, W.C. The budget of carbon in the farming of the Amazon river prawn and tambaqui fish in earthen pond monoculture and integrated multitrophic systems. *Aquac. Rep.* **2020**, *17*, 100340. [CrossRef]
79. Flickinger, D.L.; Dantas, D.P.; Proença, D.C.; David, F.S.; Valenti, W.C. Phosphorus in the culture of the Amazon River prawn (*Macrobrachium amazonicum*) and tambaqui (*Colossoma macropomum*) farmed in monoculture and in integrated multitrophic systems. *J. World Aquac. Soc.* **2020**, *51*, 1002–1023. [CrossRef]

Disclaimer/Publisher's Note: The statements, opinions and data contained in all publications are solely those of the individual author(s) and contributor(s) and not of MDPI and/or the editor(s). MDPI and/or the editor(s) disclaim responsibility for any injury to people or property resulting from any ideas, methods, instructions or products referred to in the content.

fishes

Article

Comparative Description and Analysis of Oyster Aquaculture in Selected Atlantic Regions: Production, Market Dynamics, and Consumption Patterns

Johannes A. Iitembu [1,*], Daniel Fitzgerald [2,†], Themistoklis Altintzoglou [3], Pierre Boudry [4], Peter Britz [5], Carrie J. Byron [6], Daniel Delago [7], Sophie Girard [4], Colin Hannon [8], Marcia Kafensztok [9], Francisco Lagreze [10], Jefferson Francisco Alves Legat [11], Angela Puchnick Legat [11], Adriane K. Michaelis [12], Ingelinn Eskildsen Pleym [3], Simone Sühnel [9,13], William Walton [12] and Åsa Strand [14,*]

1 Department of Fisheries and Ocean Sciences, University of Namibia, Henties Bay 13005, Namibia
2 Ecole Supérieure d'Agricultures, 55 Rue Rabelais, 49000 Angers, France; daniel@smallaxepeppers.com
3 Nofima, Muninbakken 9-13, 9291 Tromsø, Norway; themis.altintzoglou@nofima.no (T.A.); ingelinn.eskildsen.pleym@nofima.no (I.E.P.)
4 Département Ressources Biologiques et Environnement, Ifremer, 29280 Plouzané, France; pierre.boudry@ifremer.fr (P.B.); sophie.girard@ifremer.fr (S.G.)
5 Department of Ichthyology and Fisheries Science, Rhodes University, Makhanda 6140, South Africa; p.britz@ru.ac.za
6 School of Marine and Environmental Programs, University of New England, 11 Hills Beach Road, Biddeford, ME 04005, USA; cbyron@une.edu
7 Ocean Era Inc., 73-970 Makako Bay Drive, Kailua-Kona, HI 96740, USA; dan@ocean-era.com
8 Marine and Freshwater Research Centre, Atlantic Technological University, Dublin Road, H91 T8NW Galway, Ireland; colin.hannon@atu.ie
9 Primar Aquacultura, Estrada RN 03, km 10, Sítio São Félix, Piau, Tibau do Sul 59178-000, RN, Brazil; primarorganica@gmail.com (M.K.); ssuhnel@gmail.com (S.S.)
10 Center of Marine Research, Federal University of Paraná, Av. Beira-Mar, s/n, Pontal do Sul, Pontal do Paraná 83255-976, PR, Brazil; lagreze@ufpr.br
11 Brazilian Agricultural Research Corporation, Embrapa Coastal Tablelands, Av. Beira Mar 3250, Aracaju 49025-040, SE, Brazil; jefferson.legat@embrapa.br (J.F.A.L.); angela.legat@embrapa.br (A.P.L.)
12 Virginia Institute of Marine Science, College of William and Mary, Andrews Hall 440, Gloucester Point, VA 23062, USA; amichaelis@vims.edu (A.K.M.); walton@vims.edu (W.W.)
13 AquaInspiration and Innovation, Rua Túlio de Oliveira, 195, Armação, Florianópolis 88066-303, SC, Brazil
14 Department of Environmental Intelligence, IVL Swedish Environmental Research Institute, Kristineberg 566, 45178 Fiskebäckskil, Sweden
* Correspondence: jiitembu@unam.na (J.A.I.); asa.strand@ivl.se (Å.S.)
† Current address: Small Axe Peppers, New York, NY 10001, USA.

Citation: Iitembu, J.A.; Fitzgerald, D.; Altintzoglou, T.; Boudry, P.; Britz, P.; Byron, C.J.; Delago, D.; Girard, S.; Hannon, C.; Kafensztok, M.; et al. Comparative Description and Analysis of Oyster Aquaculture in Selected Atlantic Regions: Production, Market Dynamics, and Consumption Patterns. *Fishes* **2023**, *8*, 584. https://doi.org/10.3390/fishes8120584

Academic Editor: Wagner C. Valenti

Received: 20 October 2023
Revised: 21 November 2023
Accepted: 24 November 2023
Published: 29 November 2023

Copyright: © 2023 by the authors. Licensee MDPI, Basel, Switzerland. This article is an open access article distributed under the terms and conditions of the Creative Commons Attribution (CC BY) license (https://creativecommons.org/licenses/by/4.0/).

Abstract: In the face of an increasing world population and a subsequent need for an increase in sustainable and healthy food production, low trophic species, such as oysters, emerge as a promising alternative. However, regional variations in oyster production techniques, market dynamics, and consumption patterns create challenges for both the global and local industry's growth. In this study, a descriptive qualitative analysis of oyster markets across seven Atlantic regions was carried out. The Pacific oyster (*Crassostrea gigas*) was found to be farmed in most Atlantic regions except the US but is classified as invasive in Sweden and potentially invasive in South Africa. Other farmed and/or harvested species include native species (*C. gasar and C. rhizophorae*) in Brazil, the American cupped oyster (*C. virginica*) in the US, and the European flat oyster (*Ostrea edulis*) in France, Sweden, and the US. In Irish farms, Pacific oysters are primarily for export to European markets. The marine aquaculture sectors of Sweden, South Africa, and Namibia, as well as Brazil's farming for *C. gasar*, were found to be underdeveloped. This study also observed a variation in licensing, property rights, and regulatory frameworks. Financial challenges for small businesses, ecological implications of seed production techniques, biosecurity risks, and public health considerations are emphasized as critical areas for attention. This study offers valuable insights into the selected markets and can serve as a useful resource for policymakers, aquaculture practitioners, and stakeholders in optimizing global shellfish industry strategies.

Keywords: Atlantic regions; oysters; production; markets; aquaculture; low trophic

Key Contribution: Qualitative analysis of oyster markets in seven Atlantic regions highlights variations in the species farmed, production techniques, and market dynamics. Key challenges identified include financial hurdles for small businesses, ecological concerns related to seed production, biosecurity risks, and public health issues. These findings could inform policymakers and stakeholders in refining oyster industry growth strategies at both global and local levels.

1. Introduction

Humans have consumed shellfish since ancient times. Archaeological studies provide evidence of its consumption dating back to the initial migrations of *Homo sapiens* out of Africa, approximately 100,000 to 200,000 years ago [1]. Oysters, among shellfish species, have been prized as aphrodisiacs from the time of Roman emperors to contemporary Casanovas. For millennia, oysters have sustained civilizations from the Ancient Romans [2] to various Chinese dynasties [3] up to modern consumers. Few types of seafood can boast such a storied history. This popularity has resulted in overexploitation globally [4]. In Europe, overfishing led to the depletion of European flat oyster (*Ostrea edulis*) beds starting from the 18th century [5–7]. In a similar vein, eastern American oyster (Crassostrea virginica) reefs in the United States experienced a decline by the 1940s [6]. Globally, oysters represent the leading farmed molluscan species in terms of quantity produced, with increasing production since 1950 [8]. Global oyster production is dominated by China, while the next five countries with significant oyster aquaculture production include France, the US, South Korea, Japan, and the Philippines [8]. Asia dominates the global production and consumption of oysters [9] and production in other regions has also seen an increase since 1950 [8].

Oysters also play significant ecological roles in addition to their importance as a food source. They are recognized as ecosystem engineers due to their direct or indirect control over resource availability for other organisms [10]. Oysters trap organic-enriched particles with their shells and create three-dimensional structures, thereby facilitating the presence of additional fauna and seaweed, particularly in soft-sediment environments [11,12]. As filter feeders, oysters contribute to water clarity through their filtration process. This increased clarity allows for greater light penetration, which benefits adjacent vegetated habitats such as seagrass beds [13]. These vegetated habitats are vital as nursery areas and for carbon sequestration [13]. Furthermore, oyster filtration helps reduce the occurrence of harmful algal blooms, which have significant ecological and economic impacts [13,14].

As food, oysters are a highly nutritious source, containing essential nutrients not provided by land-based proteins [15,16]. They are considered one of the most nutrient-dense seafood options [17] while offering a high-protein, low-fat product that contains good levels of healthy Ω-3 fatty acids [9]. Oyster aquaculture, recognized as low-trophic, is also valued for its reduced environmental impacts, lower feed requirements [18], and potential significant contributions to circular economics [19]. Various research initiatives have focused on expanding the products and processes of low-trophic species in marine aquaculture value chains across the Atlantic [20,21]. However, globally, oyster production has faced recurring setbacks due to disease and parasitic outbreaks [8]. Furthermore, translocations of farmed oysters between geographical areas and the introduction of new species for culture have contributed to the dispersal of invasive species, pathogens, and parasites over time [22].

Despite these challenges, the global production of oysters has shown an upward trend, indicating a sustained increase in oyster aquaculture production for various markets [8]. In terms of regional and country/region-specific productions, it has stagnated in most countries and the increase is primarily driven by the Asian market [8]. While oyster production and consumption have been increasing in some regions [8,23], there are also areas where oysters are underutilized as a resource [24]. The underutilization of oysters in some regions may be attributed to the marketing of oysters, which is influenced by factors such as growth, mortality, quality, price [25], and existing sanitation programs (food safety) [26]. Additionally, there is a lack of knowledge about the comparative characteristics of various regions regarding production, consumption patterns, and market dynamics. This study, therefore, aimed to conduct a qualitative description of oyster production, consumption patterns, and market dynamics in seven regions (the US, Brazil, France, Sweden, Ireland, South Africa, and Namibia) along the Atlantic Ocean, to enhance understanding of these markets. The research questions included understanding how oyster production methods differ among the seven Atlantic regions, the implications of these differences for the industry's sustainability and growth, the unique consumption patterns of oysters in each region, and how market dynamics and strategies vary across these regions. Comparing these similarities and differences among the regions can serve as a crucial resource for understanding and optimizing global and local oyster production, marketing, and consumption strategies. Additionally, the findings from this study can also be used as a basis for detailed investigations of oyster farming strategies and practical adoption of farming methods in specific markets.

2. Methods and Materials

2.1. Study Areas

The geographical regions included in this study encompassed North America, South America, Europe, and Africa, all of which border the North and South Atlantic Oceans (Figure 1). These regions were selected based on their participation in or association with the AquaVitae project, to provide a snapshot of variability among nations/regions based on overall market trends. Due to the large geographical scale and diversity of the Brazilian and American markets, these areas were further subdivided into northern US, southern US (including the Gulf of Mexico), northeast Brazil, and southern Brazil to facilitate more detailed analyses of these regions (Figure 1). The chosen markets within these regions included (a) markets with high levels of production and consumption (mature): France, the US, and southern Brazil; (b) markets known for low levels of oyster consumption and production (immature): Sweden and northern Brazil; and (c) markets that did not comfortably fit into either category (a) or (b) (mixed): Namibia, South Africa, and Ireland.

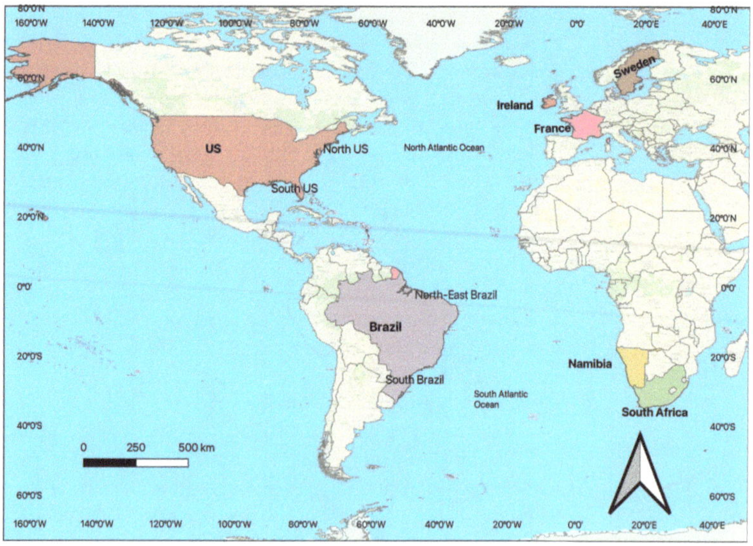

Figure 1. A world map illustrating the geographical regions (shaded in colours) examined in this study.

2.2. Data Collection and Analysis

This study employed questionnaires (open-ended questions) and personal communications with academic experts from the selected markets (Table 1) and literature searches. The experts were selected based on their substantial knowledge of the selected markets and their participation in or association with the AquaVitae project. The purpose was to conduct a qualitative description of the selected markets to gain insights into the historical trends in oyster aquaculture, consumer perceptions of oyster aquaculture, and current market trends related to oyster consumption and production. Qualitative descriptive analysis was used because this study drew from the perspectives of the selected experts, employed a purposeful selection of experts, and involved content analysis of various literature and reports [27,28]. The questionnaires, given to the experts, included a current overview of oyster production and consumption in each market. The data extracted from the questionnaires and expert interviews were utilized to profile the selected markets using the PEST (political, economic, social, and technological) analysis framework to provide insights into market characteristics [29]. For the application of PEST, the political factors (P) account for the political situation. In this study, they included governance, licensing, and regulatory measures related to health considerations. The economic factors (E) cover issues that may influence business operations and profitability. This study included issues related to the challenges and opportunities for financing. Social (S) factors cover aspects related to consumers and producers, with this study incorporating aspects of the roles of producer organizations and cooperatives. The technological (T) factors cover issues related to operations and production, which are critical for the long-term future of oyster farming in various markets. Technological considerations in this study included seed production techniques and regulation, the role of hatcheries in oyster seed production, the advantages of triploid oysters, production techniques, postharvest processing, the importance of depuration facilities, and food safety through traceability. The data used included production numbers, production methods, aquaculture cultivation, wild harvest, native and exotic species, import and export markets, consumer profiles, safety standards, and value-added products relevant to the chosen markets. Other parameters for each market considered included licensing structures, governance, food safety requirements, financing/investment availability, insurance availability, hatcheries/seed production, triploid production, water monitoring programs, producer cooperatives and organizations, and native oyster cultivation/restoration (Table 2).

Table 1. List of experts selected from each market considered.

Market	Expert ID	Affiliation
US (north)	1	University of New England
US (north)	2	University of New England
US (north)	3	University of New England
US (south)	4	Virginia Institute of Marine Science
US (south)	5	Virginia Institute of Marine Science
Brazil (north)	6	Brazilian Agricultural Research Corporation (EMBRAPA)
Brazil (north)	7	Primar Aquacultura
Brazil (south and north)	8	Primar Aquacultura, AquaInspiration and Innovation
Brazil (south and north)	9	Universidade Federal do Paraná
France	10	Ifremer
France	11	Ifremer
Sweden	12	IVL Swedish Environmental Research Institute
South Africa	13	Rhodes University
Namibia	14	University of Namibia

Table 2. Summary of factors from PEST (political, economic, social, technological) analysis framework and market characterization.

	Northern US	Southern US	Northern Brazil	Southern Brazil	France	Ireland	Sweden	South Africa	Namibia
Market characterization	Mature	Mature	Immature	Mature	Mature	Mixed	Immature	Mixed	Mixed
Species	*Crassostrea virginica*	*C. virginica*	*Crassostrea gasar*, *C. rhizophorae*, and *C. gigas*	*Crassostrea gasar*, *C. rhizophorae*, and *C. gigas*	*C. gigas* and *O. edulis)*	*C. gigas* and *O. edulis)*	*C. gigas* and *O. edulis*	*C. gigas* and *Striostrea margaritacea)*	*C. gigas*
Fishery category	Aquaculture (A) and capture fishery (C)	Aquaculture (A) and capture fishery (C)	Aquaculture and capture fishery	Aquaculture	Aquaculture	Aquaculture	Aquaculture and capture fishery	Aquaculture and capture fishery	Aquaculture
Production statistics (2019) (Aquaculture = A and capture fishery = C)	148,000 t (A) and 38,000 t (C)		1700 t (A) and 1000 t (C)		86,000 t(A)	10,700 t (A)	21 t (A)	383 t (A) and 23 t (A)	350 t (A)
Political (P), Economic(E), Social (S), and Technological (T): PEST considerations									
Leasing of seabed/licensing (P)	x	x		x	x	x	x	x	x
Oyster sanctuaries/protected areas (P)	x	x				x	x		
Economic incentives (loans, tax breaks) (E)	x	x		x	x			x	x
Disaster relief/crop insurance (E)	x	x			x				x
Efficient and timely permitting process (P)	x	x	x	x	x			x	x
Extensive seed production (wild seed collection/pond production) (T)		x	x		x	x	x		x
Hatchery seed production (T)	x	x	x	x	x	x	x		x
Use of triploids (T)	x	x		x	x	x			
Producer cooperative (S)	x	x	x	x	x	x	x	x	x

Table 2. *Cont.*

	Northern US	Southern US	Northern Brazil	Southern Brazil	France	Ireland	Sweden	South Africa	Namibia
Political (P), Economic(E), Social (S), and Technological (T): PEST considerations									
Water monitoring programs (P)	x	x		x	x	x	x	x	x
Food safety regulation (P)	x	x	x	x	x	x	x	x	x
Postprocessing facilities (T)	x	x	x	x	x	x	x	x	x
Traceability (T)	x	x		x	x	x	x	x	x

3. Results and Discussions

3.1. General Description of Production, Market, and Consumption Patterns of the Selected Markets

3.1.1. The United States

The native *C. virginica* is the primary oyster species on the eastern seaboard of the United States and is the most valuable marine aquaculture species in the US [30]. In 2019 Atlantic Ocean oyster landings, including aquaculture and wild harvest, were valued at approximately USD 192 million, with over 148,000 metric tons of farmed oysters and almost 38,000 metric tons of wild-harvested oysters (Table 2) [31]. Various production techniques are employed for oyster production, ranging from harvesting wild oysters from wild grounds to on-bottom culture using shell deployment to collect naturally produced seed, or young oysters transplanted to private, leased grounds. More intensive methods used involve off-bottom, floating, and other types of container culture, with oyster seed obtained from commercial hatcheries. Equipment commonly used varies by location largely depending on local system dynamics (e.g., water depth, tidal fluctuation, and accessibility) and US state regulations. In the US, nearshore aquaculture adheres to federal regulations and is managed at the state level. In some cases, states allow for even more localized management of oyster aquaculture, and it is managed and permitted at township or county levels.

In terms of markets, wild oysters are typically sold at lower prices compared to aquaculture-farmed oysters and have traditionally been associated with the canning market. However, there is a shift in consumer preferences towards fresh, raw oysters. In 1970, 76% of oysters produced were consumed as fresh and raw, increasing to 92% by 1994 [32]. Shellfish consumers, according to national seafood surveys, tend to have higher education and household income than average [33,34]. They are also more likely to be over 50 years of age [35]. Experienced oyster consumers show a preference for farmed oysters, while infrequent oyster consumers are more inclined towards the concept of wild-caught oysters [36].

Certification under the National Shellfish Sanitation Program (NSSP) is mandatory for all oyster dealers in the US to market oyster products across state lines. However, the proportion of oyster production exported is relatively small, partly due to the 10-year ban on shellfish trade with the EU, although this was recently lifted [37]. This shift may be linked to changes in the typical oyster consumer. Additionally, the US imports oysters from both China and South Korea, including both farmed and wild species [38].

3.1.2. Brazil

Oyster production in Brazil primarily revolves around the *Crassostrea* genus, which includes two native species (*Crassostrea gasar* (=*Crassostrea brasiliana*) and *C. rhizophorae*) and one exotic species (*C. gigas*, introduced in Brazil in 1974), with the latter being the most produced and consumed. In 2019, Brazil's total oyster production reached 2700 tons, consisting of 1700 tons from aquaculture and 1000 tons from capture fisheries, with a farm oyster value estimated at USD 1.8 million (Table 2) [31]. The native species (*C. gasar* and *C. rhizophorae*) are widely distributed across Brazilian estuaries and have been harvested by low-income fishing communities in various mangrove areas [39]. However, in several regions, oyster growers expressed concerns over the low market value of wild oysters, which imposes limitations on both market prices and farmed oysters. Sometimes, juveniles are harvested from the mangrove and used for grow-out in fisheries-based aquaculture systems. Some producers also employ seed collectors to capture native species of *Crassostrea* ssp. seeds. Oyster hatcheries play a crucial role in producing seeds for *C. gigas* farming in southern Brazil, where no wild seed collection occurs. In 2019, the estimated production of hatchery seeds reached about 46,000 seeds, although there are limited hatcheries for the native species [40], with the seed of *C. gasar* being produced in a private hatchery in the northeast of Brazil and in a public hatchery in southern Brazil.

Regarding production techniques, the majority of producers utilize suspended lanterns in longline systems and bags for grow-out. In the southern regions, suspended systems, such as longlines with lantern nets (off-bottom culture) are common, particularly in shallow and protected areas with minimal tidal variation [40], with one case in on-bottom culture in Paraná. In contrast, bottom-based systems, particularly fixed tables with plastic bags, are prevalent in other states, especially in the northeast and north. There are also cases of pond-based production of oysters in the northeast. The variation in production techniques across states is primarily influenced by the specific grow-out areas. For instance, in Santa Catarina and Rio de Janeiro, cultivation occurs in bay areas near the coast, while in the northeastern and northern states, estuarine areas are primarily utilized for oyster grow-out, with Paraná and São Paulo in both conditions, bay and estuarine areas. The duration of the grow-out cycle of *C. gigas* and *C. gasar* ranges from 6 to 12 months, depending on the harvest size and area of production.

The consumption of native oysters from wild capture is a longstanding tradition in all coastal states of Brazil, although it lacks comprehensive documentation. On the other hand, the consumption of *C. gigas* oysters, which is well documented, is based on aquaculture production. The majority of oyster products are consumed locally by Brazilians and tourists [41], particularly during the summer months (December to February), when the majority of the population goes on vacation and there is an increase in the flow of tourists to coastal regions. In the northeast region, there is also an increase in consumption in the month of July, as it is also a vacation period. The Brazilian oyster market exhibits segmentation between high-end and low-end markets. Cultured oysters sold in restaurants or stores are typically associated with the high-end market, while oysters collected by artisanal fishermen and sold in public seafood markets or by street vendors on the beaches cater to the low-end market. Oyster products in Brazil can be marketed as whole, freshly shucked, steam-cooked, boiled, or frozen [39]. In terms of regulatory measures, as of 2023, only one state in southern Brazil, Santa Catarina, has implemented a monitoring program for oyster production, with Geographical Indication (GI) for *C. gigas* from Florianópolis. Additionally, Alagoas in the northeast region of Brazil possess a government-certified shellfish depuration facility. Besides initiatives to export oysters (such as to Singapore, the US, and China) there are no exports from, or imports to, the Brazilian oyster market yet.

3.1.3. France

France holds the position of the largest oyster producer in Europe. In 2019, the total oyster production amounted to approximately 86,000 tons (Table 2) [31], with the Pacific oyster (*C. gigas*) accounting for 93% and the native European flat oyster (*O. edulis*) comprising only 1165 tons (7%) [22]. The value of oyster cultivation in the same year reached USD 446 million, with the Pacific oyster contributing USD 438 million and the flat oyster generating USD 8 million [22]. The production of flat oysters remains limited due to the presence of parasites of the genus *Bonamia* and *Marteilia* [7]. While wild harvest plays a minimal role in production, the cultivation of farmed oysters involves the traditional seasonal collection of oyster seeds during summer. Oyster hatcheries of Pacific oyster (*C. gigas*) and European flat oyster (*O. edulis*) now meet approximately 30–50% of the seed demand, supplying seed almost year-round. Notably, 90% of the hatchery-produced seed consists of triploid oysters [42]. In areas where local seed collection is feasible, many oyster farmers utilize both sea-collected and hatchery-produced seeds.

The primary production method in France involves the use of racks and trestles located in the intertidal zone. Oysters are cultivated from the seashore to deeper waters, primarily in protected shallow bays at depths ranging from 10 to 15 m [42]. In the Mediterranean Sea, where the tidal range is limited, longline suspended culture is predominantly employed [42]. Deep-water culture (on-bottom) is practiced in south Brittany, where the seed is deployed in large concessions and subsequently harvested through dredging. The duration of the grow-out cycle ranges from 2 to 4 years, depending on the specific area, applied techniques, oyster species, and the type of seed used. The transfer of developing oysters between production areas is a common practice to leverage the advantages offered by different regions, such as seed collection, on-growing, and finishing processes.

Oysters have traditionally been considered a luxury product, primarily sold live and consumed raw [7,43]. Oyster consumers tend to be older than average and mainly belong to higher income brackets [7]. Consumption patterns involve at-home consumption, which is highly seasonal, and sales through super/hypermarkets [7]. In France, oyster production follows the European Union Food Hygiene Regulations, which establish standards for end products. However, Marennes-Oléron oysters stand out with a Protected Geographical Indication (PGI), certifying the refinement of Pacific oysters in "claires" (or ponds originally used to produce sea salt) within the Marennes-Oléron basin. This certification ensures the production of the highly prized fine-de-claire oysters, known for their exceptional quality [44]. France's oyster exports remain relatively low, primarily directed towards Italy, China, Hong Kong, the Netherlands, and Spain. Import-wise, France receives just over 5100 tons of oysters from Ireland, primarily as juveniles to finish growing in French waters.

3.1.4. Ireland

Ireland's total oyster production reached approximately 10,700 tons (2019) (Table 2) [31], with the majority consisting of Pacific oysters (*C. gigas*) at 10,460 tons and flat oysters (*O. edulis*) at 256 tons. The total value amounted to USD 51 million for Pacific oysters and nearly USD 1.4 million for flat oysters [31]. Oyster production is spread across 11 counties in Ireland, with Waterford and Donegal collectively accounting for around 60% of the country's production by tonnage [45]. While Ireland has several hatcheries, the industry heavily relies on imported seeds primarily from hatcheries in France and, to a lesser extent, from British hatcheries [46].

The main cultivation method for Pacific oysters in Ireland is through the use of bags and trestles. Ireland's oyster industry is primarily export-focused, with France accounting for 88% of Ireland's exports between 2012 and 2014, with 78.4% destined for the French market in 2021 with the increase in purchasing from other European markets. In 2021, the output volume and sales value of 10,624 tons valued at EUR 47.55 million (USD 51.8 million) [47]. The industry has been supported by French–Irish joint ventures [7]. Ireland also imports 241 tons of oysters, mainly seeds from the UK, and 72 tons of oysters from France. Oyster production follows the European Union Food Hygiene Regulations, which establish standards for end products. *O. edulis* production capacity is growing in Ireland with the redevelopment of spatting ponds by aquaculture enterprises and the further redevelopment of native oyster production beds at sea by fishermen's co-ops through the deployment of cultch and coupelles [48].

3.1.5. Sweden

Sweden's oyster production is relatively small, with a total of 21 tons (Table 2) [31]. Aquaculture production in Sweden primarily focuses on the native flat oyster (*O. edulis).* The Pacific oyster (*C. gigas)* has been invasive in Sweden since its establishment in 2006 [49], and its cultivation is prohibited. For wild harvest, approximately half of the oyster catch consists of Pacific oysters while the other half comprises European flat oysters with a small portion of farmed *O. edulis*. The aquaculture production is constrained by seed availability and seeds are sourced mainly using sea-based collectors. A hatchery established in 2006 produces insufficient seed for sale to the rest of the industry [50]. The harvest of wild *O. edulis* (fisheries) remains relatively stable over time. In contrast, the harvest of wild *C. gigas* has seen an increase in recent years, with production rising from 1 ton to 8 tons between 2009 and 2019.

Wild oysters are harvested through diving (*O. edulis* and *C. gigas*) or handpicking (*C. gigas*) [51]. Farming of *O. edulis* is carried out using surface-based floating cages or suspended cages from longline systems or other structures. Sea-based seed capture is the primary method for oyster farming. Oyster consumption patterns in Sweden align with Scandinavian per capita consumption of shellfish, which remains relatively small compared to many other developed countries in Europe and North America [52]. The oyster market is primarily local, with oysters mainly consumed in restaurants and as half-shells. Grilled and gratinated wild-harvested Pacific oysters are gaining popularity. However, the demand from Swedish consumers exceeds the supply, resulting in a steady increase in oyster consumption. Imported oysters, mainly from the Netherlands and France, meet the demand, while there are no exports from Sweden. Import volume has increased significantly, from 30 tons in 2002 to 350 tons in 2010 [52]. The native flat oyster commands a premium price compared to locally harvested Pacific oysters, which are priced higher than imported Pacific oysters.

3.1.6. South Africa

Aquaculture in South Africa's marine environment is underdeveloped due to the limited presence of estuaries or sheltered bays suitable for culture operations. In 2019, the total oyster production in South Africa was 406 tons, with 383 tons from aquaculture, mainly the Pacific oyster (*C. gigas*), and 23 tons from native rock oyster (*Striostrea margaritacea*) captured for subsistence consumption and commercial use (Table 2) [53]. The aquaculture value amounted to USD 5 million [31]. The introduction of the Pacific oyster to South Africa occurred in 1973, while the current market for native oysters relies entirely on wild harvesting [54]. Wild oysters are either collected by subsistence fishers in the Eastern Cape province for their own consumption or are sold for income [55], or by licensed commercial harvesters in the southern Cape for sale to the restaurant trade. Early efforts to farm native rock oysters were abandoned due to their slower growth compared to Pacific oysters and limited cultivation knowledge [53]. Commercial oyster culture in South Africa is predominantly located in the Western Cape, specifically in the Saldanha Bay area, where the Pacific oyster is farmed in baskets suspended on longlines [53]. Currently, it takes three to four months to grow oysters to a marketable size, with the industry relying on imported oyster seed from Chile, England, and local hatcheries [53].

The South African oyster market is a niche market primarily supplying restaurants, with higher sales during December to March. South African oysters are mainly sold fresh (80–85%), with the wild rocky oyster sought after for its rich, wild taste. It commands a price premium over farmed Pacific oysters, with cultured oysters and rock oysters fetching a 20–30% premium. Oysters are sold shucked, whole frozen, or frozen in half shells, with a small portion value-added with sauce and crumbs. Smoked and canned oyster products in South Africa are imported, as well as the majority of frozen oyster products, as imports dominate the postharvest value-added market [56]. Exporting oysters from South Africa faces challenges due to the lack of international health certification standards, limiting access to most countries. Currently, oysters are exported to southeast Asian countries like Hong Kong, where health standards are less stringent. Efforts are underway to implement a shellfish water quality monitoring program to obtain EU certification and expand market access. However, high production costs and low volumes tend to make South African oyster producers uncompetitive in export markets [57].

3.1.7. Namibia

In 2019, Namibia's oyster production reached 350 tons, exclusively from aquaculture, with a value of USD 1.9 million (Table 2) [31]. The main shellfish farming areas are found around Lüderitz, Walvis Bay, and Swakopmund [58]. The Pacific oyster (*C. gigas*) is the primary cultured species in Namibia, known for its relatively shorter growth period of 9 to 15 months compared to competitors in Europe, Japan, and the US [59].

Longlines are the predominant production method in Namibia, although some pond-based production also exists. Oyster (*C. gigas*) seed is provided by a hatchery established in Walvis Bay. The local Namibian market for oysters is limited due to the absence of traditional seafood consumption and the relatively small population of around 2.8 million inhabitants. Growers in Namibia primarily sell to wholesalers, as direct selling to restaurants for small orders requires significant effort [60]. The main markets for Namibian oysters are South Africa and southeast Asia, with only a small proportion sold in the local market [61]. Namibia is a significant exporter of oyster products to South Africa and a major competitor for local producers there [56]. However, the expansion of the oyster aquaculture industry in Namibia is hindered by limited access to other export markets such as the EU, the United States, Russia, and Asian countries. Compliance with approved sanitation standards is crucial for Namibian oysters and other high-quality shellfish to enhance access to these export markets [62].

3.2. Political, Economic, Social, and Technological (PEST) Considerations in Oyster Aquaculture

3.2.1. Political Considerations (P)

Governance and Licensing

The aquaculture sector faces various challenges related to governance, including legal insecurity, cumbersome administrative procedures, and weak participatory approaches [63]. Good governance and a sound regulatory framework are crucial to address these constraints, along with effective decision-making, planning, and monitoring tools [63]. In this study, licensing requirements for oyster aquaculture were found to vary among the studied markets. For example, the US operates under well-established leasing programs that vary by state, while Brazil has no licensing for on-bottom culture, and Sweden requires culture permits [64]. The duration of leases and permits also differs, with some countries granting 3–5-year approvals (South Africa and Sweden) and France leasing for 30 years [65]. Lease costs are a factor influencing aquaculture development in the US, and high fees may hinder growth and discourage entry into the sector in some states [8].

Moreover, the balance of multiple-use conflicts among maritime actors can cause significant delays and obstacles in permit issuance [66]. Different approaches are employed to address these challenges in the studied markets. As noted above, leasing regulations vary by state, but as one example in the Gulf of Mexico of the US (Alabama), prospective farmers obtain oyster riparian rights either through waterfront property ownership or leasing from other waterfront property owners, ensuring control over designated harvesting areas [67]. In Sweden, fishing laws regulate ownership of natural oyster stocks, with oysters belonging to land/water owners up to 200 m from the shore [52], yet harvest is only allowed in designated production areas for food safety reasons. Leasing rights and culture licensing should also protect ecologically valuable public oyster grounds and consider the natural carrying capacity of bays and inlets, as shellfish production relies on complex ecological processes [52]. In Namibia, licenses are granted based on a positive environmental assessment report, and when no significant environmental risks are identified [58]. Consequently, clear licensing and ownership rights are essential to ensure the sustainability of natural oyster reef exploitation and aquaculture activities. Uncertainty arising from the absence of clear regulations may reduce financial institutions' and investors' willingness to provide necessary funding for expansion in production.

Regulatory Measures Related to Health Considerations

Consuming oysters carries significant health risks due to potential contamination with algal biotoxins, fecal-associated viruses, bacteria, and heavy metals [68]. Consequently, maintaining high food safety standards is essential to protect consumers and ensure the growth of the oyster industry. Hence, to ensure the safety of raw oysters, public health controls and monitoring programs are necessary. The occurrence of such structures appears to be less related to the maturity state of the market than many of the other aspects evaluated in this study. In the United States, the National Shellfish Sanitation Program (NSSP) regulates molluscan shellfish food safety, with oyster-growing waters classified as approved, restricted, or prohibited based on water quality parameters. State shellfish control authorities map areas open or closed for harvesting, and shellfish sanitation programs are funded by permit and license fees [69]. Similarly, the European Union (EU) has stringent regulations for shellfish water classification, with different security measures required based on classifications [70]. EU directives mandate quality standards, health conditions, and maximum levels of contaminants in shellfish [52]. The European Commission has a National Marine Biotoxin Monitoring Program implemented to detect toxins, and shellfish harvesting and sale are prohibited if unsafe levels are found [71]. The same structures apply to both mature and evolving markets, regardless of developmental stage. In contrast, South Africa and Namibia have no general regulations; however, oyster aquaculture permit conditions prescribe regular testing for heavy metals, biotoxins, and bacterial contamination. Namibia and South Africa have both established shellfish sanitation programs to meet international standards for access to export markets but have not yet achieved EU approval [59]. In Brazil, regulations mainly focus on testing for *E. coli*, toxins (in bivalve meat), and harmful algae concentration, with enforcement challenges in some regions. Brazil has new legislation for the shellfish sanitation program (National Safety Bivalves Mollusks Program MoluBis; Portaria SDA/MAPA 884, 06/09/2023), with standards based on US and EU sanitary legislation, where other contaminants will be included, and area classification will be used. The above shows that the implementation of regulatory measures, such as the Shellfish Sanitation Program and the Marine Biotoxin Monitoring Program across various regions, can potentially contribute to optimizing oyster industry growth at both global and regional levels.

3.2.2. Economical Consideration (E)

Challenges and Opportunities for Financing

Securing financing was pointed out as a challenge in oyster aquaculture in all considered markets, except in the US. Small businesses in the industry could benefit from capital investment or low-interest loans provided by provincial and national governments, as recommended in South Africa [57]. In the US, financing options are also available through the Farm Service Administration, aimed at small operators and potential entrants who do not qualify for conventional lending [72]. Moreover, insurance plays a crucial role in ensuring economic security at a larger scale, but it is rare in aquaculture, especially for low-trophic species. The absence of ISO standards for infrastructure dimensioning further complicates the situation. There are, however, examples of how these challenges can be addressed. Aquaculture insurers in North America provide coverage for intensive and semi-intensive production systems, and as such, hatchery and nursery operations qualify for insurance [73]. Nonetheless, the number of insurance policies in effect for oyster crop protection in the US is estimated to be less than 100, covering only 2.5% of the industry [73]. In Brazil, although aquaculture insurance exists for other species, there is demand for this service by the oyster culture. Finding innovative solutions to address financing and insurance gaps is essential for the sustainable growth of the oyster aquaculture sector. In Sweden, the European Maritime, Aquaculture, and Fisheries Fund combined with national funds to implement the national food strategy has been used to channel funding to investments and R&D in the aquaculture sector. In Namibia, financial institutions are assessing ways to provide support to existing shellfish farms, and expertise in aquaculture is being encouraged to assist financial institutions in assessing loan proposals [59].

3.2.3. Social Considerations (S)

The Roles of Producer Organizations and Cooperatives

In emerging oyster sectors, the lack of facilitative structures for investment in production-related infrastructure poses challenges for farmers. However, the presence of social networks and producer cooperatives has been found to enhance production and performance within the industry [74]. Small-scale producers benefit from these organizations, as they allow for resource sharing and a unified voice in advocating for the sector's interests. In Ireland, collective efforts in areas such as depuration facilities, food safety standards, logistics, marketing, and insurance against adverse events have been recognized as potential benefits of sector collaboration [45]. Similarly, in Sweden and Brazil, producer organizations and co-operatives provide support in governance, strategic development, day-to-day management, and ensuring food safety standards [41]. The establishment of these collaborative structures enables information exchange, adaptation to volatile conditions, and the empowerment of oyster growers as primary income sources [74].

3.2.4. Technological Considerations (T)

Seed Production Techniques and Regulation

Extensive oyster seed collection relies on natural variations in abiotic conditions, such as temperature, salinity, and substrate availability. This approach presents challenges in terms of control during the production cycle but offers advantages in robustness and lower resource use [75]. Sea-based seed collection methods, including hand picking, cultch deployment, and collector deployment, are practiced in different regions, indicating a dependence on wild oyster beds (several regions of the United States to varying extents, northern Brazil, eastern South Africa, Sweden, and Ireland). The most extensive collection noted in the studied markets was seed production in Brazil, where most traditional oyster farmers collect naturally settled oyster seed from the roots of the mangrove or purchase such seed from fishermen. Larger oysters are conserved as "mother oysters" to sustain larval production [76]. However, unmanaged seed harvest can deplete natural resources and is therefore sometimes regulated. For instance, Sweden prohibits the harvest of oysters smaller than 6 cm [77], while some US Gulf States restrict seed oyster collection from public reefs to private leases [61]. Louisiana manages public oyster grounds for seed production, allowing oyster producers to obtain seed for cultivation on private leases [78]. In this area, transplanting seeds from public to private areas is considered a restoration activity following natural disasters, as seen after Hurricane Rita and Hurricane Katrina [79]. Cultch deployment is practiced in Ireland and the US but is limited in areas where shallow-water dredging is prohibited, such as Sweden. Transplanting oyster seeds between areas, however, carries significant biosecurity risks, as demonstrated by the spread of oyster herpesvirus (OsHV-1) resulting in mass mortality events in France [41]. Prior to the outbreaks, Marennes-Oleron Bay supplied seed to multiple regions, potentially facilitating the transfer of the virus [80].

Role of Hatcheries in Oyster Seed Production

Hatcheries are vital to oyster aquaculture, serving as a reliable seed source, especially in areas where natural settlement is uncertain or species are locally depleted or extinct, while also aiding genetic improvement and boosting market potential. Namibia's oyster industry exemplifies the advantages of establishing a hatchery. By transitioning from importing oyster seed to domestic production, Namibia achieved self-sufficiency, reduced the risk of importing diseases and invasive species, and developed seeds better adapted to local environmental conditions, resulting in significant production advantages [59]. Hatchery production requires advanced technological and scientific expertise, making it a costly endeavor that benefits from economies of scale [52]. During production, oyster larvae are settled on shell fragments, cultch, or can be produced as a single seed without the addition of cultch. Moreover, recently, larvae have also been produced from remote setting techniques, enhancing the hatcheries' role in oyster production and diversification [81]. Hatcheries can also impact the marketability of oysters by offering desirable traits like shell color patterns and shapes that are heritable traits [82,83]. This will ensure that market-specific attributes are met, for example, the US east coast prefers the lighter shell and mantle colors found in eastern oysters [84].

Hatchery-Produced Seed and the Advantages of Triploid Oysters

Hatcheries play a pivotal role in improving oyster production by selecting broodstock and utilizing genetic advancements. In the Pacific Northwest of the US, British Columbia, Canada, and France, the production of triploid oysters, with one extra set of chromosomes, has revolutionized the oyster industry, as triploids exhibit faster growth rates, consistent quality and texture compared to diploids [85,86]. It is important to note that triploid or tetraploid shellfish are not genetically modified organisms but possess the same genomes as diploid oysters, with one or two additional sets of chromosomes. Polyploidy can occur naturally in wild oyster populations. Their sterility reduces energy investment in reproduction, allowing for year-round consumption and shorter production cycles. Initially, concerns were raised by producers regarding the consumer acceptability of triploids, but the exceptional performance and quality of triploid oysters have resulted in a booming demand for triploid seed [43]. The development of tetraploid oysters has also greatly facilitated the production of triploid seed, obtained by crossing tetraploid males with diploid females. For example, in the US and France, triploid production has become a significant component of shellfish hatchery output [84]. In order to fully capitalize on the benefits of triploid oyster production and to overcome traditional seasonality perceptions, innovative labelling and consumer education strategies may be necessary. This would shift the perception of triploid oysters as a year-round food item, emphasizing their more consistent quality and flavor [85].

Production Techniques

Oyster cultivation encompasses a range of intensities, from extensive wild harvest-based seed to intensive floating or suspended culture using hatchery-produced seeds. Mature markets have shifted away from relying on the harvest of wild populations while emerging markets still have a mix of wild harvest and aquaculture. Harvest of wild oyster beds results in a product that is less refined than cultivated oysters, showing variations in shape, size, and growth alignments (e.g., cluster oysters)

On-bottom culture farms are particularly susceptible to market price fluctuations and oyster volume sold [87]. Similar practices are seen in one region in Brazil and in Ireland, albeit with different end products. In France, off-bottom culture on trestles has become the predominant technique, accounting for 60% of total production, while suspended culture (5%) involves hanging oysters on ropes or cages in Mediterranean lagoons or open sea lines [81]. In Brazil, various on-growing systems have been developed, including fixed tables for areas with tidal variation, longline suspended systems (the most used system) for deeper waters, and intertidal longlines. In Sweden, most farmers have developed locally adapted versions of different on-growing systems. Deep water methods like longlines can mitigate conflicts near the shore and address capacity issues in bays, allowing for greater production area utilization and expansion in total volume [43].

Postharvest Processing and the Importance of Depuration Facilities

Since adherence to food safety standards is crucial for consumer trust and market expansion [41], various processing methods are employed to enhance the shelf life, quality, and usability of oysters. In addition to traditional methods (i.e., hand sorting and grading), oysters can undergo postharvest processing (PHP) techniques. Depuration plants are the most commonly used PHP method, known for their effectiveness in reducing pathogens while being low-maintenance and free from hazardous byproducts. Other techniques to add value and improve product quality are irradiation, cryogenic individual quick-freezing (IQF), cool pasteurization, and hydrostatic pressure [84]. The establishment of reliable depuration facilities is essential for public health and increasing the value and acceptability of oysters in the market. Brazil has seen the positive impact of partnering with international agencies to establish depuration facilities, resulting in increased sales volume, higher income for growers, and enhanced consumer confidence [41]. Depuration or live storage facilities are also commonly utilized in other regions across the Atlantic, including the United States and Europe (France, Ireland, and Sweden), and in some places in Brazil. The establishment of reliable depuration facilities and setting standards for these facilities may therefore increase the value and acceptability of oysters in the market.

Food Safety through Traceability

Traceability plays a crucial role in ensuring food safety and mitigating risks in the oyster industry. It allows for the tracking of oysters from farm to consumer, particularly during product recalls, which can have significant health and economic implications [30]. Effective traceability practices contribute to maintaining the reputation of businesses and industries while minimizing negative impacts. In the United States, shellfish tags are vital for traceability, containing essential information such as harvester address, harvest location, and date of harvest. These waterproof tags are required by law to be stored at the final point of sale for 90 days [30]. Additionally, wholesalers play a key role by assigning lot numbers to shipments and maintaining metadata about the product. While digital traceability technology has been developed and piloted in some regions, its widespread adoption remains limited [88].

Similarly, in the European Union, traceability regulations mandate that all fisheries and aquaculture products be traceable throughout the production, processing, and distribution stages. Adequate labelling with essential information, including species identification, production method, catch or farming area, and fishing gear category, is required [89]. These measures ensure transparency and facilitate effective traceback in the event of issues or recalls. Compliance with traceability requirements enhances consumer confidence, safeguards public health, and supports the responsible and sustainable development of the oyster industry.

4. Conclusions and Recommendations

Oyster production and consumption across Atlantic regions show marked differences in production methods, species farmed, and market dynamics. The US, France, and Brazil have robust oyster industries, while Namibia and South Africa show growth potential despite challenges. Ecological concerns arise from seed production techniques, though innovations such as hatcheries and triploid oysters offer benefits, despite resource and acceptance issues. Public health measures and a range of postharvest processing methods, focusing on depuration facilities and traceability, are vital for consumer protection and industry growth. To tackle financial hurdles, particularly for small businesses, policy reforms, innovative solutions, and government support are necessary. Enhanced insurance coverage is needed for risk mitigation and industry stability. Regulatory frameworks and property rights need strengthening to foster responsible oyster exploitation and sustainable aquaculture. The establishment of hatcheries is recommended for stable seed supply, marketability, and reducing dependence on imported seeds. Improvement in postharvest processing methods and robust traceability systems are essential for food safety and market value. While continued research is essential to produce innovative practices, enhance cultivation techniques, and ensure responsible market dynamics, this study concludes that to achieve significant growth at both global and regional levels, identified variations may need to be harmonized. While this study produced important findings, it is vital that more comprehensive research on future market dynamics and consumer preferences across regions be conducted for effective market strategies. Additionally, assessing the impact of regulatory changes on the oyster industry's growth and sustainability, as well as exploring advanced risk management strategies and insurance models tailored to the specific needs of the oyster industry, may be crucial steps for future research.

Author Contributions: Conceptualization: Å.S., D.F., S.S. and J.A.I.; Methodology: Å.S., D.F., I.E.P., S.S. and J.A.I.; Validation: Å.S., D.F., J.A.I., P.B. (Pierre Boudry) and C.H.; Analysis: D.F., Å.S. and J.A.I.; Investigation, D.F. and Å.S.; Data Curation: Å.S., C.J.B., D.F.; C.H. and J.A.I.; Writing—Original Draft Preparation: D.F., J.A.I. and Å.S.; Writing—Review and Editing: Å.S., D.F., C.J.B., J.A.I., T.A., P.B.(Peter Britz), D.D., C.H., I.E.P., W.W., S.G., M.K., F.L., J.F.A.L., A.P.L., S.S. and A.K.M.; Supervision, Å.S.; Funding Acquisition, Å.S. and S.S. All authors have read and agreed to the published version of the manuscript.

Funding: This research was financially supported by the EU-funded project AquaVitae (project number 818173).

Informed Consent Statement: Informed consent was obtained from all subjects involved in this study.

Data Availability Statement: Data is contained within the article.

Acknowledgments: This paper is based on the MSc thesis by Daniel Fitzgerald and works from AquaVitae WP 6.

Conflicts of Interest: The authors declare no conflict of interest.

References

1. Moore, D.; Heilweck, M. Aquaculture: Prehistoric to Traditional to Modern. In *Aquaculture: Ocean Blue Carbon Meets UN-SDGS*; Sustainable Development Goals Series; Springer: Cham, Switzerland, 2022. [CrossRef]
2. MacKenzie, C.L.; Burrell, V.G., Jr.; Rosenfield, A.; Hobart, W.L. The History, Present Condition, and Future of the Molluscan Fisheries of North and Central America and Europe. Volume 1: Atlantic and Gulf Coasts. 1997. Available online: https://repository.library.noaa.gov/view/noaa/3011/noaa_3011_DS1.pdf (accessed on 30 May 2023).
3. Tantcheva-Burdge, E.; Lei, Z.; Cheung, V. Enquiry into the colours of the MoGao murals at DunHuang from the Sui Dynasty, the Tang Dynasty and the Five Dynasties period. *Cult. E Sci. Color. Color Cult. Sci.* **2023**, *15*, 57–68. [CrossRef]
4. Menzel, R.W. *Cultural Practices and Production of the American Oyster with Emphasis on the South Atlantic and Gulf of Mexico. Estuarine and Marine Bivalve Mollusk Culture*; Menzel, W., Ed.; CRC Press: Boston, MA, USA, 1991; pp. 25–34.
5. Héral, M. Traditional oyster culture in France. In *Barnabé Aquaculture*; Solbe, J.F., Ed.; Ellis Horwood: Chichester, UK, 1990; pp. 342–387.
6. Woods, H.; Hargis, W.J.; Hershner, C.H.; Mason, P.A.M. Disappearance of the natural emergent 3-dimensional oyster reef system of the James River, Virginia, 1871–1948. *J. Shellfish. Res.* **2005**, *24*, 139–142.
7. Pouvreau, S.; Lapègue, S.; Arzul, I.; Boudry, P. Fifty years of research to counter the decline of the European flat oyster (*Ostrea edulis*): A review of French achievements and prospects for the restoration of remaining beds and revival of aquaculture production. *Aquat. Living Resour.* **2023**, *36*, 13. [CrossRef]
8. Botta, R.; Asche, F.; Borsum, J.S.; Camp, E.V. A review of global oyster aquaculture production and consumption. *Mar. Policy* **2020**, *117*, 103952. [CrossRef]
9. Zhu, Y.; Li, Q.; Yu, H.; Kong, L. Biochemical Composition and Nutritional Value of Different Shell Color Strains of Pacific Oyster *Crassostrea gigas*. *J. Ocean. Univ. China* **2018**, *17*, 897–904. [CrossRef]
10. Jones, C.G.; Lawton, J.H.; Shachak, M. Positive and negative effects of organisms as physical ecosystem engineers. *Ecology* **1997**, *78*, 1946–1957. [CrossRef]
11. Petersen, J.K.; Holmer, M.; Termansen, M.; Hasler, B. *Nutrient Extraction through Bivalves. Goods and Services of Marine Bivalves*; Springer: Cham, Switzerland, 2019; pp. 179–208. [CrossRef]
12. Ruesink, J.L.; Lenihan, H.S.; Trimble, A.C.; Heiman, K.W.; Micheli, F.; Byers, J.E.; Kay, M.C. Introduction of non-native oysters: Ecosystem effects and restoration implications. *Annu. Rev. Ecol. Evol. Syst.* **2005**, *36*, 643–689. [CrossRef]
13. Newell, R.I.; Koch, E.W. Modelling seagrass density and distribution in response to changes in turbidity stemming from bivalve filtration and seagrass sediment stabilization. *Estuaries* **2004**, *27*, 793–806. [CrossRef]
14. Cerrato, R.M.; Caron, D.A.; Lonsdale, D.J.; Rose, J.M.; Schaffner, R.A. Effect of the northern quahog *Mercenaria mercenaria* on the development of blooms of the brown tide alga *Aureococcus anophagefferens*. *Mar. Ecol. Prog. Ser.* **2004**, *281*, 93–108. [CrossRef]
15. Venugopal, V.; Gopakumar, K. Shellfish: Nutritive value, health benefits, and consumer safety. *Compr. Rev. Food Sci. Food Saf.* **2017**, *16*, 1219–1242. [CrossRef]
16. Wright, A.C.; Fan, Y.; Baker, G.L. Nutritional value and food safety of bivalve molluscan shellfish. *J. Shellfish. Res.* **2018**, *37*, 695–708. [CrossRef]
17. Hallström, E.; Bergman, K.; Mifflin, K.; Parker, R.; Tyedmers, P.; Troell, M.; Ziegler, F. Combined climate and nutritional performance of seafoods. *J. Clean. Prod.* **2019**, *230*, 402–411. [CrossRef]
18. Krause, G.; Le Vay, L.; Buck, B.H.; Costa-Pierce, B.A.; Dewhurst, T.; Heasman, K.G.; Nevejan, N.; Nielsen, P.; Nielsen, K.N.; Park, K.; et al. Prospects of low trophic marine aquaculture aontributing to food Security in a net zero-carbon world. *Front. Sustain. Food Syst.* **2022**, *6*, 875509. [CrossRef]
19. Thomas, J.B.E.; Sinha, R.; Strand, Å.; Söderqvist, T.; Stadmark, J.; Franzén, F.; Ingmansson, I.; Gröndahl, F.; Hasselström, L. Marine biomass for a circular blue-green bioeconomy? A life cycle perspective on closing nitrogen and phosphorus land-marine loops. *J. Ind. Ecol.* **2022**, *26*, 2136–2153. [CrossRef]
20. James, P.; Evensen, T. Live transport of the green sea urchin (*Strongylocentrotus droebachiensis*) in air and immersed in seawater and the impact on subsequent roe enhancement after in-water transport. *Aquac. Res.* **2022**, *53*, 5205–5213. [CrossRef]
21. Hughes, A.D. Defining Nature-Based Solutions Within the Blue Economy: The Example of Aquaculture. *Front. Mar. Sci.* **2021**, *8*, 711443. [CrossRef]
22. Pernet, F.; Lupo, C.; Bacher, C.; Whittington, R.J. Infectious diseases in oyster aquaculture require a new integrated approach. *Philos. Trans. R. Soc. B Biol. Sci.* **2016**, *371*, 20150213. [CrossRef]
23. Peng, D.; Zhang, S.; Zhang, H.; Pang, D.; Yang, Q.; Jiang, R.; Lin, Y.; Mu, Y.; Zhu, Y. The oyster fishery in China: Trend, concerns and solutions. *Mar. Policy* **2021**, *129*, 104524. [CrossRef]
24. Lawley, M.; Birch, D. Exploring Point of Sale Strategies for Improving Seafood Retailing: The Case of the Australian Oyster Industry. *J. Food Prod. Mark.* **2016**, *22*, 792–808. [CrossRef]
25. Askew, C.G. A generalized growth and mortality model for assessing the economics of bivalve culture. *Aquaculture* **1978**, *14*, 91–104. [CrossRef]
26. Weir, M.; Rajić, A.; Dutil, L.; Uhland, C.; Bruneau, N. Zoonotic bacteria and antimicrobial resistance in aquaculture: Opportunities for surveillance in Canada. *Can. Vet. J.* **2012**, *53*, 619–622. [PubMed]

27. Kim, H.; Sefcik, J.S.; Bradway, C. Characteristics of qualitative descriptive studies: A systematic review. *Res. Nurs. Health* **2017**, *40*, 23–42. [CrossRef] [PubMed]
28. Sandelowski, M. What's in a name? Qualitative description revisited. *Res. Nurs. Health* **2010**, *33*, 77–84. [CrossRef] [PubMed]
29. Carruthers, H. Using PEST analysis to improve business performance. *In Practice* **2009**, *31*, 37–39. [CrossRef]
30. Love, D.C.; Lane, R.M.; Kuehl, L.M.; Hudson, B.; Harding, J.; Clancy, K.; Fry, J.P. Performance and conduct of supply chains for United States farmed oysters. *Aquaculture* **2020**, *515*, 734569. [CrossRef]
31. FAO. FAO Yearbook. *Fishery and Aquaculture Statistics 2019/FAO Annuaire. Statistiques des Pêches et de L'aquaculture 2019/FAO Anuario*; Estadísticas de pesca y Acuicultura: Rome, Italy, 2019. [CrossRef]
32. Lipton, D.; Kirkley, J. A Profile of the Oyster Industry Northeastern United States. 1994. Available online: https://scholarworks.wm.edu/cgi/viewcontent.cgi?article=2297andcontext=reports (accessed on 30 May 2023).
33. Posadas, B.; Posadas, R.; Andrews, L. Consumer Preferences for Postharvest-Processed Raw Oyster Products in Southern California. 2011. Available online: https://repository.library.noaa.gov/view/noaa/38154/noaa_38154_DS1.pdf (accessed on 30 May 2023).
34. Jahns, L.; Raatz, S.K.; Johnson, L.K.; Kranz, S.; Silverstein, J.T.; Picklo, M.J. Intake of seafood in the US varies by age, income, and education level but not by race-ethnicity. *Nutrients* **2014**, *6*, 6060–6075. [CrossRef]
35. Zhang, X.; House, L.; Sureshwaran, S.; Hanson, T.R. At-home and away-from-home consumption of seafood in the United States (No. 1364-2016-108069). In Proceedings of the Southern Agricultural Economics Association (SAEA) 2004 Annual Meeting, Tulsa OK, USA, 14–18 February 2004. [CrossRef]
36. Kecinski, M.; Messer, K.D.; Knapp, L.; Shirazi, Y. Consumer preferences for oyster attributes: Field experiments on brand, locality and growing method. *Agric. Resour. Econ. Rev.* **2017**, *46*, 315–333. [CrossRef]
37. Food and Drug Administration (FDA). Equivalence Determination Regarding Implementation by Spain and the Netherlands of the European Union System of Food Safety Control Measures for Raw Bivalve Molluscan Shellfish with Additional Controls. 2020. Available online: https://www.govinfo.gov/content/pkg/FR-2020-09-24/pdf/2020-20755.pdf (accessed on 12 May 2023).
38. Wilbon, A.; Bundy, M.; Clark, K. Entrepreneurship in the Chesapeake Bay Oyster Industry. 2011. Available online: https://www.researchgate.net/profile/Anthony-Wilbon/publication/295547389_Entrepreneurship_in_the_Chesapeake_Bay_Oyster_Industry/links/56cb23e408aee3cee5415c1c/Entrepreneurship-in-the-Chesapeake-Bay-Oyster-Industry.pdf (accessed on 30 May 2023).
39. Barroso, G.F. Assessing the Potential for Mangrove Oyster Aquaculture in an Estuarine System of the Southeastern Coast of Brazil: A Geographic Information System Approach. 2004. Available online: https://dspace.library.uvic.ca/bitstream/handle/1828/648/barroso_2004.pdf?isAllowed=yandsequence=1. (accessed on 30 May 2023).
40. Valenti, W.C.; Barros, H.P.; Moraes-Valenti, P.; Bueno, G.W.; Cavalli, R.O. Aquaculture in Brazil: Past, present and future. *Aquac. Rep.* **2021**, *19*, 100611. [CrossRef]
41. de Souza Sampaio, D.; Tagliaro, C.H.; Schneider, H.; Beasley, C.R. Oyster culture on the Amazon mangrove coast: Asymmetries and advances in an emerging sector. *Rev. Aquac.* **2019**, *11*, 88–104. [CrossRef]
42. Robert, R.; Sanchez, J.L.; Pérez-Parallé, L.; Ponis, E.; Kamermans, P.; O'Mahoney, M. A glimpse on the mollusc industry in Europe. *Aquac. Eur.* **2013**, *38*, 5–11.
43. Buestel, D.; Ropert, M.; Prou, J.; Goulletquer, P. History, status, and future of oyster culture in France. *J. Shellfish. Res.* **2009**, *28*, 813–820. [CrossRef]
44. Mariojouls, C.; Paquotte, P. *Labelling and Product Differentiation in the French National Market for Oysters and Mussels. Quality of Fish from Catch to Consumer: Labelling, Modelling and Traceability*; Wageningen Academic Publishers: Wageningen, The Netherlands, 2003; p. 441.
45. Renwick, A. Regulatory failure and risk in Aquaculture: A case study of the Irish Oyster Industry. In Proceedings of the Agricultural Economics Society AES 90th Annual Conference, Warwick University, Coventry, UK, 4–6 April 2016. [CrossRef]
46. Fox, M.; Christley, R.; Lupo, C.; Moore, H.; Service, M.; Campbell, K. Preventing and mitigating farmed bivalve disease: A Northern Ireland case study. *Aquac. Int.* **2020**, *28*, 2397–2417. [CrossRef]
47. BIM. The Economic Contribution of the Aquaculture Sector. Across Ireland's Bay Areas. Report. 2022. Available online: https://bim.ie/wp-content/uploads/2022/05/BIM-Collective-Bay-Areas-Report-SPREADS.pdf (accessed on 2 October 2023).
48. zu Ermgassen, P.S.E.; Strand, Å.; Bakker, N.; Blanco, A.; Bonačić, K.; Boudry, P.; Brundu, G.; Cameron, T.C.; Connellan, I.; Costa Fd Debney, A.; et al. Overcoming *Ostrea edulis* seed production limitations to meet ecosystem restoration demands in the UN decade on restoration. *Aquat. Living Resour.* **2023**, *36*, 16. [CrossRef]
49. Wrange, A.L.; Valero, J.; Harkestad, L.S.; Strand, Ø.; Lindegarth, S.; Christensen, H.T.; Mortensen, S. Massive settlements of the Pacific oyster, Crassostrea gigas, in Scandinavia. *Biol. Invasions* **2010**, *12*, 1145–1152. [CrossRef]
50. Paisley, L.G.; Ariel, E.; Lyngstad, T.; Jónsson, G.; Vennerström, P.; Hellström, A.; Østergaard, P. An overview of aquaculture in the Nordic countries. *J. World Aquac. Soc.* **2010**, *41*, 1–17. [CrossRef]
51. Hanstén, M. Harvesting Wild Pacific Oysters, Crassostrea Gigas–Developing Methods for Locating and Live-Storing Marketable Individuals. 2017. Available online: https://www.theseus.fi/bitstream/handle/10024/129818/Hansten_Magnus.pdf?sequence=1 (accessed on 30 May 2023).

52. Joyce, A.; Kjerulf, J.; Petersen, D.; Murphy, C. Strategic Approaches for Aquaculture Industry Development: Flat Oyster Cultivation in Scandinavia. DTU Libr. 2013. Available online: https://backend.orbit.dtu.dk/ws/portalfiles/portal/103297945/Publishers_version.pdf (accessed on 30 May 2023).
53. Haupt, T.M.; Griffiths, C.L.; Robinson, T.B.; Tonin, A.F.; De Bruyn, P.A. The history and status of oyster exploitation and culture in South Africa. *J. Shellfish. Res.* **2010**, *29*, 151–159. [CrossRef]
54. Keightley, J.; von der Heyden, S.; Jackson, S. Introduced Pacific oysters Crassostrea gigas in South Africa: Demographic change, genetic diversity and body condition. *Afr. J. Mar. Sci.* **2015**, *37*, 89–98. [CrossRef]
55. Branch, G.M.; Hauck, M.; Siqwana-Ndulo, N.; Dye, A.H. Defining fishers in the South African context: Subsistence, artisanal and small-scale commercial sectors. *Afr. J. Mar. Sci.* **2002**, *24*, 475–487. [CrossRef]
56. Olivier, D.; Heinecken, L.; Jackson, S. Mussel and oyster culture in Saldanha Bay, South Africa: Potential for sustainable growth, development and employment creation. *Food Secur.* **2013**, *5*, 251–267. [CrossRef]
57. Britz, P.J.; Lee, B.; Botes, L. *AISA 2009 Aquaculture Benchmarking Survey: Primary Production and Markets*; AISA Report; Enviro-Fish Africa (Pty.) Ltd.: Grahamstown, South Africa, 2009.
58. Iitembu, J.A.; Ndakalimwe, N.G.; Tjipute, M.; Asino, H.; Hamukwaya, J. The Governance of Aquaculture in Namibia as a Vehicle for Food Security and Economic Growth. In *Food Security for African Smallholder Farmers*; Mupambwa, H.A., Nciizah, A.D., Nyambo, P., Muchara, B., Gabriel, N.N., Eds.; Springer: Singapore, 2022; pp. 391–403. [CrossRef]
59. Mapfumo, B. History and Development of Shellfish Cultivation in Walvis Bay, Namibia. Aquaculture Compendium, (109419). 2010. Available online: https://www.sarnissa.stir.ac.uk/wp-content/uploads/2020/12/WalvisBay-Oysters-CS-Aug-09.pdf (accessed on 30 May 2023).
60. Russell, D.; Kibria, G. Oyster Farming in Namibia: A Success Story to Greater Food Security. Available online: https://www.was.org/MeetingAbstracts/ShowAbstract/47375 (accessed on 31 May 2023).
61. Chiripanhura, B.; Teweldemedhin, M. *An Analysis of the Fishing Industry in Namibia: The Structure, Performance, Challenges, and Prospects for Growth and Diversification*; International Food Policy Research Institute: Washington, DC, USA, 2016.
62. Murta, M.S.G.; Kibria, M.D.G.; An Emerging Shellfish Farming Industry in Namibia—Responsible Seafood Advocate. Global Seafood Alliance. Available online: https://www.globalseafood.org/advocate/emerging-shellfish-farming-industry-namibia/ (accessed on 14 August 2022).
63. Brugère, C.; Ridler, N.; Haylor, G.; Macfadyen, G.; Hishamunda, N. *Aquaculture Planning: Policy Formulation and Implementation for Sustainable Development*; FAO Fisheries and Aquaculture Technical Paper; FAO: Rome, Italy, 2010; Volume 542.
64. Knapp, G.; Rubino, M.C. The political economics of marine aquaculture in the United States. *Rev. Fish. Sci. Aquac.* **2016**, *24*, 213–229. [CrossRef]
65. Dosdat, A.; De la Pomelie, C. Regulation and monitoring of marine aquaculture in France. *J. Appl. Ichthyol.* **2000**, *16*, 157–162. [CrossRef]
66. Young, O.R.; Osherenko, G.; Ekstrom, J.; Crowder, L.B.; Ogden, J.; Wilson, J.A.; Day, J.C.; Douvere, F.; Ehler, C.N.; McLeod, K.L.; et al. Solving the crisis in ocean governance: Place-based management of marine ecosystems. *Environ. Sci. Policy Sustain. Dev.* **2007**, *49*, 20–32. [CrossRef]
67. Walton, W.C.; Davis, J.E.; Chaplin, G.I.; Rikard, F.S.; Hanson, T.R.; Waters, P.J.; Swann, D.L. Off-bottom oyster farming. In *Agriculture and Natural Resources Timely Information: Fisheries and Aquaculture Series*; Alabama Cooperative Extension System: Autaugaville, AL, USA, 2012; p. 8.
68. Vieira, K.S.; Neto, J.A.B.; Crapez, M.A.C.; Gaylarde, C.; Pierri, B.d.S.; Saldaña-Serrano, M.; Bainy, A.C.D.; Nogueira, D.J.; Fonseca, E.M. Occurrence of microplastics and heavy metals accumulation in native oysters Crassostrea Gasar in the Paranaguá estuarine system, Brazil. *Mar. Pollut. Bull.* **2021**, *166*, 112225. [CrossRef]
69. Supan, J. Extensive culture of Crassostrea virginica in the Gulf of Mexico Region (p. 4). Southern Regional Aquaculture Center. 2002. Available online: https://shellfish.ifas.ufl.edu/wp-content/uploads/Extensive-Culture-of-Crassostrea-virginica-in-the-Gulf-of-Mexico-Region.pdf (accessed on 31 May 2023).
70. Kallis, G.; Butler, D. The EU water framework directive: Measures and implications. *Water Policy* **2001**, *3*, 125–142. [CrossRef]
71. Commission Regulation (EC) No 1881/2006 of 19 December 2006 Setting Maximum Levels for Certain Contaminants in Foodstuffs (Text with EEA Relevance). Available online: https://eur-lex.europa.eu/LexUriServ/LexUriServ.do?uri=OJ:L:2006:364:0005:0024:EN:PDF. (accessed on 31 May 2023).
72. Grant GR Wood, S.D.; Wright, W.J. Black farmers united: The struggle against power and principalities. *J. Pan Afr. Stud.* **2012**, *5*, 3–22.
73. Olin, P.G.; Smith, J.; Nabi, R. *Regional Review on Status and Trends in Aquaculture Development in North America: Canada and the United States of America—2010*; FAO Fisheries and Aquaculture Circular No. 1061/2; FAO: Rome, Italy, 2011; p. 84.
74. Cole, A.; Chen, X. Off-farm employment in aquaculture: A case study of New England's oyster growers. *Can. J. Agric. Econ./Rev. Can. D'agroeconomie* **2021**, *69*, 369–382. [CrossRef]
75. Strand, A.; Suhnel, S. *Overview of Culture Systems for Low Trophic Species*; AquaVitae: Tromsø, Norway, 2022. [CrossRef]
76. Haque, C.E.; Deb, A.K.; Medeiros, D. Integrating conservation with livelihood improvement for sustainable development: The experiment of an oyster producers' cooperative in Southeast Brazil. *Soc. Nat. Resour.* **2009**, *22*, 554–570. [CrossRef]
77. Thorngren, L.; Bergström, P.; Dunér Holthuis, T.; Lindegarth, M. Assessment of the population of Ostrea edulis in Sweden: A marginal population of significance? *Ecol. Evol.* **2019**, *9*, 13877–13888. [CrossRef]

78. MacKenzie, C.L., Jr. The Commercial Oysters of the Gulf of Mexico and the Caribbean Sea: Ecology, Biology, and Fisheries. 2005. Available online: https://aquadocs.org/bitstream/handle/1834/29832/gcfi_56-50.pdf?sequence=1 (accessed on 31 May 2023).
79. Banks, P.; Beck, S.; Chapiesky, K.; Isaacs, J. Louisiana Oyster. 2016. Available online: https://static1.squarespace.com/static/5c3 11daff2e6b1bd0e619a75/t/5c688be1ee6eb079bf19da35/1550355435448/finaldraftoysterfmp11-23-16.pdf (accessed on 31 May 2023).
80. Guillotreau, P.; Le Bihan, V.; Pardo, S. Mass mortality of farmed oysters in France: Bad responses and good results. In *Global Change in Marine Systems*; Routledge: London, UK, 2017; pp. 54–64.
81. Boeuf, G. Present status of the French aquaculture. *Aquac. Sci.* **2000**, *48*, 243–248.
82. Han, Z.; Li, Q. Mendelian inheritance of orange shell color in the Pacific oyster *Crassostrea gigas*. *Aquaculture* **2020**, *516*, 734616. [CrossRef]
83. Jourdan, A.; Morvezen, R.; Enez, F.; Haffray, P.; Lange, A.; Vétois, E.; Allal, F.; Phocas, F.; Bugeon, J.; Degremont, L.; et al. Potential of genomic selection for growth and quality traits in mixed-family breeding designs for the Pacific oyster *Crassostrea gigas*. *Aquaculture* **2023**, *576*, 739878. [CrossRef]
84. Cheney, D.P. Bivalve shellfish quality in the USA: From the hatchery to the consumer. *J. World Aquac. Soc.* **2010**, *41*, 192–206. [CrossRef]
85. Nell, J.A. Farming triploid oysters. *Aquaculture* **2002**, *210*, 69–88. [CrossRef]
86. Gagnaire, B.; Soletchnik, P.; Madec, P.; Geairon, P.; Le Moine, O.; Renault, T. Diploid and triploid Pacific oysters, *Crassostrea gigas* (Thunberg), reared at two heights above sediment in Marennes-Oleron Basin, France: Difference in mortality, sexual maturation and hemocyte parameters. *Aquaculture* **2006**, *254*, 606–616. [CrossRef]
87. Engle, C.; van Senten, J.; Parker, M.; Webster, D.; Clark, C. Economic tradeoffs and risk between traditional bottom and container culture of oysters on Maryland farms. *Aquac. Econ. Manag.* **2021**, *25*, 472–503. [CrossRef]
88. Miller, A.; Heggelund, D.; McDermott, T. *Digital Traceability for Oyster Supply Chains: Implementation and Results of a Pilot*, 222nd ed.; Gulf States Marine Fisheries Commission: Ocean Springs, MS, USA, 2014.
89. D'Amico, P.; Armani, A.; Gianfaldoni, D.; Guidi, A. New provisions for the labelling of fishery and aquaculture products: Difficulties in the implementation of Regulation (EU) n. 1379/2013. *Mar. Policy* **2016**, *71*, 147–156. [CrossRef]

Disclaimer/Publisher's Note: The statements, opinions and data contained in all publications are solely those of the individual author(s) and contributor(s) and not of MDPI and/or the editor(s). MDPI and/or the editor(s) disclaim responsibility for any injury to people or property resulting from any ideas, methods, instructions or products referred to in the content.

MDPI AG
Grosspeteranlage 5
4052 Basel
Switzerland
Tel.: +41 61 683 77 34

Fishes Editorial Office
E-mail: fishes@mdpi.com
www.mdpi.com/journal/fishes

Disclaimer/Publisher's Note: The statements, opinions and data contained in all publications are solely those of the individual author(s) and contributor(s) and not of MDPI and/or the editor(s). MDPI and/or the editor(s) disclaim responsibility for any injury to people or property resulting from any ideas, methods, instructions or products referred to in the content.

www.ingramcontent.com/pod-product-compliance
Lightning Source LLC
Chambersburg PA
CBHW041355210926
51850CB00013B/3087